Applied Quantitative Analysis for Real Estate

To fully function in today's global real estate industry, students and professionals increasingly need to understand how to implement essential and cutting-edge quantitative techniques.

This book presents an easy-to-read guide to applying quantitative analysis in real estate aimed at non-cognate undergraduate and masters students, and meets the requirements of modern professional practice. Through case studies and examples illustrating applications using data sourced from dedicated real estate information providers and major firms in the industry, the book provides an introduction to the foundations underlying statistical data analysis, common data manipulations and understanding descriptive statistics, before gradually building up to more advanced quantitative analysis, modelling and forecasting of real estate markets.

Our examples and case studies within the chapters have been specifically compiled for this book and explicitly designed to help the reader acquire a better understanding of the quantitative methods addressed in each chapter. Our objective is to equip readers with the skills needed to confidently carry out their own quantitative analysis and be able to interpret empirical results from academic work and practitioner studies in the field of real estate and in other asset classes.

Both undergraduate and masters level students, as well as real estate analysts in the professions, will find this book to be essential reading.

Sotiris Tsolacos is Professor of Real Estate Investment at Cass Business School and the Programme Director for the MSc Real Estate Investment. Previously he was with Henley Business School. He earned his PhD in economics at the University of Reading. He has over 25 years of experience in the quantitative analysis of real estate markets. He has over 40 publications in this field and is the co-author of the book *Real Estate Modelling and Forecasting*. He has long industry experience. He was with JLL in London and the Director of European Research in Property & Portfolio Research/CoStar. He is the co-founder of KappaSigma Partners, a regional property consultancy firm. He works closely with the industry on a range of topical themes in real estate. His industry experience and continual engagement with the profession has greatly inspired the content of this book.

Mark Andrew is Senior Lecturer in Real Estate Finance and Investment and Programme Director for the MSc Real Estate at Cass Business School, City, University of London. He received his PhD from the University of Reading in 2001 and has taught non-cognate and cognate real estate students for over 15 years. He has published in highly ranked real estate and non-real estate journals, such as *Real Estate Economics, Urban Studies* and *Environment and Planning A*. He has undertaken a number of government commissioned reports on his research specialty in residential markets. He currently acts as a referee for a number of leading academic real estate journals and is a serving member on the IPF Research Steering Group.

Applied Quantitative
Analysis for Real Estate

Sotiris Tsolacos and Mark Andrew

Routledge
Taylor & Francis Group

LONDON AND NEW YORK

First published 2021
by Routledge
2 Park Square, Milton Park, Abingdon, Oxon OX14 4RN

and by Routledge
52 Vanderbilt Avenue, New York, NY 10017

Routledge is an imprint of the Taylor & Francis Group, an informa business

British Library Cataloguing-in-Publication Data
A catalogue record for this book is available from the British Library

Library of Congress Cataloging-in-Publication Data
Names: Tsolacos, Sotiris, author. | Andrew, Mark, author.
Title: Applied quantitative analysis for real estate / Sotiris Tsolacos and
    Mark Andrew.
Description: Abingdon, Oxon ; New York, NY : Routledge, 2020. | Includes
    bibliographical references and index.
Identifiers: LCCN 2020013923 (print) | LCCN 2020013924 (ebook) |
    ISBN 9781138561328 (hbk) | ISBN 9781138561335 (pbk) |
    ISBN 9780203710876 (ebk) | ISBN 9781351359016 (adobe pdf) |
    ISBN 9781351358996 (mobi) | ISBN 9781351359009 (epub)
Subjects: LCSH: Real estate investment—Statistical methods. | Real property—
    Statistical methods. | Quantitative analysis.
Classification: LCC HD1382.5 .T79 2020 (print) | LCC HD1382.5 (ebook) |
    DDC 332.63/240721—dc23
LC record available at https://lccn.loc.gov/2020013923
LC ebook record available at https://lccn.loc.gov/2020013924

ISBN: 978-1-138-56132-8 (hbk)
ISBN: 978-1-138-56133-5 (pbk)
ISBN: 978-0-203-71087-6 (ebk)

Typeset in Baskerville
by Apex CoVantage, LLC

Visit the eResources: www.routledge.com/9781138561335

Contents

7 Regression diagnostics

8 Stationarity

9 Forecast evaluation

10 ARMA models

11 Vector autoregressions   273

12 Epilogue   295

Figures

Figures

Tables

Preface

Motivation

This book aims to support and accompany the teaching of quantitative education in the real estate field. Over the last few years, the quantitative analysis of real estate markets has grown in importance. Real estate is a truly multidisciplinary field reflecting the characteristics of the built environment, which itself is a factor of production, a component of fixed capital investment, a means of accumulating and holding wealth and an investment asset class. Within this wide-ranging realm there are important everyday tasks performed in the industry, such as valuations, cash flow predictions, loan underwriting and asset allocations to mention a few. The greater sophistication of the industry has generated education demands on students and analysts. Graduates and apprentices joining valuation teams, capital market, investment research and debt teams would need to carry out basic and more sophisticated quantitative analysis. The industry now requires a more analytical skill set.

Both undergraduate and postgraduate courses in real estate, real estate finance and investment include basic and advanced quantitative analysis in their curriculum. In our experience, real estate students come from a variety of disciplines. Some may not have studied mathematics or statistics since leaving school, while others may have studied quantitative techniques but applied in different contexts and therefore are unfamiliar with real estate data and theories. There are numerous econometrics textbooks with good applications in the field of economics and finance. Our motivation for this book was to cover topics, essential and advanced, that are most in demand by employers in the real estate market. The idea was to produce a book containing extensive examples of applications in real estate, so that the reader becomes familiar with real estate data, and the issues and peculiarities encountered when applying quantitative methods in its analysis, yet at the same time ensure that the level is accessible readers from a non-numerical and statistical background. Empirical and practical applications in this book are based on extensive use of real estate data, which should additionally aid the understanding of the theoretical concepts underlying the explanations of phenomena and outcomes in the market. We further illustrate the analyses using Excel and EViews, commonly used software in the real estate industry.

The origin of this book is derived from the content of lectures delivered in modules taught on the MSc Real Estate and MSc Real Estate Investment degree programmes at Cass Business School, and our frustration at the lack of an appropriate textbook to cater for our students' needs. Introductory textbooks are appropriate for students with limited backgrounds in mathematics and statistics but seldom contain any content related to real estate. They also tend to focus too much on the derivations of formulae of estimators along with proofs of their desirable and undesirable properties. On the other hand, the more advanced textbooks require a significant step up in mathematical and statistical ability to comprehend their content, and are geared more toward courses lasting more than one or two terms. Many introductory and advanced textbooks very often provide little guidance on how to actually execute a quantitative analysis. We believe that our book will help real estate students acquire the skills to work confidently and independently and provide the basis to carry out further statistical work addressing real estate issues.

Both authors of this book have had several years of experience of teaching quantitative methods to real estate students in the classroom and have published research in highly reputable real estate journals. Sotiris Tsolacos has over 16 years of experience working in various real estate research environments in the industry. Mark Andrew has been involved in housing research projects funded by the UK government and also undertaken private consultancy work. Our experience of working with the industry and government helped to shape the content of the book, emphasising elements that are often not be flagged in more mainstream econometrics textbooks. We feel that between us we have the sufficient expertise and experience to achieve our objectives.

Intended audience

The intended audience is undergraduate and postgraduate students in real estate and disciplines incorporating real estate who require a broad knowledge of modern applied statistical techniques employed in analysing its markets, and professionals in the industry who would like to refresh or extend their knowledge. An important feature of this book is its aim to address the needs of the industry and make the book a reference book for post-education. Although the applications and motivations for modelling in this book are drawn from real estate, the empirical procedures adopted in testing of theories are relevant to other disciplines, such as economics, business studies, finance and management, and should prove useful to these students.

Online resource

The content of the book is supported by online material. It contains accompanying notes for each chapter with further examples and data files in Excel and EViews so that the reader can replicate the empirical analysis undertaken throughout the book. Further, the online resource includes additional chapters:

Chapter I: building empirical regression models
Chapter II: regression analysis: a cross-section model
Chapter III: cointegration
Chapter IV: panel analysis

There is also an instructor's section that comprises a suite of lecture slides.

Prerequisites for understanding the material in this book

We assume little knowledge of quantitative techniques, although the reader should have basic knowledge mathematics and statistics at secondary (high) school. To refresh memory, we devote a chapter to cover these prerequisites. The focus of our book is on the application of the most commonly used techniques to analyse real estate data and markets. Throughout the book we refer to the theories commonly applied in real estate analyses since many readers may not have this background. The emphasis in our book is on an application of various methods to analyse real estate data and to deal with common issues related to it.

Acknowledgements

We made use of several databases provided by the private sector. We would like to express our appreciation to the CoStar Group, JLL, RCA, Knight Frank, MSCI and GPR for data provision. We have also used websites and data publicly available on the internet in our examples.

We would like to acknowledge the input of a number of individuals for their involvement in various tasks with this book. We are grateful to Nigel Almond, Mark Stansfield, Sophia Roberts and Nicole Lux for their assistance and Ervi Liusman for help with Hong Kong data. In particular, we would like to thank Cleo Flokes and Fergus Hicks for their comments and suggestions in Chapters 2 and 11.

1 Introduction

The focus of this book is the application of quantitative techniques to real estate. The material presented takes the reader from basic statistical analysis to more advanced topics addressing a range of quantitative methods employed both in real estate education and the workplace. The themes in this book are presented in an applied manner. Throughout the book we provide examples to illustrate the statistical concepts in the context of the real estate market. The online resource for this book contains further applications. The aim of the book is to make the reader confident with the application of quantitative techniques to real estate.

1.1 Motivation and rationale for this book

We highlight four inter-related trends in the real estate field that necessitate the existence of a textbook on the applied quantitative analysis of real estate markets.

(i) The application of statistical tools to analyse data is a much-sought skill in the real estate business. It is increasingly becoming an integral part of real estate education. Students in real estate degree programmes are expected to have at least a fair knowledge of basic statistical techniques. A good grasp of statistical analysis is helpful for the study of other subjects such as real estate investment, appraisals and portfolio management.

(ii) It is recognised that the real estate market interacts with the economic and broader investment environments. Quantitative analysis will assist us to quantify relationships and test them empirically. Such analysis varies from simple descriptions of features of the data, examining correlations to constructing a variety of econometric models. It opens up a greater range of options in the empirical investigation of real estate markets.

(iii) The recognition of real estate as a mainstream asset class poses challenges to how analysis in real estate markets is conducted. Investors in other asset classes are accustomed to the application of quantitative analysis and would expect similar practices in the real estate market.

(iv) The availability of data is growing. Databases are getting larger and becoming more readily available. Universities are increasingly gaining access to more databases, including proprietary data from firms and organisations. A good background of quantitative analysis enables students, researchers, analysts and others to utilise the growing availability of data.

Universities have incorporated quantitative research techniques into their real estate programmes. Modules covering simple data analysis and statistical modelling are part of the curricula, depending of course on the nature of the programme. Real estate investment and finance programmes will incorporate more advanced analysis whereas the more traditional real estate programmes will contain essential quantitative techniques. This book is motivated by the needs of both groups of students either at the undergraduate or postgraduate levels. A number of these programmes are conversion courses, usually taken by non-cognate students who may come from a subject area with little statistical background. This book is intended to bring these students up to speed with the application of quantitative techniques to real estate data analysis. The book aims to facilitate the development of these skills. Students will become familiar with the most commonly used techniques in practice and will be well equipped to directly carry out empirical work.

The work in this book is also inspired by the quantitative needs of real estate analysts in the industry. Analysts have access to large datasets. Some of them may have to brush up their quantitative knowledge. A quantitative background gives them flexibility to undertake their own work and investigate relationships using the wealth of data available. Ability to make sense of the data and carry out empirical analyses is a valuable skill for a professional in the real estate field.

1.2   Broad themes covered in the book

Quantitative analysis comprises a large set of mathematical and statistical procedures and tools for factual analysis. This book is not intended to present any branch of quantitative analysis that is potentially relevant to real estate. This would be an onerous task. The book focuses on themes most prevalent in real estate and presents them in detail and in a practical way. The book helps the reader build the necessary background for further quantitative analysis. In this section we give a short outline of four themes covered in this book. They are data manipulation, probability and inferential statistics in real estate, applied econometric modelling and forecasting.

1.2.1   Data manipulation and summarising information

Understanding and summarising data to reveal systematic patterns is informative. Information in its raw form may not be appropriate to conduct analysis or make comparisons and can be difficult to comprehend. Consider, for example, rents recorded in different units (e.g. local currency per square metre per year vis-à-vis in local currency per square foot per month). Or for that matter, rents obtained from countries with low and high inflation. Data transformations help us to understand the information they contain about the market. Rents should be expressed in similar units, or one should produce an index or adjust them for inflation. It is valuable to understand the distribution of the data, obtain statistics such as the mean and standard deviation and use techniques such as correlation analysis to study relationships between data series. Think of a situation in which we find that rents in market A are more strongly associated with the economy than in market B. What are the reasons behind this?

1.2.2   Probability analysis, inferential statistics and applications

Analysing the market entails dealing with uncertainty that arises from the complex nature of the real world. Real estate is subject to uncertainty of the same nature and intensity, as for instance in the stock market. Following data transformations and rudimentary statistical analysis, we introduce the concepts of probability, expected values and hypothesis testing.

They enable the analyst to draw conclusions reflecting the most probable outcomes. Probability distributions are the basic statistical framework for undertaking such analyses. Suppose when we study house price growth we observe that the most common growth rate values lie between 4% and 6% per annum. There are occasions, however, when house price growth can be negative, say −10%, or strongly positive, +20%. These are rare occasions, but they still occur reflecting major economic events in the market. By studying the distributional properties of the data, we are able to make useful inferences about expected values and probabilities of them occurring.

Related to this topic is sampling. Usually we work with samples from a larger and often unknown population or have a preconceived idea about the data generating process (DGP). This concept is not easy to grasp at first, a fact we acknowledge. Results reported from calculations are sample-based results with which we infer conclusions for the unknown or unmeasurable population. For example, when we work with 40 observations of rent data, we assume that this is either a subset of the unknown population of rents or the unknown DGP. If our objective is to estimate, for example, the mean value of house price growth out of sample estimates, we can utilise the concepts of a sampling distribution to determine the likelihood of the sample estimates reflecting the population or DGP value. Again, any inferences will involve the language of probability. These concepts are additionally used in hypothesis testing and modelling relationships.

1.2.3 Model building and applied econometric analysis

The book proceeds to employ and present formal techniques to study relationships within the real estate market and between the real estate market and the economy and investment environment. The complexity of the real estate market, its linkages to the economy and the importance of real estate in credit and investment markets have necessitated closer study of the dynamics of the real estate market. This market has taken a major part in causing economic and financial crises; an example is the global financial crisis in 2008–2009. As a result, we have seen an explosion in the use of quantitative analysis to explore how adjustments take place within the real estate market and measure its linkages to the external environment.

The relationships can be complex. We try to establish the factor or factors that influence a real estate variable. Housing construction can be influenced not just by economic growth but also by other influences including interest rates, construction costs and other. The empirical investigation of relationships will give evidence about prevailing theories and a priori arguments. It will be based on what happened in the past, will identify systematic influences and will use it to explain the relationship and forecast as we will discuss. These systematic influences are best uncovered by statistical techniques, as the human brain cannot work out exact systematic relationships. We will address a range of questions, for example, if interest rates rise by a percentage point, what is the expected impact on mortgage rates and eventually house prices? Such questions can be answered by regression analysis. We discuss the diagnostics tests to run to confirm the validity of the model. This background is used to extend the analysis to study relationships with more advanced tools.

1.2.4 Forecasting

Finally, we recognise that a definitive goal in the quantitative analysis of real estate is to make predictions and forecasts about the market. Explicitly addressing real estate forecasting is necessitated by the growing need for forecasts in the industry. The book exposes the reader to a variety of econometric models, forecasting techniques and assessment procedures that can help the researcher carry out forecasting work effectively. In the forecast section of the book, we highlight the limitations both of quantitative (or model-based) and qualitative (or judgemental forecasting) and discuss how they can be combined. We also discuss how forecasting takes place in practice and offer our views about the process.

1.3   Book online resource

The online resource for this book contains accompanying notes to chapters with further examples. It also contains chapters on additional topics. Most of the statistical procedures in this book can be carried out in Excel. In the online resource we illustrate the procedures in the econometric package EViews, a software package common in the real estate industry. Datafiles are posted so that the reader can replicate the analysis and practice EViews and Excel.

2 Real estate data

2.1 Introduction

This chapter provides an overview of real estate data. Before dipping into the statistical analysis of data, it is appropriate to describe the main data series available to study and analyse the real estate market. It contains definitions of data, brief explanations of how key measures are calculated, sources and a number of graphical illustrations. We present the raw data. Data transformation is the subject of subsequent chapters. In data analysis and the model building process, the particular features of real estate data should be well understood so that quantitative analysis is not conducted in a vacuum or with ignorance of the meaning of real estate data. Appropriate interpretations can then be made.

The chapter is intended to be a primer for real estate data. It introduces data and data sources in the real estate market to both undergraduate and postgraduate students previously unfamiliar with real estate. An explicit coverage of real estate data is valuable for their studies, pursuit of research topics and triggering the use of additional data not reported in this chapter. A further objective is to illustrate the plurality of data to a broader audience from other fields (e.g. other real asset classes) who would like to monitor and study the real estate market or perform analytics with data from this market. This chapter is a short version of a fuller data document available in the online resource of this book.

In order to provide structure to this chapter, we break down the real estate market into its main components: user or occupier, development (building construction) and investment sub-markets or segments. Distinctive data describe trends in each of these submarkets. Data may be produced by different sources and the method of obtaining/compiling data and constructing metrics is not identical in each segment. To complicate things further, different series exist describing the same aspect of the real estate market. For example, there are different data series for rents.

We briefly discuss key data in the direct or private market (also known as the underlying market) and the unlisted market. An introduction to real estate investment trusts (REITs) and debt market data is contained in the book's online resource for this chapter.

2.2 Segments of the real estate market

A simple framework of the real estate market proposed by Keogh (1994) is useful for our purposes (also see Tsolacos *et al.*, 1998). It breaks down the market into three major parts: the user or occupier market, the investment market and the development market (Figure 2.1). The segments of the markets are also analysed in DiPasquale and Wheaton (1992), Ball *et al.* (1998), Jowsey (2011, ch. 17) and Pirounakis (2013). Each division in Figure 2.1 is

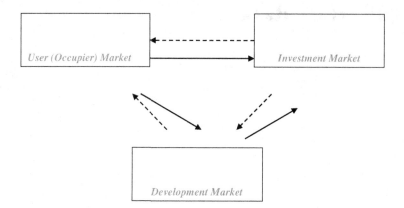

Figure 2.1 Segments of the real estate market

characterised by distinctive datasets – distinctive in terms of me thods of collection, sources and, quite importantly, properties of the data themselves.

2.2.1   User/occupier market

In this segment we discuss metrics and data relating to the tenant/leasing market. The key measures in this sub-market are demand, vacancy and rents.

2.2.2   Development market

The development market encompasses new building development as well as major refurbishments. Data on current and future building development, both under construction and planned, are closely watched. The development market supplies new space to the user market (for owner-occupation and leasing) and the investment market.

2.2.3   Investment market

In the investment market, capital values and yields are determined by demand and supply of investment interests in real estate. The investment market reflects current and forecast trends in the occupier market, development segment and the economy. It also incorporates wider asset market influences (e.g. the bond market or other real asset markets) and liquidity (investable capital in real estate).

2.3   User/occupier market data

In this segment we discuss metrics and data relating to the tenant/leasing market.

2.3.1   Demand data

Data reporting on levels of demand for space indicate the occupational requirements of firms, how much office, retail, industrial or other space they need, where, and of what type/specification. This section focuses on measures which quantify demand for business space only. There are a number of specific demand measures.

(i)   *Take up*

This is the most common reported demand series. Data are mostly collected from the agency teams of real estate service providers. However, there are real estate information and data firms collecting take up and other demand data. According to BNP Paribas, *take up* in Europe represents the total floor space known to have been let or pre-let, sold or pre-sold to tenants or owner-occupiers during the survey period, which is usually a quarter or a year. In other countries like Australia, lease renewals might be included in take-up figures.

Take up is categorised by area and different level of market aggregation (e.g. markets and submarkets). Office submarkets can be Central London or the City of London, in-town or out of town locations. For retail sub-categories are property types (high street stores, retail parks, department stores, etc.). Definitions of take up may differ.

- **See online note #2.1 on take up measurement**

Take up does not necessarily represent new demand; it could just reflect relocations within the boundaries of the market (firms churning space). An occupier moving from one building into another building of the same size counts as take up.

Data on take up is pretty accurate, although area definitions and other definitions are not universally accepted amongst agents and other data providers. The data are subject to some revisions, especially quarterly data, as some transactions can be recorded with a delay. The series can also be erratic. Consider a leasing agreement planned to be signed in mid-September (i.e. quarter 3), but for some non-economic or market reason it is signed in early October (quarter 4). Hence this transaction will be recorded in the fourth quarter instead of the third quarter. Also note that in smaller markets a few big leasing deals can have a huge impact on take-up levels in a particular time.

Take-up data can be quite informative and are broken down by size, type of business or occupier, and specification of buildings. Take up is reported in square meters or square feet, although for local market purposes a local measure of area can also be used (e.g. *ping* in Taiwan or *tsubo* in Japan).

To illustrate take up with real numbers, in the third quarter of 2018 office take up in Central London was 3.74 million square feet (sq ft) according to Knight Frank. This is approximately 347,454 square metres (sq m). The conversion factor is 1 square metre = 10.764 square feet. In the City of London, where take up was 1.67 million sq ft, the TMT (technology, media, telecoms) sector accounted for 30%, followed by the financial sector (21%) and the flexible/serviced office sector 17%. Therefore we get a pretty good idea about demand trends by sector.

Take up cannot be negative. If there are no new lettings, take up will be zero. But what happens when, say, a firm vacates a building in one area and moves to a different market? Take up is not negative, but net absorption will be impacted. We come back to net absorption more fully next.

Figure 2.2 shows take up data for the logistics sector in the European Union. We observe that European logistics take up has followed an upward general trend. This reflects an expanding logistics sector and growing demand for space. As the logistics market becomes larger, there is a corresponding higher level of take up as more firms enter the market and

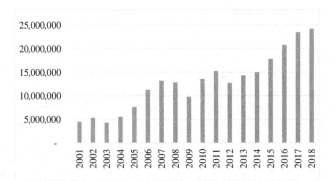

Figure 2.2 European logistics take up
Source: JLL.

churning space. Take up is a flow variable as it is not cumulative of past values and it is measured over an interval of time (usually quarter or year).

(ii)   *Net absorption*

Net absorption is the second main demand measure popular in academic research and in practice, even though across Europe and Asia data on this is often limited. This measure of
demand is different from take up. A naïve example illustrates this. There are four equally
sized buildings in a market and one building is vacant, whereby all three buildings are occupied by a different occupier. Suppose now that one occupier moves to the vacant building,
leaving behind a vacant building. Take up is one building. Net absorption is zero. The occupied stock remains three buildings. There is no new demand, just firms relocating within the
market.

Net absorption is the change in occupied stock between two points in time. Assume that at
the end of December 2018 total stock in a small office market is 600,000 sq m, out of which
550,000 sq m is occupied. A quarter later (end of March 2019) we observe that the occupied
stock increases to 580,000 sq m. Net absorption is 580,000 sq m − 550,000 sq m = 30,000 sq
m over the quarter. Note that stock may have risen to 650,000 sq m due to new supply. Often
markets see space added over time, but it does not matter in the estimate of net absorption;
we just focus on the occupied stock, which in our example increased to 580,000 sq m.

For clarity and the purpose of understanding, net absorption takes into account an expansion or decrease of space that existing firms occupy, new firms moving into the market (new
fresh demand) and existing firms in the market moving out of the market (leakage). According to CoStar Group, 'Net absorption is defined as the net change in occupied space over a
given period of time, calculated by summing up all the positive changes in occupancy (move
ins) and subtracting all the negative changes in occupancy (move outs)'. Net absorption can
be negative, for example a firm leaves the market at a time when no firm leases more space
or enters the market.

Since net absorption represents new demand, it is an appealing metric to gauge demand
conditions in the market. It is a useful metric for investors and real estate analysts. Net
absorption can be estimated or observed directly.

Table 2.1 Net absorption calculations (City of London offices)

	3Q2017	4Q2017	1Q2018	2Q2018	3Q2018
Vacancy	6.6%	7.1%	6.9%	6.9%	6.6%
Occupied stock	93.4%	92.9%	93.1%	93.1%	93.4%
Stock (sq m)	8,659,000	8,691,750	8,724,500	8,757,250	8,790,000
Occupied stock (sq m)	8,087,506	8,074,636	8,122,510	8,153,000	8,209,860
Net absorption (sq m)		−12,870	47,874	30,490	56,860
Take up (sq m) Knight Frank		216,462	196,953	171,869	155,147

Net absorption can be worked out from the vacancy (rate) and the stock that will establish the level (and rate) of occupied stock. We discuss vacancy and occupied stock in sections 2.3.2 and 2.3.3. Table 2.1 illustrates an indirect calculation estimate.

In this example there are two variables we need data on: vacancy and stock in the market. Vacancy data are commonly reported by real estate service firms and should be easy to get. Firms also have estimates for stock. From the vacancy rate we can get an estimate of the occupied stock. In 3Q2018 vacancy was 6.6%, and hence the occupied stock was 93.4% in the City of London office market. We now work out the occupied stock in square meters. In 3Q2018 93.4% of 8,790,000 sq m was occupied. The occupied stock was therefore 8,209,860 sq m in 3Q2018. By taking the absolute differences of the occupied stock between quarters we obtain net absorption. For comparison purposes we also report the take-up figures from Knight Frank. Net absorption figures will nearly always be lower than take-up figures. In 4Q2017 net absorption was negative whereas take up was positive. It means that take up was the result of firms moving around. In 3Q2018 just over a third of take up was net absorption and new demand in the City of London. In the first three quarters of 2018 the City of London office market saw positive net absorption, hence additional space was demanded and leased.

- **See online note #2.2 on further calculations of net absorption and take up**

(iii) Active demand

Active demand or active requirements is a measure of potential demand and it is certainly worth monitoring as it carries advance signals about demand conditions in a market. It refers to firms (potential occupiers) which have registered a requirement to lease space and wish to satisfy in the foreseeable future. It therefore records the number of enquiries. Agency teams of real estate firms have a good grasp of these declared enquiries. As in the case of the two demand measures discussed earlier, active demand databases contain information about sectors, size, property characteristics and other. Active demand is measured in sq m (or sq ft). This metric is a precursor of take up and net absorption, and in our view it is a useful series not only to gauge trends in the occupier market but also of the general business environment.

(iv) Employment to floor space ratio

In local markets, in particular office markets, the employment/floor space ratio (also known as the floor space utilization ratio or employment density), is useful in assessing demand trends and expected occupied space. More specifically it is the average floor space per full-time equivalent member of staff. The occupied stock is the product of the level of employment multiplied by the amount of space each employee uses. Hence if the equivalent of full-time employment is 60,000 employees in an office market and each employee on average utilises 15 sq m, the occupied stock is $60,000 \times 15 = 900,000$ sq m. If firms plan to increase employment by 5,000, the extra space required will be $5,000 \times 15 = 75,000$ sq m. This assumes that there is no shadow vacant space (defined in section 2.3.3) in the buildings. It represents net lettable area. The net lettable area is the net usable internal area for the purpose of the business. Definitions of net and gross internal areas of buildings can be found in RICS (2015).

Employment densities differ by sector. For example, in offices net lettable space per employee could be around 12 sq m, in call centres the density is about 8 sq m per employee, in data centres it is 47 sq m and in warehouses 90 sq m). Densities differ by type of jobs too. Managers require more space than junior staff. A detailed analysis with examples including the impact of hot desking and energy efficiency can be found in Home and Communities Agency (2015, 2010). The main method of generating employment density data is surveys (see Home and Communities Agency, 2010, p. 13). Information about employment needs, type of jobs and associated floor space/employee ratios have provided the basis to predict demand. An early and pioneering study is that of Kelly (1983) in the Manhattan office market.

2.3.2   Stock and supply data

(i)   Stock or inventory

Stock or inventory is the total accommodation (office, retail, industrial, other) in the market. It is the size of the market expressed in sq ft or sq m. New building development adds to stock whereas the demolition of buildings reduces the stock. In some definitions obsolete buildings that have no prospects of being occupied are also removed from the registers. Change of use also affects stock statistics. If an office building is converted into a residential block it will be subtracted from the office stock statistics. The space is added to the residential stock. A building which undergoes refurbishment will usually be removed from the statistics and re-enter when it is completed or is close to completion and marketed for occupation. Stock figures in a particular market may differ by data source. Such discrepancies are not uncommon and usually relate to unstandardised definitions and data compilation method.

- **See online note #2.3 on stock data**

(ii)   Occupied stock and occupancy rate

This is the proportion of the stock that is occupied. If the stock of distribution warehouses is 800,00 sq m and 720,000 sq mt is occupied, the occupancy rate is 720,000 sq m/800,000 sq m = 90%. That is 90% of the warehouse stock is occupied.

(iii) Supply data

NEW CONSTRUCTION AND REFURBISHED SPACE

New construction or development represents newly built structures or extensively refurbished buildings ('major refurbishments'). New construction is a flow series. It records newly built or refurbished space placed in the market. For example, a building of 3,000 sq m was completed and put onto the market in the first quarter whilst 6,000 sq m of refurbished space enters the market in the following quarter.

DEVELOPMENT PIPELINE

This metric has two components. First, *under construction*, which comprises buildings that are in the process of being constructed or extensively refurbished (to rent or for owner occupation). Data on space *under construction* gives a good proxy of new space coming on the market. Second, it includes schemes which have received planning permission and have the potential to be built in the future. The future could be the next number of years until the outstanding planning consent expires (say 5 years). Not all of the schemes that have the potential (and planning permission) to be built are actually developed.

PRE-LET AND SPECULATIVE DEVELOPMENT

Space under construction or building undergoing refurbishment could have or not have secured a tenant ahead of completion. If a firm has agreed to lease space when the building is completed, we say that part of the building is *pre-let*. This provides security to developers (and lenders). Development undertaken without pre-securing a tenant is called *speculative* development. Developers expecting good leasing conditions (a strong occupier market) might decide to consider achieving actual lettings only after completion.

BUILDING STARTS

Data on building starts refer to the value or volume of buildings for which development commences. It is a lead series of how much space will be completed. There are methodological issues with series (see Ball and Tsolacos, 2002 for a relevant discussion in the UK). At the local level, real estate firms possess accurate information and data for building works about to start. At the aggregate level, new orders represent a good series to proxy the position on the building construction cycle. It provides an aggregate picture of the building cycle whilst ironing out local market development spikes.

SUPPLY FROM PRE-EXISTING BUILDINGS

The discussion on the supply of commercial space took place in the context of new construction. Supply has a second component, that of supply of space for owner occupation or leasing from existing buildings. It is the result of lease terminations, subletting, tenants downsizing accommodation requirements, space previously hoarded by property owners and bankruptcies.

2.3.3 *Vacancy and rents*

In the previous sections we discussed data that measure the level of demand and supply of space in the real estate market. As in elementary economics, demand and supply will determine prices, i.e. rental levels in the occupier market. The real estate market is more complicated, however. It is not just the price (rent) that is determined in the occupier market by demand and supply, but also the availability of space and vacancy. Vacancy and rents are significant measures reflecting market conditions. They are part of the fundamentals of the market.

(i)   Vacancy/availability

VACANCY

Vacancy represents the total floor space (part of buildings or the whole building) which is physically vacant and ready for occupation over the next few months (for instance in the next 3 months) at the time of the survey. At any point in time it represents the volume of space that is empty and ready for occupation. As a definition it sounds similar to supply. The difference is that it incorporates both new and existing space supplied and the level of actual vacancy in the market also reflects the strength of demand. Vacancy is as useful metric to monitor as vacancy swings through time and variation among property sectors and markets are the result of demand/supply balance. There are a few variants of the definitions of vacancy.

• **See online note #2.4 on vacancy measurement and data**

VACANCY RATE

Vacancy rate is the total vacant floor space including sublettings divided by the total stock at the survey date. Hence is it expressed as a percentage. A 6% vacancy in an office market means that 6% of the total stock is vacant. The remaining 94% of the stock is occupied, hence the occupancy rate we defined earlier.

AVAILABILITY

Availability refers to all space available for (1) immediate occupation, which means the space is physically vacant and (2) space under construction that will be completed within a certain period of time (usually 6 to 12 months) and is not let. In some definitions availability may include space which is currently occupied but the tenant will vacate it within 3 or 6 months. Since this space will become available it is being marketed. Availability is mostly reported in physical units (sq m or sq ft).

SHADOW VACANCY

A useful statistic extensively monitored in the US market is shadow vacancy. This refers to space that is leased but not utilised. This could be the result of firms restructuring their accommodation needs or layoffs in difficult business conditions. In such a case firms will have spare accommodation which they might hoard rather than market it. Consider now what happens when business conditions improve. Since the firm has a high level of shadow vacancy it will not take up new space (absorption) but it will utilise the vacant space kept idle. The importance of this statistic is clear, but given the nature of this type of vacancy it is difficult to get accurate data.

FINANCIAL VACANCY

This is a metric reported in the MSCI database and requires familiarization with the rent definition presented in the next section. The *financial vacancy rate* is total market rental value in vacant units divided by total market rental value. To give the reader an idea of actual financial vacancy data, according to MSCI figures the financial vacancy in the UK was 7% for all property, ranging from 2.9% in hotels to 12.6% in offices as per mid-2018.

NATURAL VACANCY RATE

This is a metric mostly used in real estate research (as in our empirical analysis later in this book). The US real estate literature theorises that at any point in time, some of the stock will be vacant for a number of reasons. First, space takes time to be let. Also, landlords might deliberately keep space vacant in rising markets in expectation of higher rents. Buildings which were vacated due to a tenant defaulting or a lease termination may be undergoing some minor refurbishment. Or there may be a mismatch between what tenants look for and what is available. A review of the theory of the natural vacancy rate can be found in Wheaton (1990), Clapp (1993), and FRSBF (2001), among many studies primarily in the US.

(ii)   Rents

Firms pay rent for space they lease. The description of rent data sounds like a straightforward task. This is not the case, however. What might be confusing for students and new analysts is the existence of several definitions of rents in the commercial market. Before proceeding it is appropriate to make some preliminary points to provide context for the discussion of rent definitions:

(a)   Rents are reported in different ways in international markets – in local currency and area metrics. In the UK rents are reported in GBP per sq ft per annum, similarly in the US in USD per sq ft per annum. In Germany rents are in EUR per sq m per month, in Japan in JPY per tsubo per month, in China in RMB per sq m per month.

(b)   Rent data are obtained from valuations or transactions. In the former case, the rent is estimated from an external valuer at the time of the valuation. The frequency of the valuations for institutionally owned real estate tend to be monthly, quarterly or annual. In the case of transaction rents the rent is obtained from an actual lease transaction. The transaction-based rent is the market determined rent. Traditionally appraisal-based rents have been the most common measure for rents internationally. But databases for transaction-based rent data are now available and are developing further.

(c)   When an occupier rents space, the lease will typically be in force for a number of years depending on the geography and local market practices. It could be for 1, 3, 5, 9, 10, 15 years or longer. A term of the lease is the period between rent reviews. Although a lease may expire in 15 years, rent reviews might be agreed, say, every 3 years. The rent will be reviewed with reference to open market rent (see following definition), inflation or other according to the terms of the lease at the end of each 3-year (or other) period.

(d)   When an occupier leases space, the landlord may grant lease incentives including rent concessions in the form of a rent-free period (no rent paid for, say, 12 months on a 10-year lease) or a reduced (stepped) rent. Suppose a firm takes up a 5-year lease and is granted a 6-month rent-free period. The firm will start paying the full rent in the seventh month. We elaborate further on this point later under the 'Headline Rent' section.

(e)   The real estate market is made up both of modern high-quality buildings and older premises. Depending on the market maturity, rent data can be available for all categories of buildings in each property sector (e.g. prime, Grade A, B or C buildings).

ASKING OR FACE RENT

This is the rent asked by landlords for space marketed for lease (e.g. USD 60 per sq ft per annum).

ESTIMATED RENTAL VALUE AND OPEN MARKET RENT VALUE

Estimated rental value (ERV) or the open market rent value (OMRV) is the external valuers' opinion of the open market rent. It is also known as the *achievable* rent. This is the rent at which the premises might reasonably be expected to let at in the open market, on the date of the valuation or at the rent review date according to the terms of the hypothetical lease. That is the ERV or OMRV is the rent estimated by valuers that the property would fetch if it was to be rented now. ERV is more commonly used in the market.

PASSING OR CONTRACT RENT

The rent that an occupier will pay will be negotiated between the owner and the tenant. The exact terms of the lease, the size of letting and the strength of the covenant also have an impact. *Passing* or *contract* rent is the actual rent payable by the tenant as stipulated in the lease agreement. Contract rent could differ from the asking rent, especially during periods of market downturns. As the contract rent can be fixed for a number of years (depending on lease terms), it can diverge from the open market rent. Passing rent can be higher or lower than the ERV or market rent. It tends to be lower than the open market rent at periods of rising rents (the property is *under-rented*) and higher at periods of weak market conditions and falling rents (*over-rented* property).

REVERSIONARY RENT

Following from the previous definition, when the passing rent is different to the estimated rental value, a *rent reversion* arises. The rent is anticipated to increase or decrease to the estimated rental value on rent reviews, new lettings of vacant space or lease renewal. Investors generally want to buy a property offering income growth possibilities, hence assets with reversionary rent potential are appealing.

HEADLINE RENT

Headline rent is the gross rent that is paid under a lease. The gross rent becomes payable after the end of any rent-free or reduced rent period and after all tenant incentives have expired. Rent free periods are most common in the office market: an occupier can take on a 10-year lease with the landlord agreeing to give rent free period for 1 or even 2 years. The length of the rent-free period or other concessions depends primarily on market conditions (in a buoyant occupier market, incentives fall) and the length of lease. The *headline rent* could be higher than what the property would have actually achieved at the time of signing the lease. This practice keeps the rent high at periods of soft market conditions such as recessions (headline rent steady but incentives rise). *Headline rents* usually form the benchmark for any 'upward only' rent review in the lease.

PRIME RENT

Prime rent is the highest rent that could be achieved for a typical building/unit of the highest quality and specification in the best location to a tenant with a strong covenant (that is a firm with low risk of going bankrupt and defaulting on rents). The prime rent is likely to exclude an extreme rent paid in a letting deal which is not considered representative of the market. The prime rent reported may not change for some time if it is due to a particular (but not extreme) deal. Landlords can keep the *headline prime rent* steady but increase incentives.

NET EFFECTIVE RENT

Net effective rent is the rent paid that takes into account any rent-free periods or other tenant incentives. If a firm takes up a 7-year lease and gets a rent-free period of 1 year, the firm will pay the contracted rent for 6, not 7, years. Dividing the total contractual rent paid for 6 years by 84 months (7 years), we get the average rent per month. It is termed effective to denote that this is the actual receivable rent by the landlord/payable by the tenant. Incentives are treated as a cost to rent. If a lease incorporates reviews every, say, 3 years, the incentives will be spread over the sub-period to the first rent review.

We give a simple example to illustrate the calculation of the net effective rent in online note #2.5 (see also RICS, 2013).

• **Online note #2.5: Net effective rent calculation**

NET OPERATING INCOME

The meaning of net operating income (NOI) is how much money a real estate asset generates after all operating expenses are accounted for. NOI is equal to the property's gross income less total expenses for the period at consideration (month, quarter, year). Hence,

NOI = Gross operating income (GOI) − operating expenses

Gross operating income is the rental income and any other income (e.g. renting out part of the car park of the building or renting space for an event). Rents represent a primary component of net operating income.

Operating expenses, which differ from one-off expenses when the building was acquired, include void costs such as advertising costs, re-letting costs, unrecoverable service charges (repairs and maintenance, property management fees, insurance, property taxes, business rates, utilities). Depending on the terms of the lease some of these costs can be borne by the tenants, e.g. insurance. The reader is advised to consult valuation textbooks for a discussion on lease arrangements (e.g. Wyatt, 2013). The reader will familiarise with the full repairing and insuring (FRI) lease terms, which is a typical lease in the UK, and other arrangements such as the triple-net (NNN) lease.

2.4   Investment market

2.4.1   Investment market context

A growing and more sophisticated real estate investment market is supported by ample performance measures and data series. As in the case of the occupier market, this section does not offer a complete guide to real estate investment data. We focus on key measures that are most commonly reported in practice and are essential in the study of this segment of the market. In order to provide context and a workable structure to the discussion, we categorise the data with respect to the so-called four quadrant investment approach. The four quadrant framework illustrates the main ways investors can take exposure to the real estate market (see CBRE, 2019; Rees and Wood, 2007). We present two versions of the four quadrant approach:

Table 2.2 Four quadrant framework I

Private (equity investments)	*Public or securitised market (equity investments)*
– Direct: investments in buildings	– REITs: real estate investment trusts (equity)
– Indirect: unlisted funds	– Listed property firms
Private debt (debt investments)	*Public debt (debt investments)*
– Bank loans	– Commercial mortgage-backed securities (CMBS)
– Mortgage trusts	– Corporate bonds
– Mezzanine debt	– Mortgage REITs

Table 2.3 Four quadrant framework II

Private or direct market (equity investments)	*Unlisted market (equity investments)*
– Direct investments in buildings	– Unlisted funds
	– Other equity vehicles
Public or securitised market (equity investments)	*Debt (private and public)*
– REITs: real estate investment trusts (equity)	– Commercial mortgage-backed securities (CMBS)
– Listed property firms	– Corporate bonds
	– Mortgage REITs
	– Bank loans
	– Mortgage trusts
	– Mezzanine debt

The difference between the two quadrant investment frameworks is the maturity of the debt market. In Europe and most of Asia for example, the real estate debt market is not well developed for investment purposes and hence emphasis is given to the direct and indirect/unlisted market. The framework in Table 2.2 is typical of the US where there is long tradition of private and securitised debt vehicles and data.

The discussion of investment market data is structured along these two frameworks: Direct (private), direct (unlisted), public equity (REITs), public debt (CMBS) and private debt. Data in the private and unlisted segments of the market are presented in this chapter, with the online resource introducing key data series in the other segments.

2.4.2   *Direct market performance indicators and data*

The direct investment market quadrant is the reference point for all other quadrants and segments. Performance in this segment is firmly linked to the performance of the direct real estate investment market.

(i)   Income return

Income return refers to net income receivable over the value of the previous period calculated net of all irrecoverable costs incurred by the investor. MSCI define income return (IR) at time t as:

$$IR_t = \frac{NI_t}{CV_{t-1} + CAPEXP_t} \times 100 \qquad (2.1)$$

where NI is the net income, which is the rent received by the investor net of property management costs, ground rent, collection costs associated with rent in arrears and other expenses that depend on the terms of the tenant lease contract. NI in the MSCI definition is a corresponding term for the net operating income we presented in the previous section. CV is the estimated *capital value* at the end of period $t - 1$ or beginning of period t (hence it is a valuation-based figure), $CAPEXP_t$ is *capital expenditure* incurred during time t (the quarter or year). Income return is calculated as a percentage (e.g. income return 5.3% in 2018). Income return data are also presented in the form of an index. Income return is the corresponding term for *direct return, net rental* (or income) *yield* or *running yield*.

It is straightforward to compute an income return. Using the following information, the income return on a property asset for 2018 (end of 2018) is:

- Property value £12,000,000 as of end of December 2017
- Net income (or NOI) in 2018: £750,000
- Capital expenditure in the course of 2018: £160,000

$$IR_t = \frac{750000}{12000000 + 160000} \times 100 = 6.17\%$$

(ii)   Capital value or price

Capital value or price is the value of the property asset as estimated in the regular valuations. Price data based on market transactions are compiled by RCA, CoStar Group and NCREIF. Real estate service firms also have such databases but in most cases data are not compiled as systematically as by the aforementioned firms.

(iii)   Capital value growth or appreciation return

Capital value (CV) growth measures the change in the value (price) of the building asset over a given period of time (month, quarter, annual) similar to a share. The calculation is net of any capital expenditure and receipts over the period. Hence the change in the market value of the building is:

$$CV\ Growth_t = \frac{CV_t - CV_{t-1}}{CV_{t-1}} \times 100 \qquad (2.2)$$

The online version of the chapter contains the definitions used by MSCI and NCREIF.

(iv)   Total return

Total return is calculated from adding up income and capital value growth. Figure 2.3 plots total returns and its components, using a subset of NCREIF data for a randomly chosen market and sector.

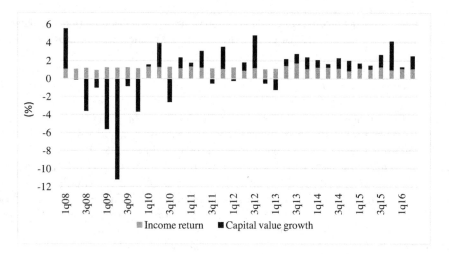

Figure 2.3 Components of total returns

Source: NCREIF.

Total returns in Figure 2.3 is the sum of the two bars representing income return and *capital growth* or *appreciation returns*. In the example from NCREIF we observe that income is a steady proportion of total returns. Capital values is the volatile component of total returns. In the second quarter of 2009, capital growth in this sample fell by 11% whereas income return remained steady. This is a representative picture for the components of total returns in any sector and any market. The series that fluctuates considerably is capital growth, with total return swings reflecting the variability in capital growth but mitigated by income returns. The stability of income returns is on the other hand one of the attractions in real estate investments. Remember, if the market fluctuates and prices experience a period of falls, the investor can weather this volatility if the tenants continue to pay the rent (assuming that there is no need for a forced sale).

VALUATION AND TRANSACTIONS CAPITAL VALUE DATA

Historically data on capital values were mainly compiled from regular building valu-ations. The valuation process is still the main source for capital value indices as well as income return and total return indices. As in the case of rents, the investment mar-ket has seen the development of transaction-based capital value databases and indices. Actual market transactions feed into the construction of these indices. Appropriate sta-tistical techniques utilizing repeat sales also apply to obtain transaction-based indices. A detailed discussion is beyond the objectives of this chapter. The interested reader can consult the methodologies used by Real Capital Analytics (RCA) and the CoStar Group. As an example, in Figure 2.4 we plot capital values for retail and industrials constructed from transactions in the Asia Pacific region compiled by RCA. We present aggregate data. RCA maintain disaggregated data too. In the particular example and for the pur-pose of comparison we present the data expressed in a common denominator: USD per sq ft per annum.

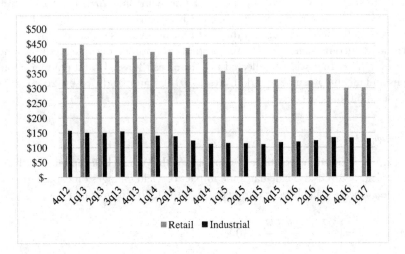

Figure 2.4 Transaction-based capital values in Asia-Pacific

Source: Real Capital Analytics.

Note: USD per sq ft per annum.

(v) Net initial yield

The *net initial yield* (NIY) is the net income received in the first year divided by the purchase price plus purchase costs. It is an important metric, as it is the yield accepted by the investor at acquisition. Assume an investor purchases a property which is let at £2,000,000 per year. If the purchase price is £35,000,000, the gross initial yield is:

$$\frac{2,000,000}{35,000,000} \times 100 = 5.7 \left(\text{or } 5.7\%\right).$$

Assume now that after deducting costs (ground rent, service charges and other property expenses), the net income is £1,900,000 per year. The investor has also incurred acquisition costs of £1,500,000 that need to be added to the purchase price. The net initial yield is:

$$\frac{1,900,000}{36,500,000} \times 100 = 5.2 \left(\text{or } 5.2\%\right).$$

There is a close relationship between the net initial yield with income return or the *running yield*. Both income return and running yield are calculated at any time during the investment period. The net initial yield reflects a potential market participant's assessment of risk reflected in the bid price for the property asset. Two investors will pay a different price for an asset if they have diverse views on rent prospects and risks attached to this asset.

(vi) Equivalent yield

The *equivalent yield* is an overall or *all-risk* yield that is used to capitalise the current and future cash flows after each rent review. More precisely, Wyatt (2013) defines it as the *overall yield* that can be used to capitalise both current and reversionary income. *Reversionary income* relates to the situation in which the rent passing is below ERV or market rent and it is likely to revert to market rent in the future, usually at the next rent review. It is essentially the internal rate of return of an income that grows. If the property is not fully occupied and becomes fully occupied over time, that will be reflected in future cash flows. Useful reading about yield definitions and meaning can be found in Wyatt (2013), Baum and Crosby (2014) and Brown and Matysiak (2000).

(vii)   Yield impact data

In Figure 2.5 we showed the volatility of capital growth. We can link capital growth volatility to shifts in yields. MSCI also calculates the impact of yield changes on capital growth. The so-called *yield impact* is calculated monthly through the following formula:

$$yield\ impact_t = \left[\frac{Syield_t - Eyield_t}{Eyield_t}\right] \times 100 \tag{2.3}$$

where *Syield* and *Eyield* are the yield at the start and end of the month, respectively. The formula is set up in a way to denote the negative effect of a rise in yield on capital growth and vice versa. In the following example, the negative sign denotes the negative impact on capital growth (i.e. prices fall). In Figure 2.5 we illustrate the relationship graphically.

$$yield\ impact_t = \left[\frac{5.2 - 5.5}{5.5}\right] \times 100 = -5.45$$

We illustrate the yield impact on capital growth by plotting annual capital growth and the equivalent yield (EY) change from mid-2002 through mid-2018. We observe a nearly perfect negative relationship. At times when the yield change is negative (yields fall), capital growth is positive and vice versa. The equivalent yield change is not identical to MSCI's yield impact data, but still it illustrates the point of the nearly perfect inverse association

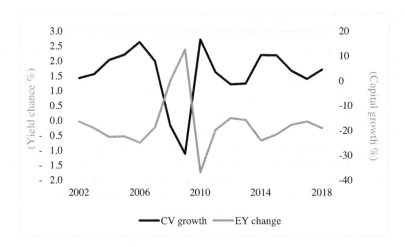

Figure 2.5 Capital growth and yield impact

Source: MSCI, Authors.

between capital values and yield changes. It follows that if we predict yield changes we have a view about expected capital values, or if we predict capital values we can make inferences about yield shifts.

(viii)   Cap rate (capitalization rate)

The *cap rate* is a measure analogous to net initial yield and its calculation is in principle similar to that for the net initial yield. In the net initial yield the net income largely reflects the current net rent. But the net rent may not be equal to the market rent or ERV. It could be lower. Hence in the initial yield formula the numerator is contractual rent (net of costs). That means rent based on existing leases, which may not be at market rental level. In the cap rate definition, the rent is the market rent or ERV. It follows that when an investor purchases a building which is let at market value, the cap rate will be similar to the net initial yield. In other versions of the cap rate it is defined as the NOI divided by the current value of the property asset.

(ix)   Valuation and transaction data for yields

Yields and cap rates can either be valuation- or transaction-based. For the latter, see datasets compiled by data and information providers such as RCA and CoStar Group.

2.5   Indirect investment – property funds

2.5.1   Background

Investors can invest in real estate indirectly via unlisted property funds. Investors need to study trends in the real estate market to decide what type of funds to invest in (in terms of sector, geography, style – core, value add, opportunistic or other). In addition, they require information about the characteristics of real estate funds. Associations of unlisted funds such as INREV in Europe with a sister organization in Asia (ANREV), NAREIM and NCREIF in the US and AREF in the UK are sources of fund level data. The construction of metrics for fund performance and risks are based on fund-level data provided by fund managers. From these data useful aggregate indices for indirect investments are constructed. This section outlines key metrics that can be found in these databases. The discussion in the next section draws much upon INREV's database (see INREV, 2018).

2.5.2   Fund performance metrics and data

Total return, capital value growth and income return

Similarly to the direct market, data on total return and the constituent components capital growth and income return are available in the unlisted indirect market. The formulas for these performance metrics are somewhat different from those used in the direct market. Key elements are the net asset value (NAV), contributions, redemptions and distributions. Net asset value is the market value of all company's assets (not limited to its properties) less all its liabilities and obligations. Net asset values are based on the valuations of the underlying assets. Investor contributions to a fund are implemented via capital calls. Capital calls (also termed drawdowns) denote raising funds from investors or limited partners to execute

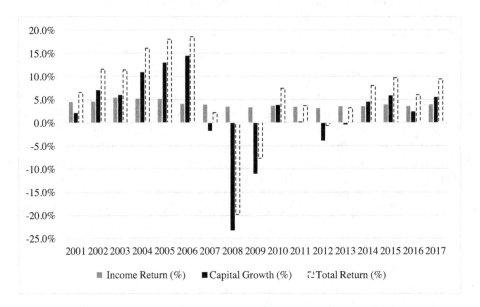

Figure 2.6 Total return components series for European unlisted property funds

Source: INREV.

an investment strategy. The commitment of the investor is not necessarily funded at once upon admission but through successive capital contributions. Distributions are the income generated from investments distributed to investors, typically monthly or quarterly. It is a key component of income returns. Finally, redemptions occur when assets are liquidated and the principal (the amount invested) is returned to investors. These fund performance data form the basis for assessing trends in the indirect market.

Figure 2.6 presents fund performance based on INREV data. As in the case with direct market, we observe the stable income return component and a volatile capital growth constituent feeding into the volatility of total returns.

Performance – yields

The Association of Real Estate Funds (AREF) defines yield as the gross yield (%), which is the gross of tax, net of expenses distribution for the last 12 months expressed as a percentage of the latest (end of year) net asset value.

IRR (internal rate of return)

INREV constructs an IRR annual index to measure the performance of the fund (closed end) since fund inception. The IRR index methodology is described in INREV (2018).

2.5.3   Other fund and investment data

In addition to performance indicators, a variety of other data is collected. For example INREV registers data for vehicle structure (open ended or closed end), target region and/or country, target sector (multi/single sector), gearing as a percentage of gross asset value (GAV) and loan information to highlight a few of the fields for data analysis. NCREIF's fund level data include performance metrics for a number of fund structures such as open ended, closed end value add, timberland and the daily priced fund. The daily priced fund index is based on a small pool of open-ended funds that are priced daily.

2.5.4 *Volume of transactions*

An area that has seen significant data improvements is that of commercial real estate investment (and increasingly residential), both domestic and international or cross border. Investors are now presented with robust statistics that has contributed to making the investment market transparent. Cushman and Wakefield's Money Into Property and MSCI's data on capital expenditure (the sum of money spent on purchases of new properties, expenditure on development and other capital expenditure), net investment and turnover (the sum of capital expenditure and receipts) have been sources of monitoring the volume of investment transactions. Property services firms have also been running and evolving their databases on real estate transactions. The databases have become more detailed with the arrival of firms such as RCA (Real Capital Analytics) and CoStar Group, with a particular focus on capital flows and transactions.

The data span across the main sectors office, industrial, retail, apartment, mixed use and alternatives. They can be broken down geographically, by source and type of investor. Definitions by source might differ. For example RCA captures sales of properties and portfolios of USD 2.5 million or greater in the US and USD 10 million or greater outside the US. JLL would record transactions over USD 5 million. The databases may further differ by the items included, e.g. alternative sectors, land transactions, property company mergers and acquisitions (M&A), REIT formations and so forth. The reader is urged to visit the websites for RCA, CoStar Group or real estate service firms for background to capital transaction data and data trends. Figure 2.7 plots the volume of transactions in Asia.

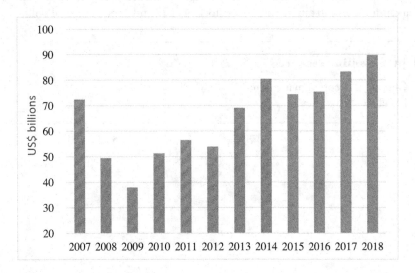

Figure 2.7 Volume of transactions in Asia-Pacific (offices)
Source: RCA.

The availability of capital transactions data in real estate has triggered significant research interest examining the impact of internationalisation on global capital flows (see Lizieri and Pain, 2014) and in particular the determinants of cross-border capital flows (Fuerst *et al.*, 2015; McAllister and Nanda, 2016a). Factors affecting domestic capital transactions are studied by Devaney *et al.* (2017), with Mauck and Price (2017) distinguishing between the contributing factors to domestic and international real estate investment. More detailed questions are asked in this research such what assets cross-border investors focus on and whether cross border investors exhibit different preferences to domestic investors (see Devaney *et al.*, 2019). The key question of capital flows and pricing is considered in a number of studies (Oikarinen and Falkenbach, 2017; McAllister and Nanda, 2015, 2016b).

- **See online note #2.6 for an introduction to data in the REIT sector and private and public debt markets**

2.6   Final remarks

The objective of this chapter is to give a flavour of the data we use and analyse in the real estate market. As an introduction, the chapter along with the online resources focus on a subset of real estate data which by no means covers the full inventory of data in each part of the market. We presented data and metrics that are prevalent in the study of the real estate market in practice and customary in empirical quantitative research.

The chapter is essential for the reader with no or little prior knowledge of real estate data in particular of the private or direct market. Some of the metrics presented are complex, and one will need time to take in the different types of real estate data. Initially a real estate analyst or student should understand the meaning of the data and metrics presented. Working with data will improve knowledge.

The production of real estate data continues and quality improves through larger samples and systematic collection. The presence of real estate data and information intelligence firms has further boosted the breadth and quality of data. The corollaries of this evolution are a requirement to work and analyse a greater stock of data and good analytical skills.

Chapter 2 online resource

- Chapter 2 accompanying notes
- Excel file: "ch2_excel"

---

3 Data, common manipulations and descriptive statistics in analysis

3.1 Introduction

This chapter has two purposes. First, it outlines the main representations of information and the usual manipulations required in preparing data for undertaking empirical research. Second, it introduces the statistical concepts underlying the standard calculations used to summarise and identify important features of a data set.

Section 2 of this chapter reviews the common representations of information through values, variables, constants, parameters and equations. Section 3 considers different data structures in which information are held. Section 4 briefly reviews the mathematical operations, rules and symbols contained in formulae, which are instructions on how to execute a calculation. Section 5 examines the important concepts involved in the construction of indices (indexes), a common method of recording and grouping heterogeneous information together, and the interpretation of their values. Section 6 emphasises the importance of understanding the data being analysed, including the recognition that there may be a requirement to prepare the data for analysis. It takes the reader through a simple example outlining the construction of a commercial property return index and percentage returns as representations of the commercial property asset market. Next, it proceeds to illustrate the preparation of total return information for performing an analysis of past investment performance. We explain the reasons for adjusting information recorded or derived from monetary values for changes in general prices and show how to deflate such variables in levels and rates.

The next part of this chapter, sections 7 and 8, introduce the basic statistical calculations and the concepts underlying them in the descriptive analysis of data. The analytical framework in applied statistical analysis is the distribution, where data is organised and then summarised to highlight its important features. Descriptive statistics is a term used to describe calculations characterising a distribution, including measures such as the arithmetic mean, median, mode, standard deviation, semi-interquartile range, range, skewness and kurtosis. We additionally consider the geometric mean, which on occasion has the advantage of being more representative than the arithmetic mean. The final section illustrates the application of descriptive statistics by investigating the past investment performance of three assets: commercial property, equities and bonds.

3.2 Data representations

3.2.1 Values, variables, constants, parameters and equations

It is convenient to use variables as a representation of a container in which the information in the form of values are stored. Latin alphabet letters are typically used to depict variables, the two most common being the letters X and Y. Let us assume we have data about three recent house prices sales in Brook Street: £250,000, £300,000 and £350,000. Instead of listing the values, we can define X to represent information of the sale prices of houses in Brook Street. Since the values vary, X is referred to as a variable. Variables are usually represented by capital letters, such as X, and the values contained within them as lowercase letters, such as x.

A constant is a special type of variable in which the values are constant. Constants are often used to facilitate calculations rather than storing information. For example, regressions applying the ordinary least squares estimator requires a constant to ensure that an estimated impact is calculated correctly.

An equation is a statement written in numbers and symbols on either side of an equality or inequality sign. Equations can be used to depict relationships between variables. Unknowns (unknown values) in equations can be solved.

A parameter is a number attached to a variable to capture its direction and magnitude in influencing the values of another variable. The relationship between variables is represented by an equation. The description of an equation is often communicated by specifying variables as being on the left-hand and right-hand side of a sign. As a simple example:

$$Y = 2 + 3X$$

The equation comprises a constant 2 and a parameter 3. The values in variable Y are determined positively by the values in variable X. Let us assume that one of the values of X is 3. At that value, the value of Y will be:

$$Y = 2 + 3(3) = 11$$

An equation informs us of the relationship between values in variables. There may be more than one variable determining the values of a variable. For example, let us assume that values in the variable Z also determine values in Y. The relationship is now represented as:

$$Y = 2 + 3X + 5Z$$

Equations and variables enable us to communicate information and relationships between entities succinctly, as it means that all the values involved in a calculation and its output do not have to be listed separately.

3.2.2   *Variable notation*

Subscripts and superscripts may be attached to variables. A subscript can be used to denote an element of the data. For example, let us assume that we wish to store information of house prices in a region over time. It is common practice to use the letter t as the reference to a point in time. A variable can be defined to represent house prices in each time period in that region, say Y_t. If we specify an explicit time period for t, say the year 2000, then the value of the variable represented by $Y_{t=2000}$ is the house price in that year. The extension to depicting a relationship between variables, for example, house prices and household incomes, denoted by the letters Y and X respectively, in a region at each point in time is straightforward:

$$Y_t = 2 + 3X_t$$

Equally, the house prices of multiple regions may be represented by a subscript. As it is common practice to use the letter i, the variable would be Y_i, where i refers to a region. $Y_{i=London}$ and $Y_{i=Southeast}$ contain information of the house prices in London and the southeast of England, respectively.

Rather than separately listing values for each region in each time period, appropriate notation can be attached to a variable as a representation, say Y_{it}. For example, the relationship between house prices and household incomes in each region over time may be represented as:

$$Y_{it} = 2 + 3X_{it}$$

When using notation in this way, we should explicitly define the variables and the subscripts attached to them.

3.2.3 Qualitative and quantitative variables

There are two basic types of variables: qualitative and quantitative. A qualitative variable is nonnumeric and the information that it contains can only be classified into distinct categories. The categories are unordered (not ranked). Examples of qualitative data include dwelling type (detached houses, semi-detached house, terraced houses and apartments) or type of commercial property (offices, shops, warehouses, hotels). A qualitative variable can be converted into a quantitative variable usually by counting, for example reporting the number of houses and flats in a postcode.

On the other hand, quantitative variables are numerical in nature. Quantitative variables are either discrete or continuous. Discrete variables can assume only certain values and are characterised by gaps between the values, for example, the number of bedrooms in a house (1, 2, 3, etc.). A house can have 3 or 4 bedrooms, but it cannot have 2.3 bedrooms. Discrete variable values are usually obtained from counting.

A continuous variable contains values within a specified interval. Examples include house prices, rents and yields. For example, the initial yield could be 5.4%, 5.36% or 5.358%. Typically, continuous variable values are obtained from measuring.

3.3   Data structures

3.3.1 Time-series data

Time-series data refers to information about a subject recorded at regular intervals in time. A time series of office prices in London shows how they vary through time. The time series can be recorded at varying frequencies (periods): daily, monthly, quarterly or annually. Statistical analysis using time-series data relies on variation in values over time in calculations. Time-series data is useful for capturing long-term trends and cycles.

3.3.2 Cross-sectional data

Cross-sectional data refers to information recorded about multiple subjects (individuals, firms, markets, countries and other entities) at a specified point in time. The key distinction is that there is no time dimension to the data. An example of a cross-section could be rents recorded in 40 office markets (locations) in the US in 2016. Statistical analysis using cross-section data relies on variation in values over the cross-section in calculations, such as location.

3.3.3 Pooled cross-sectional data

A pooled cross-section refers to a repeated number of cross-sectional data compiled over time. The values in this data structure change over time and vary through the cross-sections. Information from different subjects or different samples are recorded in each period. For example, the prices of apartments recorded each month over a year in an administrative area such as the London Borough of Richmond upon Thames will mainly comprise of different apartments, as high search and transactions costs make it unlikely the same apartment would be sold multiple times in such a short period. Statistical analysis using this type of data benefits from the variation in values over time and the cross-section.

3.3.4 Panel/longitudinal data

Panel or longitudinal data has a unique data structure in that the values record the measurement of the same unit at several points in time. Unlike pooled cross-sections, panel data are collected from sampling the same subjects repeatedly over time. A database for property fund performance could include information for 30 funds over a time horizon, say from 2000 to 2010. Panel data can be balanced, containing information on the same number of subjects over time; or unbalanced, where some subjects are measured more than others due to non-response (missing information) or attrition (disappearing) from a sample. Panel data are described in terms of the length of the time (T) and size of the cross-section (N) dimensions. Macro-panels refer to panel data having a large T and small N structure, while micro-panels refer to the panel data having small T and large N dimensions. Panel data have a number of advantages over repeated cross-sectional data. They allow for the measurement of within-sample change over time, enable the measurement of the duration of events (e.g. how long an individual resides in a house) and record the timing of various events.

3.4 Mathematical symbols, operations and rules

We briefly list and review the operations and operators required for understanding the content in our book. Some of these operations will be applied more regularly than others.

3.4.1 Mathematical symbols

There are standard mathematical symbols to represent mathematical operations and to communicate the relative magnitudes of values. Table 3.1 provides a list along with a succinct explanation of what they represent.

In addition to the operations listed, there are additional operators which are commonly employed and these are considered next.

3.4.2 Summation operator (Σ)

Sigma (Σ) is commonly used to instruct an addition operation.

Suppose: $x_1 = 5$, $x_2 = 6$, $x_3 = 4$, $x_4 = -5$

Table 3.1 Common mathematical symbols

Symbol	Name	Representation	Example
=	Equals sign	equality	$7 = 2 + 5$ 7 is equal to $2 + 5$
≠	Not equal sign	inequality	$6 \neq 5$ 6 is not equal to 5
≈	Approximately equal	approximation	$x \approx y$ means x is approximately equal to y
>	Strict inequality	greater than	$6 > 4$ 6 is greater than 4
<	Strict inequality	less than	$6 < 7$ 6 is less than 7
≥	Weak inequality	greater than or equal to	$4 \geq 3$, $x \geq y$ means x is greater than or equal to y
≤	Weak inequality	less than or equal to	$3 \leq 5$, $x \leq y$ means x is less than or equal to y
( )	parentheses	calculate expression inside first	$2 \times (3 + 5) = 16$
[ ]	brackets	calculate expression inside first	$[(1 + 2) \times (1 + 5)] = 18$
+	plus sign	addition	$4 + 5 = 9$
−	minus sign	subtraction	$3 - 2 = 1$
±	plus – minus	both plus and minus operations	$3 \pm 4 = 7$ or -1
*	asterisk	multiplication	$2 * 5 = 10$
×	times sign	multiplication	$2 \times 5 = 10$
·	multiplication dot	multiplication	$2 \cdot 5 = 10$
÷	division sign/obelus	division	$6 \div 3 = 2$
/	division slash	division	$6/3 = 2$
a^b	power	exponent	$2^3 = 8$
$a{\wedge}b$	caret	exponent	$2 \wedge 3 = 8$
$\sqrt{a}$	square root	$\sqrt{a}.\sqrt{a} = a$	$\sqrt{9} = \pm 3$
$\sqrt[3]{a}$	cube root	$\sqrt{a}.\sqrt{a}.\sqrt{a} = a$	$\sqrt[3]{8} = 2$
$\sqrt[n]{a}$	nth root (radical)	for $n = 4$	$\sqrt[4]{256} = 4$

Adding all the values together:

$$x_1 + x_2 + x_3 + x_4 = 5 + 6 + 4 - 5 = 10$$

We can use the summation operator to convey this addition operation more succinctly:

$$\sum_{i=1}^{4} x_i = 10$$

where i designates the element required in the operation, for example, $i = 1$: $x_i = x_1 = 5$.

The numbers designating the elements at the bottom and top of the summation operator inform the reader of the first and final elements in a series of numbers (values) which should be added together. However, when it is clear which values should be used, this part of the instruction is redundant, and the instruction is conveyed simply as $\sum x$.

3.4.3   Product operator (Π)

The product operator expresses a long multiplication sequence in a compact manner. Suppose $x_1 = 2$, $x_2 = 3$, and $x_3 = 5$. The definition of the product operator is:

$$\prod_{i=1}^{3} x_i = x_1 \times x_2 \times x_3 = 2 \times 3 \times 5 = 30$$

Similarly, to the summation operator, the instruction is conveyed as $\prod x$ when it is clear which values should be multiplied together.

3.4.4   Lag operator (L)

Due to the nature of the real estate market, adjustments to economic and financial shocks such as a change in interest rates are characterised by lagged effects over time. Table 3.2 illustrates the concept of lags in a time-series using CoStar's office price index for New York. We will explain what an index represents shortly. For now, treat it as a variable.

Column (ii) reports the original data at time t and columns (iii) and (iv) the data lagged by one and two periods respectively, represented by variables OFI, OFI_{t-1} and OFI_{t-2}. OFI_t is the variable containing the contemporaneous values of the office price index, the reported values pertinent to quarter for the year. The one-period lagged variable, OFI_{t-1}, which in this case is a one-quarter lag, contains the previous quarter's values of this variable. For example, the second quarter 2018 value (2018Q2) of OFI_{t-1} is 165.2, the first quarter 2018 (2018Q1) value of OFI_t. As the example illustrates, it is possible to construct variables with multiple lags, for example, OFI_{t-2}.

A formal way to denote a lagged series is to use the lag symbol or lag operator L. Consider a time-series variable X_t. Then:

$$LX_t = X_{t-1} \tag{3.1}$$

$$L^2 X_t = X_{t-2} \left(L(LX_t) = L(X_{t-1}) \right) \tag{3.2}$$

Table 3.2 Lags in data

(i)	(ii)	(iii)	(iv)
		Office price index (OFI)	
	t	$t-1$	$t-2$
	OFI_t	$OFIt{-}1$ $(LOFI_t)$	$OFIt{-}2$ $(L2OFI_t)$
2017Q4	164.6	–	–
2018Q1	165.2	164.6	–
2018Q2	166.1	165.2	166.1
2018Q3	166.1	166.1	166.1
2018Q4	165.7	166.1	165.7

and in general

$$L^s X_t = X_{t-s} \tag{3.3}$$

In the preceding example $LOFI_t = OFI_{t-1}$ and $L^2 OFI_t = OFI_{t-2}$.

3.4.5 *Laws of mathematical operations*

Formulae are sets of instructions on how to execute a calculation. Very often formulae require calculations involving several mathematical operations. There are mathematical laws governing the sequence of operations in a formula. The laws are:

Rule 1: Simplify all operations inside parentheses.
Rule 2: Simplify all exponents, working from left to right.
Rule 3: Perform all multiplications and divisions, working from left to right.
Rule 4: Perform all additions and subtractions, working from left to right.

Let us assume that we wish to execute the following calculation:

$$2^2 + 3 \times 6/2 - (3 - 1)$$

The rules inform us that we should proceed in the following manner:

(i)   Execute the calculation in brackets

$$2^2 + 3 \times 6/2 - 2$$

(ii) Execute the exponent calculation

$$4 + 3 \times 6/2 - 2$$

(iii) Execute either the division or multiplication operation

$$4 + 3 \times 3 - 2$$

(iv) Execute the remaining division or multiplication operation

$$4 + 9 - 2$$

(v)   Execute the addition or subtraction operation

$$13 - 2$$

(vi) Execute the remaining addition or subtraction operation.

$$11$$

Now let us examine the application of these laws of operation in the calculation of a sample variance formula:

$$\sum \frac{(x - \bar{x})^2}{n - 1}$$

The laws of mathematical operations inform us we should execute the calculation in brackets first $(x - \bar{x})$, take its square $(x - \bar{x})^2$ and then divide each $(x - \bar{x})^2$ term by $n - 1$ before adding all the values together.

3.4.6 Logarithms

Logarithms have special properties which are exploited in calculations and in econometric modelling. A logarithm is the power to which a base number must be raised to be equal to a stated number. If the stated number is 8 and the base is 2, then the power to which 2 should be raised to equal 8 is 3: $2^3 = 8$. Using logarithmic notation, this is represented by $\log_2 8 = 3$, since $2^3 = 8$, where *log* represents the logarithm. The main base number used in logarithmic operations is 10 or the mathematical constant e ($e \approx 2.71828$), the natural exponent. The natural exponent e is special in financial mathematics, as it is used in calculations involving continuous compounding or discounting. When the base is the natural exponent e, it is referred to as the natural logarithm (ln):

$\ln(Y) = X$ because $e^X = Y$.

Any positive (non-zero and non-negative) number can be converted into a logarithmic measurement unit. For example, $\log_{10} 5 = 0.699$ and $\ln(5) = 1.609$. The branch of statistics considered in this book employs natural logarithms. The main properties of natural logarithms are summarised in Table 3.3:

Table 3.3 Properties of natural logarithms

	Natural logarithm
Product	$\ln(XY) = \ln X + \ln Y$
Division	$\ln\left(\dfrac{X}{Y}\right) = \ln Y - \ln X$
Power	$\ln(X^n) = n\ln(X)$
Inverse	$\ln\left(\dfrac{1}{X}\right) = -\ln(X)$
Unity	$\ln(1) = 0$

We now outline two additional properties of natural logarithms to those listed in Table 3.3.

A first difference refers to the subtraction of consecutive values of a variable. First difference calculations are usually undertaken in time-series rather than cross-sectional data. Taking first differences in logarithms is approximately equal to a relative change (or a growth rate in decimals).

$$\ln(X_t) - \ln(X_{t-1}) \approx \frac{X_t - X_{t-1}}{X_{t-1}} \tag{3.4}$$

The first difference in logarithms is only approximately equal to a relative change, as this method diverges from a relative change when the differences in consecutive values of a

Table 3.4 Percentage and logarithmic changes in office prices

Year	Office Price Index	Percentage growth (%)	Natural logarithm of price index	Logarithmic change*
2014	100.0		$\ln(100.0) = 4.605$	
2015	102.5	$\left(\dfrac{(102.5-100.0)}{100.0}\right)100 = 2.5$	4.630	$4.630 - 4.605 = 0.025$
2016	104.0	1.46	4.644	$4.644 - 4.630 = 0.014$
2017	109.5	5.29	4.696	$4.696 - 4.644 = 0.052$
2018	122.0	11.42	4.804	$4.804 - 4.696 = 0.108$
Cumulative growth:		0.220		$4.804 - 4.605 = 0.199$

*logarithmic change: $= \ln\left(\dfrac{X_t}{X_{t-1}}\right) = \ln\left(X_t - X_{t-1}\right)$

time-series variable become larger. Table 3.4 shows some differences in estimating house price growth and compounded returns using the percentage (precise) and logarithm (approximation) methods.

The first column displays the year, the second the office price index, the third the percentage growth of the office price index, the fourth the natural logarithmic value of the office index, and the final column the first difference in the logarithmic values of the office index. The calculations reveal that the smaller the change in the office index values, the more accurate the logarithmic first difference is as an approximation to a relative change – the smaller change in value between 2015 and 2014 compared to a large change in value between 2017 and 2018.

Logarithmic differences in values over longer time horizons or time intervals (e.g. day) within time periods (e.g. year) can be used to derive the continuously compounded return or cumulative growth that occurred. Consider the discrete method for calculating a cumulative change in office values using rates is:

$$\prod (1+r_t) = (1+r_1)(1+r_2)...(1+r_T) \tag{3.5}$$

The cumulative change can be converted into a cumulative growth rate or compounded rate by subtracting the answer by 1. In Table 3.4, $(1 + 0.025)(1 + 0.0146)(1 + 0.01529)$ $(1 + 0.1142) = 1.220$, yielding a compounded return equal to 22.6% ($1.220 - 1 = 0.22$ or 22.6%).

If the office price at the beginning of the period is X_0 and the price at the end of period T is X_T, the cumulative growth or compounded return is: $\ln(X_T) - \ln(X_0) = \ln(122.0) - \ln(100.0) = 0.199$.

The difference in values reflects the fact that the logarithmic method computes the continuously compounded return rather a return compounded discretely.

Another useful feature of converting values into logarithmic units concerns the transformation of non-linear relationships between variables into linear relationships. Figure 3.1 depicts the relationship between the total return of US REITs with time in index and logarithmic units.

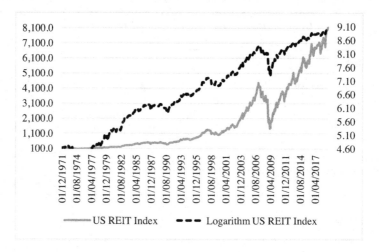

Figure 3.1 US REIT total return index

The graph displays a non-linear relationship between the US REIT index and time, initially rising slowly at the beginning of the period and then becoming steeper, reflecting the effects of compounding due to the assumption of achieved total returns being reinvested in the construction of the index. The graph illustrates a significant cycle in the period of the global financial crisis, followed by a steep trend with more fluctuations (volatility) compared to the past. The logarithmic transformation of the total return index values converts this relationship into a linear trend. The cycle reflecting the global financial crisis is still evident but other cycles are also more discernible now due to the feature of logarithmic changes in values being much smaller than index unit changes. Logarithms can be used to reduce the distortions caused by outliers (extreme values) in a series and are often used to investigate different functional forms between relationships of variables.

3.5   Indices/indexes

An index is a useful way of recording, describing and graphing information. The plural for an index is either indices or indexes.

3.5.1   *Index numbers*

An index expresses a value of a subject relative to a reference value. The values of an index are called index numbers or index units. Indices are commonly used to record time-series information, where the reference value refers to the value of the base period. Examples of time-series indices include the Consumer Price Index (CPI), FTSE All Share index and the MSCI real estate indices. Alternatively, indices can be constructed from cross-sectional data to express the value relative to a value of another element in the same time period. For example, the Index of Multiple Deprivation (IMD) is the official measure of relative deprivation for small areas in England for a particular year, such as 2015. The values for each location are expressed relative to England as a whole.

Table 3.5 Prices of items and expenditure

Item	Price (pence)	Price (pence)	Price (pence)	Expenditure (£)	Expenditure (£)	Expenditure (£)
	2016	2017	2018	2016	2017	2018
Bread	75	89	89	5,000	5,550	5,550
Butter (250 g)	120	126	126	8,000	7,900	7,900
Milk (1L)	100	96	96	9,500	9,900	12,500

Index numbers are additionally useful for summarising (grouping together) information about many (heterogeneous) items, capturing compounding effects and identifying (the relative magnitude) trends and cycles. In this textbook the focus is on indexes reporting time-series information, as this is the type of index a reader is likely to encounter most often. There are different ways to construct an index, depending on the objective. Broadly, they can be categorised as:

- Simple
- Unweighted aggregate
- Weighted aggregate.

To illustrate the important concepts, let us consider the construction of various price indices using information of prices and expenditure on three items displayed in the Table 3.5.

3.5.2 *Constructing a simple index*

A simple index contains numbers used to measure the relative change in just one item (subject), such as bread. Each index number is a ratio of two values multiplied by 100. The denominator value is fixed as it is the reference value representing the base period. The base period index value will always be 100, as it is formed by the value of a number divided by itself multiplied by 100.

Let us examine the formula used to construct a simple index. The generic formula is:

$$I_t = \frac{value_t}{value_0} \times 100 \tag{3.6}$$

$Value_0$ denotes the generic representation of the base period.

Let $p_{bread,t}$ represents the value of the contemporaneous price of bread and $p_{bread,0}$ the price of bread in the base period. The simple index I_t is constructed by:

$$I_t = \frac{p_{bread,t}}{p_{bread,0}} \times 100 \tag{3.7}$$

For example, the simple index value for 2017 adopting 2016 as the base year is calculated using the price of bread for 2016 and 2017:

$$I_{2017} = \frac{89}{75} \times 100 = 118.67$$

The index informs us that the price of bread is 1.18 times its value in 2016 or increased by 18.67% $((118.67 - 100.0)/100.0 \times 100)$.

Alternatively, it is possible to construct an index with a different base year, for example, an index with 2018 as the base period. The simple index value for bread in 2016 is:

$$I_{2016} = \frac{75}{89} \times 100 = 84.27$$

The only difference in input concerns the value for the base period. When an index number is less than 100, the base period value, it signifies that the value of the numerator is less than the denominator. In other words, the price of bread in 2016 was lower than 2018. The index value 84.27 indicates that it is 15.73% lower relative to its price in 2018.

3.5.3   Constructing an unweighted aggregate index

Additionally, an index is a convenient method to record a change in the values of a diverse group of items. The Consumer Price Index (CPI), for example, comprises of a basket of items including cars, shoes, milk, eggs, etc., where prices are expressed in different units such as pounds per kilogram, pounds per litre and so on. Converting the prices of diverse goods and services into an index enables the government and other bodies to be informed about changes to overall cost of living. An unweighted aggregate index can be constructed by:

$$I_t = \frac{\sum value_{it}}{\sum value_{io}} \times 100 \tag{3.8}$$

where the subscript i refers to an item, t the time and $t = 0$ the base period.

The summation in equation (3.8) occurs within a time period. The convention is to use the letter P rather than I to depict a price index and the items to be represented by the small letter p. Let p_{it} represent the price of an item i at time t. The formula for the unweighted index is:

$$P_t = \frac{\sum p_{it}}{\sum p_{io}} \times 100 \tag{3.9}$$

Table 3.5 reports that the prices of bread and butter have increased while the price of milk has fallen. We can construct an aggregate price index to find out what has happened to prices overall, or equivalently the cost of purchasing this basket of goods. There are three items in the table: bread, butter and milk. All we do is add the prices of each item together for each time period. The unweighted aggregate index value for 2016 for a price index with a base year in 2016 will be 100.0, as the numerator and denominator take the same values.

For 2017, however:

$$P_{2017} = \frac{(89+126+96)}{(75+120+100)} \times 100 = 105.4$$

Thus, the cost of living has increased as overall prices have risen by 5.4% since the base period, 2017.

As the prices for each item in 2018 remained unchanged from 2017, the price index value in 2018 is the same as in 2017.

$$P_{2018} = \frac{(89+126+96)}{(75+120+100)} \times 100 = 105.4$$

The values of an unweighted aggregate index are influenced by the units of measurement of its constituent items. Furthermore, an unweighted aggregate index does not take into consideration the relative importance of each item purchased. Modern aggregate indexes are usually weighted.

3.5.4   Constructing a weighted aggregate index

There are two important concepts underlying the construction of a weighted aggregate index, the Laspeyres and the Paasche methods. The Laspeyres method uses **base-period** weights while the Paasche method uses **current-period** weights.

(i)   Laspeyres price index

The *Laspeyres price index*, developed by Etienne Laspeyres in the latter part of the 18th century, is a weighted price index with base-period quantities as weights. Since then the method has evolved to employing other instruments as weights, such as the expenditure share of an item. The key feature of this method is that the values of the weights are determined by the base period. The generic formula for a Laspeyres price index is:

$$P_t^L = \frac{\sum w_{i0} p_{it}}{\sum w_{i0} p_{io}} \times 100 \tag{3.10}$$

where w_{i0} represents the value of the weights in the base period 0 for item i.

To construct the expenditure weight for an item, we calculate the total expenditure and then divide the expenditure of an item by it. In the case of a Laspeyres index, the base period values are used. If the base period is designated to be 2016, the weights for each item are:

$$\text{Let bread be item 1: } w_{1,0} = \frac{£5000}{£5000 + £8000 + £9500} = 0.22$$

$$\text{Let butter be item 2: } w_{2,0} = \frac{£8000}{£5000 + £8000 + £9500} = 0.36$$

$$\text{Let milk be item 3: } w_{3,0} = \frac{£9500}{£5000 + £8000 + £9500} = 0.42$$

In decimal form, the sum of the weights is equal to 1 $(0.22 + 0.36 + 0.42)$.

The Laspeyres price index value for the base period, in this case 2016, is equal to 100 as the numerator and the denominator take the same values. For 2017:

$$P_{2017}^L = \frac{\sum w_{i0} p_{i2017}}{\sum w_{i0} p_{i2017}} \times 100$$

$$P_{2017}^L = \frac{(0.22 * 89) + (0.36 * 126) + (0.42 * 96)}{(0.22 * 75) + (0.36 * 120) + (0.42 * 100)} \times 100 = 103.5$$

According to the Laspeyres price index, the price for this basket of goods rose by 3.5% since 2017. The Laspeyres price index reports a lower rise in prices compared to the unweighted aggregate price index, as it takes into consideration the relative importance of each item in its contributing to the overall cost of the basket of goods.

The Laspeyres price index value for 2018 is the same as 2017, as prices of each item have remained unchanged and the weights have fixed values. For 2018:

$$P_{2018}^{L} = \frac{\sum w_{i0} p_{i2018}}{\sum w_{i0} p_{i2018}} \times 100$$

$$P_{2018}^{L} = \frac{(0.22 \times 89) + (0.36 \times 126) + (0.42 \times 96)}{(0.22 \times 75) + (0.36 \times 120) + (0.42 \times 100)} \times 100 = 103.5$$

(ii) Paasche price index

The main disadvantage of the Laspeyres method lies in its assumption that the base-period weights are a realistic depiction of the relative importance of the items in each period. For example, it would assume that the relative importance of mobile phones in 1989 is the same as in 2018 or that the relative importance of buying video cassette recorders (VCRs) is the same in 2018 as in 1989.

Hermann Paasche proposed an alternative index. Instead of employing base-period weights, a Paasche index employs current-period weights, which in effect update the relative importance of items constituting an aggregate index.

The generic formula for a Paasche price index is:

$$P_t^P = \frac{\sum w_{it} p_{it}}{\sum w_{it} p_{i0}} \times 100 \tag{3.11}$$

One disadvantage is that it requires weights to be constructed for each item in each time period.

For 2016 the weights are:

$$\text{Bread, item 1: } w_{1,2016} = \frac{£5000}{£5000 + £8000 + £9500} = 0.22$$

$$\text{Butter, item 2: } w_{2,2016} = \frac{£8000}{£5000 + £8000 + £9500} = 0.36$$

$$\text{Milk, item 3: } w_{3,2016} = \frac{£9500}{£5000 + £8000 + £9500} = 0.42$$

For 2017 the weights are:

$$\text{Bread, item 1: } w_{1,2017} = \frac{£5550}{£5550 + £7900 + £9900} = 0.24$$

$$\text{Butter, item 2: } w_{2,2017} = \frac{£7900}{£5550 + £7900 + £9900} = 0.34$$

$$\text{Milk, item 3: } w_{3,2017} = \frac{£9900}{£5550 + £7900 + £9900} = 0.42$$

The 2017 expenditure share weights reveal that bread has become a more important purchase and consequently a more important item in contributing to a value in the index, while butter has declined and milk has remained unchanged in importance since 2016.

In order to illustrate a feature of the Paasche price index, we assume prices in 2018 were the same as 2017. Expenditure on bread and butter also stayed the same, but expenditure on milk rose substantially.

For 2018 the weights are:

$$\text{Bread, item 1: } w_{1,2018} = \frac{£5500}{£5500+£7900+£12500} = 0.21$$

$$\text{Butter, item 2: } w_{2,2018} = \frac{£7900}{£5500+£7900+£12500} = 0.31$$

$$\text{Milk, item 3: } w_{3,2018} = \frac{£12500}{£5500+£7900+£12500} = 0.48$$

Milk has thus become a much more important purchase compared to bread and butter.

The Paasche price index value for the base period 2016 is equal to 100 as the numerator and the denominator take the same values.

For 2017 the Paasche price index value is:

$$P_{2017}^{P} = \frac{\sum w_{i,2017}\, p_{i,2017}}{\sum w_{i,2017}\, p_{i,2017}} \times 100$$

$$P_{2017}^{P} = \frac{(0.24\times89)+(0.34\times126)+(0.42\times96)}{(0.24\times75)+(0.34\times120)+(0.42\times100)} \times 100 = 103.7$$

The Paasche price index indicates that overall prices rose by 3.7% in 2017.

Unlike the Laspeyres price index, there is a requirement to calculate a value for the Paasche price index in 2018 because the weights have changed. For 2018:

$$P_{2018}^{P} = \frac{\sum w_{i,2018}\, p_{i,2018}}{\sum w_{i,2018}\, p_{i,2018}} \times 100$$

$$P_{2018}^{P} = \frac{(0.21\times89)+(0.31\times126)+(0.48\times96)}{(0.21\times75)+(0.31\times120)+(0.48\times100)} \times 100 = 102.9$$

The Paasche price index value in 2018 is 102.9, indicating that overall prices are 2.9% higher compared to 2016 (the base period), but lower compared to 2017 as the index fell from 103.7 to 102.9. Even though prices of each item remained unchanged, the Paasche price index reports that prices in general are lower because of the change in the value of the weights.

The Paasche price index requires weights to be computed for each period. As different weights are used each year, it is impossible to attribute changes in the index to changes in price alone as the difference in index values could reflect changes in weight values. Conversely, the changes in the index values of a Laspeyres price index can be attributed to pure changes in the prices.

(iii)   Fisher price index

Irving Fisher proposed an index called *Fisher's ideal index* to offset these shortcomings. It is the geometric mean of the Laspeyres and Paasche indexes. The geometric mean is a special way of calculating an average value. We explain what a geometric mean is later in this chapter.

The generic formula for the Fisher price index is:

$$P_t^F = \sqrt{P_t^L \times P_t^P} \tag{3.12}$$

Using the calculations made for the Laspeyres and Paasche price indexes, the Fisher price index values are:

$$P_{2017}^F = \sqrt{P_{2017}^L \times P_{2017}^P}$$

$$P_{2017}^F = \sqrt{103.5 \times 103.7} = 103.6$$

$$P_{2018}^F = \sqrt{P_{2018}^L \times P_{2018}^P}$$

$$P_{2018}^F = \sqrt{103.5 \times 102.9} = 103.2$$

The Fisher price index indicates that there has been a marginal fall in overall prices in 2018.

(iv)   Chain-linking in index construction

The modern method of exploiting the desirable features underlying the Laspeyres and Paasche indexes is chain-linking. This involves constructing separate Laspeyres indexes for specified time intervals, where the weights are fixed within an interval but are allowed to vary between intervals. We illustrate the chain-linking concept using as an example monthly indexes representing returns achieved by an asset constructed using the Laspeyres method.

Table 3.6 reports the index values of four separately constructed Laspeyres return indexes. The base periods for each index are displayed in brackets. Each Laspeyres index uses a

Table 3.6 Monthly Laspeyres return indices

Year	Month	Laspeyres Return Index	Laspeyres Return Index	Laspeyres Return Index	Laspeyres Return Index
		(1980 = 100)	*(1981 = 100)*	*(1982 = 100)*	*(1983 = 100)*
1980	12	100.0			
1981	1	99.9			
	2	99.8			
	3	99.5			
	4	99.4			
	5	99.4			
	6	99.3			
	7	99.2			

	8	98.7		
	9	98.5		
	10	98.5		
	11	98.2		
	12	98.0		
1981	12	100.0		
1982	1	98.9		
	2	99.8		
	3	99.5		
	4	99.3		
	5	99.4		
	6	99.8		
	7	100.0		
	8	100.5		
	9	101.0		
	10	101.2		
	11	101.3		
	12	101.5		
1982	12		100.0	
1983	1		100.1	
	2		100.2	
	3		100.4	
	4		100.6	
	5		100.8	
	6		101.0	
	7		101.4	
	8		101.8	
	9		101.9	
	10		101.9	
	11		102.0	
	12		102.5	
1983	12			100.0
1984	1			99.8
	2			100.5
	3			100.9
	4			101.2
	5			101.8
	6			102.1
	7			103.4
	8			103.7
	9			104.5
	10			104.8
	11			104.9
	12			105.0

fixed weight, where the values are determined in the base period. The third column reports the constructed Laspeyres monthly return index for 1981 (January to December 1981), the fourth column the Laspeyres monthly return index for 1982, the fifth column the Laspeyres monthly return index for 1983 and so on. An index value represents the value of the index at the end of the month – for example, $I_{12}^{1981} = 98.0$ represents the 1981 Laspeyres index value at the end of December.

Chain-linking involves utilising the information in the separately constructed Laspeyres indexes to form a single index, thereby allowing weights to be fixed for a certain time interval (in this case a year), but to also vary between these periods (years). In the following example, we demonstrate a simple method to construct a chain-linked annual return index.

Table 3.7 summarises the information reported in Table 3.6.

The December 1981 Laspeyres monthly index value is 98.0, representing the return achieved over the course of that year. The index value for the beginning of the year (start of January) is the value at the end of December 1980, or 100.0. The table also contains similar information for the other Laspeyres monthly return indices. The construction of a chain-linked index requires a base period to be defined. To maximise the information contained in the separate monthly Laspeyres indexes, the end of (December) 1980 or the beginning of January 1981 is chosen to be base period (row 1, column 7). By calculating the relative change in each Laspeyres index, it is possible to update the 1980 chain-linked index value. For example, the relative change over the course of a year in the monthly index value for 1981 is:

$$r_{1981} = \frac{(98.0 - 100.0)}{100.0}(100) = -2.0\%$$

We can replicate the calculation to compute the relative changes for each index and stack them into a single column (column 6). The relative change in the index value between 1981 and 1982 is:

$$r_{1982} = \frac{(101.5 - 100.0)}{100.0}(100) = 1.5\%$$

Having obtained and stacked the relative (annual) changes, we can proceed to update the chain-linked 1980 index value. For 1981, the index value is:

$$I_{1981} = \left(1 + \frac{(-2.0)}{100}\right)I_{1980}$$

Table 3.7 Chain-linked index

Year	Laspeyres Return Index, 1981	Laspeyres Return Index, 1982	Laspeyres Return Index, 1983	Laspeyres Return Index, 1984	Relative Change in Index	Chain-Linked Index
	(1980 = 100)	(1981 = 100)	(1982 = 100)	(1983 = 100)	(%)	(1980 = 100)
1980	100.0					100.0
1981	98.0	100.0			−2.0	98.0
1982		101.5	100.0		1.5	99.5
1983			102.5	100.0	2.5	102.0
1984				105.0	5.0	107.1

Since the index value for 1980 is 100 ($I_{1980} = 100.0$):

$$I_{1981} = \left(1 + \frac{(-2.0)}{100}\right) 100.0 = 98.0$$

The generic formula for updating the chain-linked index is:

$$I_t = (1 + r_t) I_{t-1} \qquad (3.13)$$

For 1982, the index value is:

$$I_{1982} = \left(1 + \frac{(1.5)}{100}\right) 98.0 = 99.5$$

And for 1983:

$$I_{1983} = \left(1 + \frac{(2.5)}{100}\right) 99.5 = 102.0$$

Repeating this process can result in a chain-linked index over a long time horizon. The resulting chain-linked index has weights which are fixed for a year (a time interval) but allowed to vary between years (time intervals).

3.6 Applications: preparing data for analysing investment performance

3.6.1 Measuring investment performance

We shall illustrate the importance of understanding how data are compiled and basic data manipulations by considering an application involving an analysis of past investment performance among three assets: bonds, equities and commercial property. Investment performance is measured by the return achieved (realised) in a time period. A return on an investment measures the gain or loss generated on an investment relative to the amount of money invested (employed). The total return reflects the return on capital (change in the value of the asset) and on income (the income received from holding onto the asset). Returns can be reported as a percentage or as an index. Returns are measured over a fixed time (holding) period to enable a standard assessment in performance.

The capital return (CR_t) represents the relative change in the value of an asset over a holding (time) period:

$$CR_t = \frac{V_t - V_{t-1}}{V_{t-1}} \qquad (3.14)$$

where:
V_t = value of asset at the end of the holding period
V_{t-1} = value of asset at the start of the holding period (end of the previous holding period)

The income return (IR_t) represents all the income received from an asset over a holding period expressed relative to the value of the asset at the start of the holding period (end of the previous holding period):

$$IR_t = \frac{\sum R_t}{V_{t-1}} \tag{3.15}$$

where:

$\sum R_t$ = sum of all income received within a holding period

V_{t-1} = value of asset at the start of the holding period (end of the previous holding period)

The total return (TR_t) is the combined return from change in the asset value and the income received over a holding period, obtained by adding them together:

$$TR_t = CR_t + IR_t \tag{3.16}$$

or equivalently:

$$TR_t = \frac{(V_t - V_{t-1}) + \sum R_t}{V_{t-1}} \tag{3.17}$$

These generic formulae can then be applied to calculating returns for different assets using the relevant inputs.

(i)   Bonds

$\sum R_t$ = sum of coupon payments (interest) received within a holding period
V_t = price of bond at the end of the holding period
V_{t-1} = price of the bond at the start of the holding period

(ii)   Equities

$\sum R_t$ = sum of dividend payments received within a holding period
V_t = price of share at the end of the holding period
V_{t-1} = price of the share at the start of the holding period

(iii)   Commercial property

$\sum R_t$ = sum of rental payments received within a holding period
V_t = valuation of property at the end of the holding period
V_{t-1} = valuation of property at the start of the holding period

A broad understanding of the constituent components of a total return is informative. It reveals that the cash-flow from holding equities as an asset is the least predictable, depending on economic conditions, while bonds are the most predictable due to guaranteed cash-flow (coupon) payments if held until maturity. The cash-flow payments for commercial property is determined by the institutional context; in most countries it would lie between bonds and equities due to cash-flows being guaranteed for a limited period by lease contracts.

3.6.2 Constructing a commercial property return

Table 3.8 provides an example in greater detail of how a monthly commercial property return is constructed. The main figures of interest are the valuation of the building at the beginning and end of the month (December) and the income after costs received within that month. We substituted 'xxx' for numerical figures to focus our discussion on the key points.

The record reveals that this particular property was valued at £11,000,000 (V_{t-1} = £11,000) at the beginning of December (end of November) and £10,000 (V_t = £10,000) at the end of December, and received a net income of £500 ($\sum R_t$ = £500) in that month. Employing equation (3.17), the total return achieved in December 2017 is:

$$TR = \frac{(£10,000 - £11,000) + £500}{£11,000} = -0.045$$

Expressed as a percentage, the total return is −4.5%.

3.6.3 Constructing a portfolio return of commercial properties

Individual commercial property records can be combined to construct a portfolio of commercial properties' returns to represent the investment performance of the commercial property market. We shall illustrate this using a simple example.

Table 3.9 displays the figures representing the numerator and denominator components of a total return for five properties for 5 months, as well as the components of a portfolio total

Table 3.8 A commercial property record

	Inputs (£000)	Calculate (£000)
Valuations:		
Capital Value Dec 31st 2017	10,000	
Capital Value Nov 30th 2017	11,000	
Capital Gain:		−1,000
Revenue Account:		
Gross Rent Receivable	xxx	
Ground Rent Payable	xxx	
Irrecoverable Operating Costs	xxx	
Management Costs & Fees	xxx	
Net Income Receivable:		500
Capital Account:		
Refurbishment of Vacant Units	xxx	
Dilapidations Recovery	xxx	
Other	xxx	
Net Capital Spending:		0
Numerator of Return:		−500
Denominator of Return or Capital Employed:		11,000
Total return (%)		−4.5

Table 3.9 Commercial property portfolio total returns

	Property 1	Property 2	Property 3	Property 4	Property 5	Portfolio (Total)
	Numerators of Total Return — Appreciation + Income £					
Month 1	£1,021	£14,180	£6,310	£1,245	−£2,008	£20,748
Month 2	−£475	£34,919	£6,965	£7,790	£2,818	£52,017
Month 3	£3,026	£77,144	£12,361	£13,104	£5,668	£111,303
Month 4	£1,989	£115,243	£10,927	£5,587	£304	£134,050
Month 5	£2,265	£112,586	£6,815	£4,571	£15,983	£142,220
	Denominators of Total Return — Capital Employed £					
	Property 1	Property 2	Property 3	Property 4	Property 5	Portfolio (Total)
Month 1	£12,490	£487,910	£72,745	£75,878	£37,029	£686,052
Month 2	£11,488	£457,541	£71,418	£66,705	£30,566	£637,718
Month 3	£8,987	£445,928	£70,720	£64,598	£28,486	£618,719
Month 4	£9,982	£474,755	£76,113	£68,858	£29,573	£659,279
Month 5	£9,943	£542,082	£80,268	£66,940	£25,484	£724,717
	Individual Property Total Returns — %					Portfolio Total
	Property 1	Property 2	Property 3	Property 4	Property 5	Return (%)
Month 1	8.2	2.9	8.7	1.6	−5.4	3.0
Month 2	−4.1	7.6	9.8	11.7	9.2	8.2
Month 3	33.7	17.3	17.5	20.3	19.9	18.0
Month 4	19.9	24.3	14.4	8.1	1.0	20.3
Month 5	22.8	20.8	8.5	6.8	62.7	19.6

return comprising of these properties. The resulting percentage returns for each property and the portfolio are reported at the bottom of the table.

Let us recall that the formula for a total return (equation (3.17)) is:

$$TR_t = \frac{(V_t - V_{t-1}) + \sum R_t}{V_{t-1}}$$

The numerator and denominator in the table refers to the numerator, $(V_t - V_{t-1}) + \sum R_t$, and denominator, V_{t-1}.

The numerator component of the portfolio return can be obtained by summing the components of each individual property. For example, the numerator for the portfolio for month 1 is £1,021 + £14,180 + £6,310 + £1,245 − £2,008 = £20,748. Similarly, the denominator for the portfolio for month 1 is £12,490 + £487,910 + £72,745 + £75,878 + £37,029 = £686,052. The portfolio percentage return for month 1 is £20,748 / £686,052 (100) = 3.0%.

Table 3.10 Constructing weights: commercial property portfolio returns

	Weight					
	Property 1	*Property 2*	*Property 3*	*Property 4*	*Property 5*	*Total*
Month 1	0.02	0.71	0.11	0.11	0.05	1.00
Month 2	0.02	0.72	0.11	0.10	0.05	1.00
Month 3	0.01	0.72	0.11	0.10	0.05	1.00
Month 4	0.02	0.72	0.12	0.10	0.04	1.00
Month 5	0.01	0.75	0.11	0.09	0.04	1.00

Constructing the portfolio total return in this manner leads to returns which are value weighted, where the total return contribution of each constituent commercial property is dependent on its capital value relative to the combined capital values of all properties in the portfolio. A more valuable property makes a larger contribution to the computed portfolio total return. To demonstrate this, we can construct the portfolio's total return using a weighted average of the individual commercial property total return. The weight for each commercial property can be calculated by dividing the value of property by the value of the portfolio (values of individual properties added together). They are displayed in the Table 3.10.

The sum of the weights in each month must equal 1. The first row in the table reveals the relative importance of the contribution of each property to the construction of the portfolio return in month 1. Property 2 is the most (0.71) and property 1 (0.02) the least influential contributor to the portfolio (combined) return in month 1. The weight for property 1 is obtained by taking its value at the start of the month (£12,490) and dividing it by the combined value of all properties in the portfolio (£686,052). Similarly, the weight for property 2 is its value (£487,910) divided by the portfolio value (£686,052). Weights therefore reflect the relative value of each property in the portfolio. The portfolio total percentage return is obtained by adding together the product of the weight and total return of individual properties. The portfolio total return for month 1 is:

$$0.02 \ (8.2\%) + 0.71 \ (2.9\%) + 0.11 \ (8.7\%) + 0.11 \ (1.6\%) + 0.05 \ (-5.4\%) = 3.0\%.$$

The portfolio total percentage returns for the other months are constructed in a similar manner using the relevant weights and property returns.

3.6.4 Understanding recorded information

Prior to undertaking any analysis, it is important to make sure you understand the information recorded in a dataset. As mentioned previously, total return data is typically reported in one of two forms, as a percentage return, which is a rate, and as an index, which is a level. Other examples of information recorded as a rate include the inflation rate and economic growth rate, and their level counterparts, the Consumer Price Index (CPI) and the gross domestic product (GDP). Rates can be expressed as decimal or percentage by multiplying the decimal by 100. The choice of which format to record this information usually depends upon the objective of conveying the information, as each format has advantages.

Table 3.11 displays UK annual time-series information of returns in percentage form for commercial property recorded by JLL, equities and bonds.

The same information can be conveyed by presenting the data in levels as return indices (Table 3.12).

Although the values are different, the same information is conveyed in both tables. Before preceding to demonstrate this, we introduce some notation to aid communication. The data

Table 3.11 Nominal percentage returns

Year	Commercial Property Percentage Returns	Equities Percentage Returns	Bonds Percentage Returns
1981	18.6	13.6	1.8
1982	5.9	28.5	51.3
1983	7.2	28.8	15.9
1984	9.7	31.6	6.8
1985	7.5	20.2	11.0
1986	10.4	27.3	11.0
1987	18.3	8.7	16.3
1988	30.0	11.5	9.4
1989	19.2	35.5	5.9
1990	−5.5	−9.6	5.6
1991	−2.6	20.8	18.9
1992	−3.9	19.8	18.4
1993	20.2	27.5	28.8
1994	14.2	−5.9	−11.3
1995	3.6	23.0	19.0
1996	8.1	15.9	7.7
1997	17.3	23.6	19.4
1998	12.0	13.7	25.0
1999	14.1	23.8	−3.5
2000	11.4	−5.9	9.2
2001	8.0	−13.2	1.3
2002	12.5	−22.3	9.8
2003	11.0	20.2	1.6
2004	20.6	12.5	7.2
2005	19.9	21.6	8.4
2006	17.7	16.4	−0.1
2007	−5.6	5.1	5.2
2008	−21.2	−29.8	12.9
2009	5.9	29.0	−1.0
2010	15.2	14.1	9.4
2011	7.6	−3.4	21.4
2012	3.3	12.1	4.8
2013	11.8	20.5	−7.2
2014	18.3	1.2	18.3
2015	13.8	1.1	0.5
2016	4.2	16.4	11.5

in both tables show total returns (income plus capital return) realised by each asset in each year. We can label the variables using letters, such as r for returns, although as we shall see later, there are some variables which have special symbols reserved for them, such as π, for the inflation rate. Superscripts and subscripts can be attached to our representations of this information to aid communication. For example, let r represent total returns recorded as a

Table 3.12 Nominal return indexes

Year	Commercial Property (1980 = 100)	Equities (1980 = 100)	Bonds (1980 = 100)
1980	100.0	100.0	100.0
1981	118.6	113.6	101.8
1982	125.6	146.0	154.0
1983	134.7	188.0	178.5
1984	147.7	247.4	190.7
1985	158.8	297.4	211.6
1986	175.4	378.6	234.9
1987	207.5	411.5	273.2
1988	269.8	458.9	298.9
1989	321.6	621.8	316.5
1990	304.0	562.1	334.2
1991	296.3	679.0	397.4
1992	284.7	813.4	470.5
1993	342.2	1037.1	606.0
1994	391.0	975.9	537.5
1995	405.0	1,200.4	639.7
1996	437.9	1,391.3	688.9
1997	513.6	1,719.6	822.6
1998	575.0	1,955.2	1,028.2
1999	656.2	2,420.5	992.3
2000	730.7	2,277.7	1,083.5
2001	789.5	1,977.1	1,097.6
2002	888.5	1,536.2	1,205.2
2003	986.0	1,846.5	1,224.5
2004	1,188.6	2,077.3	1,312.6
2005	1,425.0	2,526.0	1,422.9
2006	1,677.8	2,940.2	1,421.5
2007	1,584.1	3,090.2	1,495.4
2008	1,248.7	2,169.3	1,688.3
2009	1,322.2	2,798.4	1,671.4
2010	1,523.2	3,193.0	1,828.5
2011	1,638.7	3,084.4	2,219.8
2012	1,693.1	3,457.7	2,326.4
2013	1,892.8	4,166.5	2,158.9
2014	2,239.1	4,216.5	2,554.0
2015	2,548.5	4,262.9	2,566.7
2016	2,656.1	4,962.0	2,861.9

percentage. The subscript t denotes that we are looking at returns over time and another subscript B represents bonds can be used to identify each time series, $r_{B,t}$. Thus, the variable $r_{B,t}$ informs the reader that it contains time-series information about bonds. If we include a specific year, say $r_{B,1981}$, then it informs the reader that we are referring to the bond percentage return value in 1981, which is 1.8%. Similarly, let I_t be the generic term to represent the return index of assets. A subscript to denote a particular asset, for example $I_{E,t}$, could be used to denote the return index for equities. We shall regularly use r_t and I_t as generic representations of percentage returns and a return index, respectively.

It is possible to construct a return index from information on percentage returns, as an index provides information of a value *at the end of a year* (relative to the base period) while a percentage return represents the relative *change in value* which occurred *during the year*. Any return index constructed from percentage returns assumes that achieved returns are fully reinvested. To initiate its calculation, a base period must be selected and its index value set to 100. The year 1980 is chosen to be the base period in the preceding tables. Next, the 1981 index value can be obtained by updating the starting index value by the relative change in its value during 1981, which is reflected by the percentage return. For commercial property for example, the 1981 index value is 100 plus 100*(18.6/100) = 100 plus 18.6 = 118.6. The generic instruction (formula) for undertaking this calculation is:

$$I_t = (1 + r_t) I_{t-1} \tag{3.18}$$

The preceding calculation translates to:

$$I_{1981} = (1 + r_{1981}) I_{1981-1=1980}$$

Plugging in the values for commercial property:

$$I_{1981} = \left(1 + \frac{18.6}{100}\right) \times 100 = 118.6$$

For 1982, we simply amend the formula and input values:

$$I_{1982} = (1 + r_{1982}) I_{1981}$$

$$I_{1982} = \left(1 + \frac{5.9}{100}\right) 118.6 = 125.6$$

Understanding this relationship is informative, as it reveals that a return index captures the effects of compounding as a consequence of the assumption that returns achieved each year are completely reinvested in the asset. To illustrate:

$$I_{1982} = (1 + r_{1982}) I_{1981}$$

Substituting the index value in 1980 for 1981,

$$I_{1982} = (1 + r_{1982}) I_{1981} = (1 + r_{1982})(1 + r_{1981}) I_{1980}$$

The current index value at time t reflects the multiplicative increases in the index values from the base period, in this case, 1980. The capturing of the compounding effects in the return

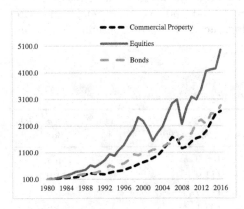 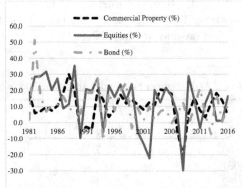

Figure 3.2 Nominal return indices and nominal percentage returns

indices can be seen in the right graph of Figure 3.2, where the asset value increases initially at a slow rate but then at a faster rate later in the investment horizon.

Equally, it is possible to derive the percentage total return by a calculation based on the information from a total return index. The annual percentage return, representing the relative change in the value of an asset, is obtained by taking the difference from its end and starting value and expressing this difference relative to the starting value:

$$r_t = \left(\frac{I_t - I_{t-1}}{I_{t-1}}\right) \tag{3.19}$$

For example, the percentage return for equities in the year 2000 can be obtained from the equity return index by inputting the relevant information into expression (3.19).

$$r_{2000} = \left(\frac{I_{2000} - I_{1999}}{I_{1999}}\right) = \left(\frac{2207.1 - 2306.6}{2306.6}\right) = -0.045$$

Reported as an annual percentage this is equal to −4.5%.

Since percentage returns only report information about changes which have occurred in a year, it does not capture any effects of compounding. But they do reveal the extent of annual fluctuations over the investment horizon, as the left-hand graph in Figure 3.2 shows.

3.6.5 *Real and nominal returns*

When assessing the past investment performance, it is important to account for changes in purchasing power, as it is conceivable that the final realised value of an asset may be insufficient to maintain a targeted standard of living due to general rises in the prices of goods and services. An individual saving via an equity fund may find that although the investment may have increased in value over time, its increase may be less than the rise in the general prices of goods and services. An analysis of an economic or financial time series usually requires an adjustment for the price level or changes to it. A price or inflation rate adjusted financial time series is measured in real terms while the unadjusted time series is referred to as nominal (as stated).

The procedure of adjusting for varying price levels or the inflation rate is referred to as deflating. It requires information about the prices of goods and services or their inflation rate over time. Table 3.13 displays the Long-Term Indicator of Prices of Consumer Goods and Services provided by the ONS (CPI).

Table 3.13 Consumer price index and annual inflation rate

Year	CPI (1974 = 100)	CPI (1980 = 100)	Inflation Rate (%)
1980	263.7	100.0	n.a.
1981	295.0	111.9	11.9
1982	320.4	121.5	8.6
1983	335.1	127.1	4.6
1984	351.8	133.4	5.0
1985	373.2	141.5	6.1
1986	385.9	146.3	3.4
1987	402.0	152.4	4.2
1988	421.7	159.9	4.9
1989	454.5	172.4	7.8
1990	497.5	188.7	9.5
1991	526.7	199.7	5.9
1992	546.4	207.2	3.7
1993	555.1	210.5	1.6
1994	568.5	215.6	2.4
1995	588.2	223.1	3.5
1996	602.4	228.4	2.4
1997	621.3	235.6	3.1
1998	642.6	243.7	3.4
1999	652.5	247.4	1.5
2000	671.8	254.8	3.0
2001	683.7	259.3	1.8
2002	695.1	263.6	1.7
2003	715.2	271.2	2.9
2004	736.5	279.3	3.0
2005	757.3	287.2	2.8
2006	781.5	296.4	3.2
2007	815.0	309.1	4.3
2008	847.5	321.4	4.0
2009	843.0	319.7	−0.5
2010	881.9	334.4	4.6
2011	927.8	351.8	5.2
2012	957.6	363.1	3.2
2013	986.7	374.2	3.0
2014	1,010.0	383.0	2.4
2015	1,020.0	386.8	1.0
2016	1,037.7	393.5	1.7

The original downloaded data is reported in column two, which has 1974 as a base year [CPI (1974 = 100)]. Column three displays the same information using a different base period [CPI (1980 = 100)]. We did this for convenience, as our information on asset returns only begins in 1980. The procedure to change the base period of an index is known as rebasing, which just involves dividing the index values in each year by the index value for the new base year. The mathematical operation to rebase is:

$$I_t^{Rebased} = \left(\frac{I_t^{Original}}{I_{new}^{Original}} \right) \times 100 \tag{3.20}$$

In our example, the new base year is 1980. For example, the rebased index value for 2002 is obtained by dividing 695.1 by 263.7 and then multiplying by 100 (695.1/263.7*100 = 263.6). Any index can be rebased in this way.

The convention is to use the letter P rather than I to denote a price index. We can convert the price index to obtain the inflation rate, the percentage change in the price level, conventionally depicted by the symbol π_t. As our price level data are annual, π_t represents the annual inflation rate:

$$\pi_t = \frac{(P_t - P_{t-1})}{P_{t-1}} \tag{3.21}$$

The inflation rate reports the relative change in prices. Let us use the rebased price index to compute the annual inflation rate for 2010. The annual inflation rate is $\pi_{2010} = \frac{(P_{2010} - P_{2009})}{P_{2009}} = \frac{(334.4 - 319.7)}{319.7} = 0.046$. As shown in Table 3.13, this can be displayed as a percentage by multiplying by 100.

We are now able to deflate the percentage return data. To deflate a total return index, which represents the value of an asset at the end of a year (a level), we employ the price index since it represents the cost of goods and services at the end of the year. A real index is obtained by dividing the nominal index by the price index and multiplying by 100:

$$I_t^r = \left(\frac{I_t}{P_t} \right) 100 \tag{3.22}$$

where I_t^r with the superscript r represents the real index.

For example, the real index value of equities in 2010 is obtained by using the values of the nominal index and price index in that year:

$$I_{2010}^r = \left(\frac{I_{2010}}{P_{2010}} \right) 100 = \left(\frac{3193.0}{334.4} \right) 100 = 954.8$$

We have just explained the generic procedure to deflate any time-series variable in levels measured in monetary units (e.g. GDP) or derived from a monetary measure, such as an asset return index. Table 3.14 displays the real return indices of commercial property, equities and bonds.

Sometimes it is more convenient to deflate a percentage return. To deflate the percentage return, which is a rate as it represents the return achieved during a year, a measure of the relative change in prices during the year, namely the inflation rate, is required. The nominal annual percentage return can be converted to a real percentage return:

$$r_t^r = \left(\frac{1+r_t}{1+\pi_t} - 1\right)$$
(3.23)

Table 3.14 Real return indices

Year	Commercial Property	Equities	Bonds
	(1980 = 100)	(1980 = 100)	(1980 = 100)
1980	100.0	100.0	100.0
1981	106.0	101.5	91.0
1982	103.4	120.1	126.8
1983	106.0	148.0	140.5
1984	110.7	185.5	142.9
1985	112.2	210.1	149.5
1986	119.8	258.7	160.5
1987	136.1	270.0	179.2
1988	168.7	286.9	186.9
1989	186.6	360.7	183.6
1990	161.1	297.9	177.2
1991	148.3	339.9	199.0
1992	137.4	392.6	227.1
1993	162.6	492.7	287.9
1994	181.4	452.7	249.3
1995	181.5	538.2	286.8
1996	191.7	609.0	301.6
1997	218.0	729.9	349.1
1998	236.0	802.3	422.0
1999	265.2	978.2	401.0
2000	286.8	894.1	425.3
2001	304.5	762.5	423.3
2002	337.1	582.8	457.2
2003	363.5	680.8	451.5
2004	425.6	743.8	470.0
2005	496.2	879.6	495.5
2006	566.1	992.1	479.6
2007	512.5	999.9	483.8
2008	388.5	675.0	525.3
2009	413.6	875.4	522.8
2010	455.4	954.8	546.8
2011	465.8	876.7	630.9
2012	466.2	952.2	640.6
2013	505.8	1,113.5	577.0
2014	584.6	1,100.9	666.8
2015	658.9	1,102.1	663.6
2016	675.0	1,260.9	727.3

The real annual percentage return from commercial property in 2011 is calculated by:

$$r^r_{2011} = \left(\frac{1+7.6/100}{1+5.2/100} - 1 \right) = 0.023.$$

Once again, this is displayed as a percentage by multiplying it by 100 in Table 3.15.

Table 3.15 Real percentage returns

Year	Commercial Property Percentage Real Returns	Equities Percentage Real Returns	Bonds Percentage Real Returns
1980	n.a.	n.a.	n.a.
1981	6.0	1.5	−9.0
1982	−2.5	18.3	39.3
1983	2.5	23.1	10.8
1984	4.5	25.4	1.7
1985	1.3	13.3	4.6
1986	6.8	23.1	7.3
1987	13.6	4.3	11.6
1988	24.0	6.3	4.3
1989	10.6	25.7	−1.7
1990	−13.6	−17.4	−3.5
1991	−8.0	14.1	12.3
1992	−7.4	15.5	14.1
1993	18.3	25.5	26.8
1994	11.6	−8.1	−13.4
1995	0.1	18.9	15.0
1996	5.6	13.2	5.2
1997	13.7	19.8	15.8
1998	8.3	9.9	20.9
1999	12.4	21.9	−5.0
2000	8.2	−8.6	6.1
2001	6.2	−14.7	−0.5
2002	10.7	−23.6	8.0
2003	7.8	16.8	−1.3
2004	17.1	9.2	4.1
2005	16.6	18.3	5.4
2006	14.1	12.8	−3.2
2007	−9.5	0.8	0.9
2008	−24.2	−32.5	8.6
2009	6.4	29.7	−0.5
2010	10.1	9.1	4.6
2011	2.3	−8.2	15.4
2012	0.1	8.6	1.5
2013	8.5	16.9	−9.9
2014	15.6	−1.1	15.6
2015	12.7	0.1	−0.5
2016	2.4	14.4	9.6

Equation (3.23) is a generic formula to deflate financial values recorded as rates. For example, it can be applied to obtaining a real rental growth rate by replacing the nominal percentage return with the nominal rental growth rate.

The preceding data manipulations have prepared the information available to us in the required format to undertake an analysis of past investment performance. The next part of this chapter explains a method can be used in executing this analysis. This method is also a starting point in more advanced empirical investigations.

3.7   Descriptive statistics

3.7.1   Background

The calculations summarising numerical information are called descriptive statistics. Descriptive statistics involves the organisation of the data to reveal its general pattern, namely the location of where most values tend to concentrate and their dispersion, and in the process exposing extreme or unusual data values and any asymmetries. We illustrate the concepts underlying descriptive statistics by applying it to the conventional analysis of past investment performance, using the annual return data prepared in the previous section.

Our objective is to assess the past investment performance of each asset and their relative (comparative) performance. The statistical analytical framework conventionally applied to assessing past investment performance as well as most economic and financial data is the normal distribution. A distribution reflects a way of organising and summarizing data. The normal distribution is extensively used due to the convenience of only having to describe two characteristics, the typical (arithmetic mean) values and the typical (standard) deviations from them. In our application, they yield the achieved typical annual return and the risk experienced respectively. The validity of adopting the normal distribution as the analytical framework can be examined by testing our return data empirically, which is undertaken in Chapter 4.

Consider the graphs in Figure 3.3.

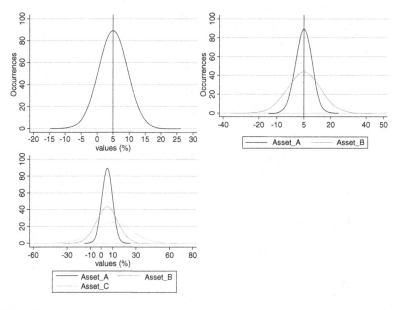

Figure 3.3 Simulated distributions of percentage total returns

The graphs depict the distribution of annual percentage returns which have been generated by software (STATA) using a (continuous) normal distribution. The normal distribution is considered in more detail in Chapter 4. The *x*-axis represents the annual percentage return data values and the *y*-axis the frequencies or occurrences of obtaining these values. To create the graphs, the data are arranged in ascending order from the lowest to the highest observed return value. The next step involves determining the number of occurrences at each value. When examining investment performance, the focus is on the centre of the distribution, as it displays the value which occurs most frequently in the data and the extent of the dispersion around it. Since this is a normal distribution, the shape and tails of the distribution can be ignored as there are relatively few observations which have values deviating away from these central values.

Consider the top left graph. The value of 5% is associated with the peak of the distribution (as shown by the vertical line on the graph), the most common return value reported for each year. However, this 5% return is not guaranteed since it is possible to obtain a higher or a lower return. Rational investors would take the risk of not achieving a typical return, in this case a 5% return, into consideration. Conventional analysis assesses this risk by considering the deviation away from this typical return value based on a standardised unit. The concern relates to deviations below the typical return value, as investors happily accept achieving returns above it. We shall explain what a standardised unit means later, but for now accept that it is equal to 2% for asset A. This means that there are a significant number of observed returns with values lying between 3% to 7%.

When analysing relative (comparative) investment performance, the analyst is in effect comparing the distribution of observed returns. The second graph (Figure 3.3, top right) compares the returns of two assets; both have the same typical return but different risk (spread around it). The distribution of return values of Asset B has a lower peak than Asset A, reflecting that it is more dispersed around the typical value, 5%. The typical return value achieved by Asset B is the same as Asset A, but the risk of not achieving that value is higher for Asset B than Asset A, as there are a significant number of observed returns with values lying between 2% to 8%.

Finally, asset C returns has a much lower peak than assets A and B, or equivalently a wider dispersion around the typical value, but the peak is associated with a return of 10%. Although asset C achieved the highest typical return, the occurrence of this value is relatively low as there are a significant number of observed return values lying between 6% to 10%. In summary, asset C had the highest typical annual percentage return (10%) but also exhibited the highest standard deviation or risk (4%); asset A and Asset B achieved the same typical annual percentage return (5%), but asset A had a lower risk (2%) compared to asset B (3%).

The preceding discussion is also pertinent to investigations involving other topics. Analysts should use the summarised information to familiarise themselves with the main features of the data and check for any anomalies or unusual values.

3.7.2 *Characteristics and moments of a distribution*

The characteristics of the data may be summarised by attaching numerical values to the features exhibited by the centre and at the tails of a distribution. These are the measures of central tendency, dispersion, skewness and kurtosis, or alternatively referred to first, second, third and fourth moments respectively. A distinction is made between population and sample distributions. A population refers to all possible observed values (observations), for example, individual information recorded by a census, whereas a sample is a subset of observations drawn from a population, for example, the General Household Survey. Population measures are always correct ('true'), whereas sample measures can be imprecise as they may be disproportionately affected by extreme values. For this reason, the formulae for the same summary statistic differ slightly between a population and a sample – the difference reflecting the need to account for the loss of accuracy when using sample data. Sample statistics are often used to draw inferences about the population. We shall examine three measures used to characterise the location of the centre of a distribution, which in our application can be used to reveal the typical annual percentage real return achieved from investing in an asset. Table 3.16 presents the data used to illustrate the descriptive statistics calculations. Due to

Table 3.16 Nominal and real de-smoothed commercial property percentage returns

Year	Commercial Property Nominal (%)	Commercial Property De-smoothed Nominal (%)	Commercial Property Real (%)	Commercial Property De-smoothed Real (%)
1981	18.59	21.21	6.01	8.35
1982	5.93	−1.26	−2.47	−9.09
1983	7.20	8.10	2.50	3.36
1984	9.70	11.32	4.49	6.03
1985	7.48	6.36	1.32	0.26
1986	10.44	12.33	6.81	8.63
1987	18.34	23.09	13.60	18.16
1988	30.02	36.98	23.95	30.58
1989	19.18	13.05	10.58	4.89
1990	−5.47	−19.62	−13.64	−26.57
1991	−2.55	−0.69	−7.95	−6.20
1992	−3.91	−4.54	−7.38	−7.98
1993	20.22	34.40	18.33	32.29
1994	14.24	10.94	11.55	8.32
1995	3.58	−2.46	0.11	−5.72
1996	8.12	10.93	5.58	8.32
1997	17.29	22.78	13.72	19.04
1998	11.97	9.04	8.26	5.43
1999	14.13	15.54	12.40	13.79
2000	11.35	9.90	8.15	6.75
2001	8.04	6.28	6.16	4.43
2002	12.54	15.32	10.69	13.43
2003	10.97	10.22	7.85	7.12
2004	20.55	26.28	17.07	22.63
2005	19.89	19.67	16.60	16.38
2006	17.74	16.66	14.10	13.05
2007	−5.59	−18.98	−9.47	−22.31
2008	−21.17	−30.05	−24.19	−32.74
2009	5.88	21.76	6.45	22.41
2010	15.20	20.78	10.12	15.45
2011	7.59	3.33	2.26	−1.79
2012	3.32	1.00	0.10	−2.14
2013	11.79	16.88	8.50	13.43
2014	18.30	22.24	15.57	19.42
2015	13.82	11.38	12.70	10.29
2016	4.22	−1.18	2.45	−2.87

'smoothing' issues resulting from using valuations rather than transacted prices in recording commercial property returns, the original JLL commercial property returns have been de-smoothed using the Blundell and Ward (1987) procedure.

- **Online note #3.1: A detailed explanation of the Blundell and Ward procedure**

3.7.3   Central tendency

The measure of central tendency or central location is a single value describing the middle point of a set of values which has been organised (arranged in ascending order and counted). The mean, median and mode are all valid measures of central tendency, and in a normal distribution they are the same value. But under different circumstances they differ and one measure may be more appropriate to adopt than the others.

(i)   Arithmetic mean

The arithmetic mean (or average) is the most commonly used measure of central tendency. It can be used for both discrete and continuous variables, although its use is most often associated with continuous variables. The arithmetic mean represents the first moment of a distribution. The mean is equal to the sum of all the values divided by the number of values in the data set. In statistics, sample and population measures have very different meanings, and these differences are very important, even if, as in the case of the mean, they are cal-culated in the same way. This is explained in more detail in the next chapter. Population measures are conventionally depicted by Greek letters and sample measures by Latin letters. To acknowledge that we are calculating the population mean and not the sample mean, the Greek lower-case letter *mu*, denoted as μ, is adopted.

The population formula is:

$$\mu = \frac{\sum x_i}{N} \tag{3.24}$$

The sample formula is:

$$\bar{x} = \frac{\sum x_i}{n} \tag{3.25}$$

The sample arithmetic mean has a bar on top of the letter to distinguish it from a data value. Note that capital N is used to denote all values in a dataset representing a population and small n for all values in a sample. In our examples, we adopt the sample measures of the characteristics of the distribution as they will be used to make inferences and test for the appropriateness of adopting the normal distribution as an analytical framework for analys-ing investment performance.

The arithmetic mean is essentially the value that is most common, although it is rarely one of the actual values you observe in your data. The mean is the only measure of central tendency where the sum of the deviations of each value from the mean is always zero. An important property of the mean is that it includes every value in your data set as part of the calculation. The conventional method for finding the typical annual percentage real return achieved by an asset is the arithmetic mean. Using the de-smoothed real return data in Table 3.16

and equation (3.25), the mean annual percentage real return achieved by commercial property is:

$$\bar{x} = \frac{\sum x_i}{n} = \frac{\left(8.35 + (-9.09) + .. + 10.29 + (-2.87)\%\right)}{36} = 6.0\%$$

The arithmetic mean has a major disadvantage in that it is particularly susceptible to the influence of outliers (extreme values), values which are unusual compared to the rest of the data set by being especially small or large.

(ii) Median

The median is a better measure of central tendency when the data contain a few outliers with very large or very small values, or when the data are skewed. Skewness refers to the situation when extreme values on one tail deviate from the mean value significantly more than those on the other tail. The median as a measure ignores the distorting effects of values at the tails of a distribution since the median is the middle value observed after the data have been arranged in an ascending order of magnitude. Column 2 in Table 3.17 demonstrates this using commercial property annual percentage real returns. The median can be found by locating the value in the middle. When there are many data values, the location of the middle data value can be found using the equation:

$$L_2 = \frac{(n+1)}{2} \tag{3.26}$$

Note that equation (3.26) does not yield the median value, just its location. When there are an even number of data values, as in our example, there is an additional step. As we have 36 values of returns, the location of the median value is the $(36+1)/2 = 18.5$th observation. A valid approach for finding the median value is to take the average of the 18th and 19th data values, that is $(7.12\% + 8.32\%)/2 = 7.72\%$.

(iii) Mode

The mode is the most frequent value found in the data and mainly applied to categorical data to identify the most common category. The main problems of adopting the mode are that it may not be unique, as there may be two or more values that share the highest frequency, and there is a tendency for every value to be different in continuous variables.

3.7.4   Measures of dispersion

Solely relying on measures of central tendency is insufficient to fully describe the data, as it is possible for two variables to have the same mean but very different concentrations (spread), as shown by the second graph in Figure 3.3. We also need to know the extent of variability, or dispersion. The range, interquartile range, and standard deviation are the three commonly used measures of dispersion.

(i) Range

The range is the difference between the largest and the smallest value in the data. It is very easy to calculate but has a lot of disadvantages. It is very sensitive to outliers and does not use

Table 3.17 Finding the median, semi-interquartile range and standard deviation

Year	Commercial Property De-smoothed X (%)	Year	Commercial Property De-smoothed X (%)	Mean X̄ (%)	Deviation from mean (x − X̄) (%)	Squared deviation from mean (x − X̄)² (%%)
				5.97		
2008	−32.74	1981	8.35		(8.35−5.97) = 2.38	(2.38)^2 = 5.68
1990	−26.57	1982	−9.09		−15.05	226.58
2007	−22.31	1983	3.36		−2.61	6.81
1982	−9.09	1984	6.03		0.07	0.00
1992	−7.98	1985	0.26		−5.71	32.58
1991	−6.20	1986	8.63		2.66	7.09
1995	−5.72	1987	18.16		12.19	148.60
2016	−2.87	1988	30.58		24.61	605.60
2012	−2.14	1989	4.89		−1.08	1.16
2011	−1.79	1990	−26.57		−32.54	1058.59
1985	0.26	1991	−6.20		−12.16	147.98
1983	3.36	1992	−7.98		−13.95	194.60
2001	4.43	1993	32.29		26.32	692.84
1989	4.89	1994	8.32		2.36	5.56
1998	5.43	1995	−5.72		−11.69	136.69
1984	6.03	1996	8.32		2.35	5.51
2000	6.75	1997	19.04		13.07	170.91
2003	7.12	1998	5.43		−0.54	0.29
1996	8.32	1999	13.79		7.82	61.22
1994	8.32	2000	6.75		0.78	0.61
1981	8.35	2001	4.43		−1.54	2.37
1986	8.63	2002	13.43		7.46	55.66
2015	10.29	2003	7.12		1.15	1.33
2006	13.05	2004	22.63		16.66	277.67
2002	13.43	2005	16.38		10.41	108.44
2013	13.43	2006	13.05		7.08	50.13
1999	13.79	2007	−22.31		−28.27	799.43
2010	15.45	2008	−32.74		−38.70	1497.98
2005	16.38	2009	22.41		16.44	270.25
1987	18.16	2010	15.45		9.48	89.93
1997	19.04	2011	−1.79		−7.75	60.12
2014	19.42	2012	−2.14		−8.11	65.71
2009	22.41	2013	13.43		7.47	55.73
2004	22.63	2014	19.42		13.45	180.98
1988	30.58	2015	10.29		4.32	18.64
1993	32.29	2016	−2.87		−8.84	78.06
					SUM	7,121.35

all the observations in a data set. It is often more informative to provide the minimum and the maximum values rather than providing the range. The range of commercial property real returns in our data is $32.29\% - (-32.74\%) = 65.03\%$. This measure of dispersion is reported in conjunction with the mode.

(ii)   *Semi-interquartile range*

Once the data have been ordered, it is possible to divide them into four groups, called quartiles. The last value in the first and third group represents the first and third quartile, respectively. The interquartile range is defined as the difference between the first and third quartile and thus describes the middle 50% of observations. If the interquartile range is large, it means that the middle 50% of observations are widely spaced. While the interquartile range is not affected by extreme values, its main disadvantage is that it is not amenable to mathematical manipulation. The interquartile range is found by identifying the value lying in the middle of the lowest value and the median, and the value lying in the middle of the highest value and the median.

To locate the first quartile value:

$$L_1 = \frac{1}{4}(n+1) \tag{3.27}$$

To locate the third quartile value:

$$L_3 = \frac{3}{4}(n+1) \tag{3.28}$$

In Table 3.17, the location of the first quartile is the $\frac{1}{4}(36+1) = 9.3$rd ordered value and the third quartile is the $\frac{3}{4}(36+1) = 27.8$th ordered value.

The 9.3rd ordered value is not observed. There are different methods involving interpolations which can be used to adjust the ninth ordered value, but the simplest approach is to take the nearest whole number as the first quartile value as an approximation. If it lies between the 9.5th to 9.9th ordered value, then it would be the tenth ordered value. In the table, the first quartile value (Q1) is -2.14%. The third quartile value (Q3) is the 27.8th observed data value, which we take to be the 28th ordered data value. The difference between the value of the third and first quartile is the interquartile range: Q3 − Q1 = $15.45\% - (-2.14\%) = 17.59\%$. Often the semi-interquartile range (SIR) is preferred for interpretation convenience, as it represents the average deviation from median value. The SIR is obtained from:

$$SIR = \frac{1}{2}(Q3 - Q1) \tag{3.29}$$

In our example:

$$SIR = \frac{1}{2}(15.45\% - (-2.14)\%) = 8.80\%$$

Thus, the average deviation from the median real percentage return is 8.8%. The larger the SIR, the greater the dispersion around the median.

(iii)   Variance and standard deviation

Variance and standard deviation are the measures of dispersion reported alongside the arithmetic mean. The variance and standard deviation measure the spread of data values around the arithmetic mean.

The population variance is:

$$\sigma^2 = \frac{\sum (x - \mu)^2}{N} \qquad (3.30)$$

The sample variance is:

$$s^2 = \frac{\sum (x - \bar{x})^2}{n-1} \qquad (3.31)$$

The population formula contains Greek letters representing the arithmetic mean and the variance, while the sample formula depicts them with Latin letters. Note also that the sample variance has an adjustment in the denominator, -1, to correct for a lost degree of freedom which will be explained in Chapter 4.

The population standard deviation is:

$$\sigma = \sqrt{\sigma^2} \qquad (3.32)$$

The sample standard deviation is:

$$s = \sqrt{s^2} \qquad (3.33)$$

A variance calculation is based on the deviation from the mean, the difference between a data value and the arithmetic mean value. Using sample notation, this is represented by $(x - \bar{x})$. Since the concept of dispersion is a measure summarizing the data, the deviation from the mean is calculated for each data value. If a data value (x) is larger than the mean value $(\bar{x})$, there is a positive deviation. A negative deviation occurs when it is smaller. The deviation from the mean is then squared $(x - \bar{x})^2$ for each observation. Squaring the deviations from the mean exaggerates the impact of large deviations and dampens the effect of small deviations from the mean. The variance is obtained when the average is taken, $\frac{\sum (x - \bar{x})^2}{n-1}$.

Table 3.17 illustrates the method of calculation. The deviation from mean for the percentage real return in the year 1981 is $(8.35\% - 5.97\%) = 2.38\%$ and its squared deviation from the mean value is $(2.38)^2 = 5.68\%\%$. The sum of the squared deviations from the mean is $\sum (x - \bar{x})^2 = 7121.35\%\%$. Equation (3.31) informs us that the variance is just the average of the squared deviations from the mean, $s^2 = \frac{\sum (x - \bar{x})^2}{n-1} = \frac{7121.35\%\%}{36-1} = 203.47\%\%$.

The unit of measurement of the variance is not the same as the mean, and consequently it is often preferable to present the standard deviation, the square root of the variance, to depict the extent of dispersion around the mean. The standard deviation $s = \sqrt{s^2} = \sqrt{203.47\%\%} = 14.26\%$. The standard deviation has a particular significance in investment performance analysis as it represents the risk of investing in an asset. More generally in an analysis involving time-series data, the standard deviation is an indication of volatility.

3.7.5   Coefficient of variation

The coefficient of variation (CV) is a relative measure of dispersion within a distribution. It reveals the extent of variability in relation to the arithmetic mean. The coefficient of variation (CV) is the ratio of the standard deviation to the arithmetic mean.

The population coefficient of variation measure is:

$$CV = \frac{\sigma}{\mu} \tag{3.34}$$

The sample coefficient of variation measure is

$$CV = \frac{s}{\bar{x}} \tag{3.35}$$

The CV is independent of the unit in which the measurement has been taken, as the unit in the numerator is the same as the denominator and they cancel each other out. The CV is used for making comparisons between data sets with different units or widely different arithmetic means. For example, it is used as a measure of income inequality in economics.

In our application of examining investment performance, the arithmetic mean represents the realised typical return and the standard deviation the risk. The CV therefore represents the risk incurred per unit of mean return:

$$CV = \frac{14.26\%}{5.97\%} = 2.39$$

The higher the CV, the more risk experienced in achieving a unit of mean return.

3.7.6   The shape of a distribution

Further characterisations of the data include measures of skewness and kurtosis as they reveal the shape of the distribution.

(i)   Skewness

Skewness is a measure of the degree of asymmetry of a distribution and represents the third moment of a distribution. If the left tail (at the end of the distribution containing smallest values) is more pronounced than the right tail (at the end of the distribution containing the largest values), the distribution exhibits negative skewness or is left skewed. By left skewed, we mean that the left tail is longer relative to the right tail. If the reverse is true, it has positive skewness or is right skewed, the right tail being longer relative to the left tail. If the two are equal, it has zero skewness or is symmetrical around the typical value (arithmetic mean, median, mode). The graphs displaying symmetry (SK = 0), negative skewness (SK = −0.5) and positive skewness (SK = 0.5) are displayed in Figure 3.4.

All graphs have the same mean, variance and kurtosis. The first depicts symmetry (SK = 0), so the mean, median and mode (location of the peak) have the same value, 5.9; the second is a distribution with a negative skew (SK = −0.5), where the mean gets 'dragged down' by having very small extreme values and resulting in the median being larger than the mean and the

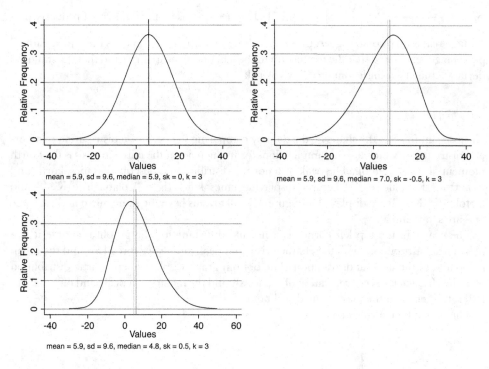

Figure 3.4 Graphical illustration of skewness

mode having the highest central value; and the third graph is a positively skewed distribution, where the mean is larger than the median as it gets 'boosted' by having very large extreme values and the mode has the smallest central value. Skewness raises the issue of which of the summary measures (mean, median and mode) should be adopted as a central value.

The population skewness measure is:

$$SK = \frac{1}{N}\sum\left(\frac{x-\mu}{\sigma}\right)^3 \tag{3.36}$$

The sample skewness measure calculated by Excel is:

$$SK = \frac{n}{(n-1)(n-2)}\sum\left(\frac{x-\bar{x}}{s}\right)^3 \tag{3.37}$$

A step-by-step example of how the sample measure of skewness can be calculated for commercial property percentage real returns can be found in the online resource on manual skewness calculation. The formulae reveal that the calculation can be split into two terms. The first term calculation involves the sample size n. For example, in our data, $\frac{n}{(n-1)(n-2)} = 0.030$. The second term is simply the summed cubed standardised deviation from the mean $\sum\left(\frac{x-\bar{x}}{s}\right)^3 = -24.485$. Multiplying them yields Excel's skewness measure,

$SK = \dfrac{n}{(n-1)(n-2)} \sum \left(\dfrac{x-\bar{x}}{s}\right)^{3} = -0.771$. The negative skew in commercial property per-

centage real return could suggest that the arithmetic mean under predicts the typical annual percentage return and that the median might be a better measure for it and that the standard deviation may be a misleading measure of risk.

(ii)   Kurtosis

Kurtosis summarises the shape of the data by comparing the peak of a distribution to a normal distribution, which also has implications for frequencies at the tails. Kurtosis is the fourth moment of a distribution. The peak of a normal distribution is described as mesokurtic and its kurtosis (K) value is 3. The graphs displaying a mesokurtic (K = 3), platykurtic (K < 3) and leptokurtic (K > 3) are displayed in Figure 3.5. All graphs have the same mean and variance and are symmetrical.

The graph in the top left corner is a normal distribution. The graph alongside it is a platykurtic distribution, with the characteristic flat peak and wider spread around the mean compared to the normal distribution. The bottom graph displays a leptokurtic distribution with the characteristic concentration of values around the mean (high peak) and tails that are fatter (thicker) than the mesokurtic distribution.

The population kurtosis measure is:

$$K = \frac{1}{N} \sum \left(\frac{x-\mu}{\sigma}\right)^{4} \tag{3.38}$$

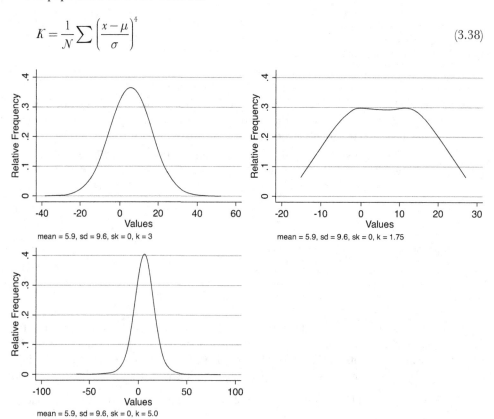

Figure 3.5 Graphical illustration of kurtosis

The sample kurtosis measure is:

$$K = \frac{n(n+1)}{(n-1)(n-2)(n-3)} \sum \left(\frac{x-\bar{x}}{s}\right)^4 \tag{3.39}$$

A step-by-step example of how the sample measure of kurtosis is calculated for commercial property percentage real returns can be found in the online notes.

- **Online note #3.2: Calculation of Kurtosis using Excel**

Manual kurtosis calculation. This calculation can be split into two. The first involves the sample size n. In our data, $n = 36$ so that $\dfrac{n(n+1)}{(n-1)(n-2)(n-3)} = 0.034$. The second term is simply the summed fourth power of the standardised deviation from the mean, $\sum \left(\dfrac{x-\bar{x}}{s}\right)^4 = 127.113$. Multiplying it yields $K = \dfrac{n(n+1)}{(n-1)(n-2)(n-3)} \sum \left(\dfrac{x-\bar{x}}{s}\right)^4 = 4.312$.

Since this value is greater than 3, it implies that commercial property returns are leptokurtic. That is, it has a higher peak at the centre and fat tails (a higher frequency of extreme values) compared to a normal distribution. This would imply that the standard deviation as a measure of risk does not fully describe the extent of the risk of investing in the asset as it ignores the possibility of making very low returns.

The kurtosis measure calculated in Excel is excess kurtosis. Excess kurtosis is kurtosis minus 3, $(K-3)$, and represents the extent of the deviation away from the normal distribution. $(K-3) = 0$ implies a characterisation equivalent to a normal distribution, $(K-3) > 0$ a leptokurtic distribution and $(K-3) < 0$ a mesokurtic distribution. The Excel formula for the sample measures of excess kurtosis is:

$$(K-3) = \frac{n(n+1)}{(n-1)(n-2)(n-3)} \sum \left(\frac{x-\bar{x}}{s}\right)^4 - \frac{3(n+1)^2}{(n-2)(n-3)} \tag{3.40}$$

The value of the additional term is $\dfrac{3(n+1)^2}{(n-2)(n-3)} = \dfrac{3(36+1)^2}{(36-2)(36-3)} = 3.275$, confirming that commercial property percentage returns are leptokurtic, $(K-3) = 4.312 - 3.275 = 1.037$.

There are alternative distributions with very different characteristics which we have not covered. A more advanced statistics textbook provides examples of alternative distributions to the ones considered here.

3.7.7 Geometric mean

While the arithmetic mean is the preferred measure of a typical (central) value, the geometric mean can sometimes be the better method to adopt, especially in cases when the values of a variable are not independent of each other, exhibit large fluctuations, are skewed, or non-normally distributed. It considers all the observations in a data set without being unduly influenced by the extreme values and is less sensitive to their 'pull' than an arithmetic mean in smaller samples. Strictly speaking, the geometric mean is not a characteristic of distribution. By the nature of its calculation, the geometric mean of data values will always be less than or equal to the arithmetic mean and requires all data values to be non-zero and positive. The general formula for calculating a geometric mean is:

$$\overline{x}^{GM} = \sqrt[n]{x_1 x_2 .. x_n} \tag{3.41}$$

In an application examining investment performance, the geometric mean provides an indication of the annualised return, also known as the compounded rate of return over an investment horizon or the typical return achieved over multiple holding periods. Why have an additional measure of the typical return achieved? The annualised return measures the cumulative change in investor wealth and informs us of the actual investment performance experienced by an investor fully reinvesting any positive return earned. Readers familiar with financial mathematics will understand that it reports the benefits of compounding. The arithmetic mean of the annual percentage returns only captures the typical return achieved each year rather than the typical return achieved each year over n years (the investment horizon).

Let us assume that an individual initially invests £100 in an asset. In the first year there is a 100% return, so that the value of the investment doubles to £200. Suppose this amount is reinvested in the following year, making the total value of the investment £200. Unfortunately, in that year the investment loses its value by 50% and consequently the investment is worth only £100. The annual percentage mean return is 25%, but this is misleading as the individual's wealth remained unchanged – he or she started with £100 and ended up with £100. The return achieved over 2 years is zero, which is clearly not the mean of the two annual returns.

A geometric mean calculation can be applied to returns recorded in either levels or in rates. Recall that returns can be recorded as an index (levels) or as a percentage return. We adopt conventional notation and use $r_1, r_2, \ldots, r_n$ to denote annual percentage return values as decimals, and I_n and I_0 to represent the end and starting index values respectively. Equation (3.41) is not directly applicable as we are interested in finding the typical growth in values rather than the typical level of values in this application. Suppose our return data are recorded as a percentage. The annualised percentage return or the percentage geometric mean return is calculated by:

$$r^{GM} = \left((1+r_1)(1+r_2)..(1+r_n) \right)^{\frac{1}{n}} - 1 \tag{3.42}$$

The annualised percentage return may also be obtained from returns recorded as an index. In this case, the annualised percentage return is calculated by:

$$r^{GM} = \left(\frac{I_n}{I_0} \right)^{\frac{1}{n}} - 1 \tag{3.43}$$

Both calculation methods yield numbers in decimals. A percentage can be obtained by multiplying them by 100. Note that equations (3.42) and (3.43) are derived from equation (3.41).

For convenience we use the level method, equation (3.43), to demonstrate the calculation of the annualised return for commercial property. We use the index values of commercial property real returns to extract the rate of compounding that took place over the investment horizon. The data can be found in Table 3.14. The annualised return over the time horizon 1980 to 2016 spans 36 years. Hence:

$$r^{GM} = \left(\frac{I_n}{I_0} \right)^{\frac{1}{n}} - 1 = \left(\frac{I_{2016}}{I_{1980}} \right)^{\frac{1}{36}} - 1 = \left(\frac{675.0}{100.0} \right)^{\frac{1}{36}} - 1 = 0.054$$

The annualised percentage real return of commercial property is 5.4% (0.054 × 100), the rate of return achieved after reinvesting any positive returns received in each year.

3.8 Applications: analysing investment performance

We now illustrate how the concepts covered so far can be used in analysing past investment performance. The implicit assumption in this analysis is that returns are normally distributed. Table 3.18 displays the summary measures reported using Excel, along with the annualised return calculation.

There are differences among the mean and median return values for each asset. Real return values are unique as there is no mode. The arithmetic mean indicates that typical annual percentage return achieved by equities over the investment horizon was 8.4%, the highest. Bonds and commercial property achieved similar typical returns at 6.1% and 6.0%, respectively. This is rather unexpected as the cash-flow properties for the assets, as outlined earlier, suggest that the mean real return for bonds should be lower as they are the least risky investment. The annualised real percentage returns confirm that bonds achieved a high real rate of return over the investment horizon. Their standard deviations, however, reveal that commercial property is riskier. The median reveals the expected pattern, as equities (13.0%) achieved the highest typical real percentage return, followed by commercial property (7.7%) and then bonds (4.9%). Similarly, the SIR indicates that the risk for commercial property lies in between bonds and equities. When we examined the data further, we discovered that the financial crash in 2007 and 2008 had a greater adverse effect on commercial property investment performance than bonds. Until then, commercial property achieved a higher average and annualised rate of real return than bonds.

The CV allows us to rank investment performance of the assets by combining the risk and mean return measures. Over the investment horizon, the CV reveals that the risk per mean unit of return is highest for commercial property, followed by equities and then bonds. In other words, commercial property performed worst and bonds best. For completeness, we also report the range in the table.

The descriptive statistics reveal that real returns of commercial property and equities are negatively skewed while bond real returns are positively skewed. The excess kurtosis

Table 3.18 Investment performance – key summary statistics

Measure	Returns (%)		
	Commercial Property	*Equities*	*Bonds*
Mean	6.0	8.4	6.1
Median	7.7	13.0	4.9
Mode	#N/A	#N/A	#N/A
Standard deviation	14.3	14.9	10.4
SIR	8.8	9.4	6.4
Range	65.0	62.2	52.7
CV	2.383	1.774	1.705
Annualised return	5.4	7.3	5.7
Skewness	−0.8	−1.0	0.9
Excess kurtosis	1.0	0.5	1.9

measures indicate that all three asset returns are leptokurtic, especially bonds. One implication of this is that the standard deviation may be insufficient in describing risk, as the kurtosis measures imply that tail-risk (experiencing large negative returns) may be present. Examining the shape of a distribution formally might shed further light on these anomalies, which we undertake in Chapter 4.

3.9   Concluding remarks

This chapter reviewed the common data representations, data structures and mathematical operations and rules required for the usual manipulations in preparing data for an analysis, and the analytical framework underlying applied statistical analysis, the distribution. The requirement to understand the differences in data structures will become apparent in later chapters on the topics of regression. A lot of information about the economy and the real estate market is reported as indexes, which is why we devoted a section to it. Understanding the different types of units of measurements reported as values helps in selecting the correct manipulations in data preparation. For example, assessing past investment performance requires return information to be adjusted for differences in general prices over time and the correct deflation procedure depends on whether the values of a return variable are in levels or rates. The data manipulations considered in this chapter are general in nature and can be applied to a wide variety of disciplines. We also explained how information can be organised and summarised by a distribution, the calculations to obtain measures of the central (typical) value, the spread around it, as well as features associated with extreme values. The collective application of these calculations in data interpretation is known as descriptive statistics. We illustrated their application by assessing past investment performance of commercial property, equities and bonds.

Chapter 3 online resource

In the online resource supporting this chapter, a brief guidance document and a dataset are provided so that you can replicate the calculations using Excel.

- Chapter 3 accompanying notes
- Excel file: "ch3_excel"

---

4 Random variables, correlation, estimation and hypothesis testing

4.1 Introduction

The main objective of this chapter is to introduce the mathematical and statistical concepts and calculations underlying the analysis of data. This chapter is broadly divided into three parts: random variables and probability distributions, correlation analysis, and estimation and hypothesis testing. The concepts covered will aid the reader in understanding the method of regression presented in subsequent chapters.

Section 2 introduces the concept of a random variable, the foundation of applied statistical analysis. By treating historic (realised) data as information derived from a data generating process (DGP) and stored in a random variable, it can be used to make statements about observed outcomes in markets. Due to its importance and widespread application, section 3 proceeds to examine the properties of the normal and the standard normal distributions. Our examples illustrating the relationship between values and their associated probabilities should provide a good base for understanding the construction of confidence intervals and hypothesis testing considered later in this chapter. Section 4 extends the concept of a random variable to explaining the relationships between them through linear measures of association known as covariance and correlation. We demonstrate an application of correlation analysis by considering the construction of a portfolio of assets to enhance their expected investment performance. The final parts of this chapter explore the branch of statistics known as inferential statistics. Sampling distributions are the foundation of the techniques employed in estimation and making statistical inferences – the execution of statistical calculations on data to make general inferences about the population (beliefs). The fifth section explains the concept of sampling distributions in estimation. A point estimate can provide an unbiased estimate of a population parameter but it does not address the issue of precision, while a confidence interval estimate incorporates a margin of error. We explain how to construct confidence intervals for the arithmetic mean and the standard deviation and in doing so, introduce the reader to the t and chi-squared probability distributions. The application of these methods is demonstrated using an example analysing future investment performance among three assets. The final part of this chapter explains hypothesis testing. We focus on the significance level and p-value approaches to hypothesis testing. A broad review of the steps of hypothesis testing is presented before going into detail, using as examples two- and one-tailed tests of beliefs concerning the risk premium of commercial property. The required adjustments for small-sized samples are also considered. There are other types of hypothesised sampling distributions in addition to the standard normal and t. The normality test of the distribution of percentage returns permits us to illustrate a hypothesis test based on the chi-squared sampling distribution when we examine the validity of adopting estimates of typical percentage return and risk using the arithmetic mean and standard deviation in investment performance analysis. Finally, we extend our consideration of hypothesis testing to correlation, testing propositions of the presence of significant linear relationships between variables.

4.2   Random variables and probability distributions

The concept of a random variable forms the basis of undertaking empirical analyses using financial, economic and real estate market information. It allows us to treat information stored in variables as outcomes of market processes. Examples include the use of historic data to assess current and future investment performance which utilises the concepts of expected return and variance, and the modelling of the statistical relationship between house prices and household incomes presumes a systematic (probabilistic) relationship between them.

4.2.1   The concept of probability

An experiment (trial) refers to the process that leads to the occurrence of one of several possible observed values (observations) or outcomes, for example, throwing a die or the realised (achieved) market rent in a period. The outcome refers to the result or value – observing a six after throwing a die or an achieved market rent in 2018 of £65 per square foot. We may also describe it as an event, which refers to a collection of one or more outcomes, for example, throwing a 6 and then throwing a 3, or throwing a value above 3 or market rents being above £25 per square foot.

A probability indicates the likelihood of an event or outcome. The convention is to use P(.) to describe it. For example, the probability of an event A occurring is P(A). An event which is guaranteed has a probability equal to 1, whilst an event which is certain never to materialise has a probability equal to 0. The sum of all possible events/outcomes must equal 1, for example, in the case of two outcomes, A or B, their probabilities must sum to 1: P(A) + P(B) = 1. Probabilities may be expressed as percentages, fractions or decimal points.

4.2.2   The concept of random variables

A random variable, usually depicted by a capital letter such as X, is a variable which can assume a set of possible values, for example, a return, but there are probabilities attached to each possible value that it can assume. A generic representation of the probability that a random variable X takes a value, represented by a small letter such as x, is denoted by $P(X = x)$. Random variables can be discrete or continuous. A discrete random variable has values which are countable and may be therefore represented by precise values. On the other hand, a continuous random variable is uncountable as outcomes are infinite and they must be represented by intervals of values. We will mainly focus on explaining concepts associated with continuous random variables but will, when appropriate in elucidation, examine discrete random variables.

4.2.3   Probability density function

A continuous random variable is a random variable with a set of possible values that are infinite. The possible values a continuous random variable can take is depicted by the probability density function (PDF). The PDF is like a frequency distribution except that the values on the vertical axis reflect probabilities rather than frequencies. The probabilities of a continuous random variable (X) are defined as the area under the curve of its PDF for intervals of values rather than specifying a precise value. Probabilities in continuous PDFs can be equivalently written using weak or strict inequalities in defining an interval of values, that is: $P(X > x) = P(X \geq x)$ and $P(X < x) = P(X \leq x)$. Finally the probability of a random variable X taking a value lying within an interval between the values x_1 and x_2 can be expressed as $P(x_1 < X < x_2)$ or $P(x_1 \leq X \leq x_2)$.

The PDF in Figure 4.1 plots the relationship between possible return values and probabilities of a random variable. The height of the curve is called the probability density. The entire area under the curve equals 1, as it represents all the possible values a random variable could assume. The area under the curve between the interval 4 and 6 (dashed vertical lines) represents the probability. Suppose the area of this interval is 0.20; therefore, the probability that the random variable X takes a value between 4 and 6 is 0.20, or more succinctly, $P(4 < X < 6) = 0.20$. The most likely value that a random variable will take is its expected value, represented by the term $E(X)$ in the figure.

Let $f(x)$ represent the probability density (probabilities) associated with values. The expected value is:

$$E(X) = \mu = \int x f(x) \, dx \tag{4.1}$$

where the integral $\int dx$ denotes the summation of probabilities for an interval of return values. For readers who are unfamiliar with integration, it refers to finding the area under the probability distribution by multiplying the relevant values with their associated probabilities and adding them together. In effect, it is the same mathematical operation as calculating the expected value of a discrete random variable, which we will consider later. As the PDF in Figure 4.1 is symmetrical, the expected value is the value associated with the highest point in the probability density.

The expected variance is:

$$\mathrm{Var}(X) = \sigma^2 = \int [x - E(X)]^2 \, f(x) \, dx \tag{4.2}$$

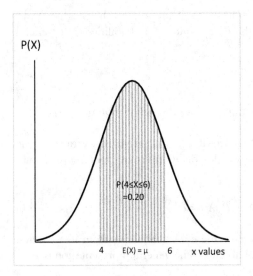

| Key Features |
| An area under the curve represents the cumulative probabilities |
| The total area under the curve represents the sum of probabilities of all possible values and is equal to 1. |
| In the figure, the area under the curve between the dashed lines represent the probability that a random variable X takes a value between 5 and 6, $P(4 < X < 6)$. |
| If the area under the curve between the dashed lines is 0.2, this can be restated as: $P(4 < X < 6) = 0.2$. |

Figure 4.1 A continuous probability density function

An analysis may focus on the first two moments of the distribution of a random variable provided the third moment (skewness) exhibits symmetry and the fourth moment (kurtosis) is mesokurtic.

The expected skewness is:

$$SK = \int \left[x - E(X) \right]^3 f(x) dx \tag{4.3}$$

And finally, the expected kurtosis is:

$$K = \int \left[x - E(X) \right]^4 f(x) dx \tag{4.4}$$

An observant reader would have noted that the expected value is equivalent to the population arithmetic mean and the expected variance and expected standard deviation to the population variance and population standard deviation. This is also the case for expected skewness and expected kurtosis.

4.2.4   The cumulative distribution function (CDF)

The cumulative distribution function (CDF) calculates the cumulative probability up to a specified variable value. The CDF is used to determine the probability that a value from a random variable will be less than or equal to a certain value. For example, a cumulative distribution function can show the proportion of returns with a value less than 10%, that is, $P(X < 10\%)$.

4.2.5   Probability mass function

A discrete probability distribution, also known as a *probability mass function*, describes the probability of a random variable taking a precise value. For example, suppose that the probability of a discrete random variable X assuming a value of 5 is equal to 0.20. This can be written as $P(X = 5) = 0.20$. A *probability mass function* is usually represented as a table but can also be depicted as a bar chart or histogram.

The expected value of a random variable is its mean, which represents the weighted average of its possible values, the weights being the corresponding probabilities. The general representation is:

$$E(X) = \sum x P(X = x) = \mu \tag{4.5}$$

It can be clearly seen that the expectations operator implies the application of probabilities as weights to obtain a weighted average value.

The expected variance of a discrete random variable is the weighted average of the squared deviations from its expected value, the weights being the corresponding probabilities. The general representation is:

$$\text{Var}(X) = E\left[\left(x - E(X) \right)^2 \right] = \sum \left[x - E(X) \right]^2 P(X) = \sigma^2 \tag{4.6}$$

The expected skewness is the weighted average of the cubic deviations from its expected value, the weights being the corresponding probabilities. The general representation is:

$$SK = E\left[\left(x - E(X) \right)^3 \right] \tag{4.7}$$

The expected kurtosis of a random variable is the weighted average of the deviations from its expected value raised to the power of 4, the weights being the corresponding probabilities. The general representation is:

$$K = E\left[\left(x - E(X)\right)^4\right]$$ (4.8)

To facilitate the understanding of these concepts, we shall consider an example of investment returns for a hypothetical asset under three scenarios with three possible outcomes. Since there are only three possible outcomes, it is a discrete probability distribution.

4.2.6   An example of probability mass function: future investment performance

Table 4.1 displays three possible scenarios for an asset's future return. Let the discrete random variable X represent the asset return. The asset will a return a value of 12.0% if there is a boom, 7.0% if there is normal economic growth and −8.0% if there is a recession.

Associated with each scenario is a probability. The probability of the asset achieving a 12.0% return is 0.20, which can be more succinctly written as $P(X = 12.0) = 0.20$. There is a 20% likelihood that the asset will achieve a return of 12.0%. The reader is left to interpret the outcomes associated with the other scenarios.

From equation (4.5), the expected return is:

$$E(X) = \sum xP(X = x)$$
$$E(X) = 12.0\%(0.20) + 7.0\%(0.35) + (-8.0\%)(0.45) = 1.25\%$$

However, as the probabilities indicate, this return might not be achieved. We can apply the expected variance to obtain the expected risk of not achieving the expected return. From equation (4.6), the expected variance is:

$$Var(X) = \sigma^2 = \sum [x - E(X)]^2 P(X)$$

The easiest way to calculate this is to set up a table. The deviation from mean, $[x - E(X)]$, is obtained by subtracting the return value for a scenario from the expected value. The next step is to obtain its square, $[x - E(X)]^2$. The third step is to weight each squared deviation by the respective probability.

Table 4.1 Random variable discrete probabilities

Scenario (S)	State of the Economy	Probability P(X)	Return (%) X
1	Boom	0.20	12.0
2	Normal	0.35	7.0
3	Recession	0.45	−8.0

Table 4.2 Random variable discrete variance

Scenario	State of the Economy	Probability P(X)	Return (%)	E(X) (%)	x − E(X) (%)	(x − E(X))² (%)(%)	P(X) (x − E(X))² (%)(%)
1	Boom	0.20	12.0	1.25	10.750	115.563	23.113
2	Normal	0.35	7.0		5.750	33.063	11.572
3	Recession	0.45	−8.0		−9.250	85.563	38.503

For example, scenario 1:

$$x - E(X) = [12.0\% - 1.25\%] = 10.750\%$$
$$[x - E(X)]^2 = [10.750\%]^2 = 115.563\%\%$$
$$P(X)[x - E(X)]^2 = 0.20(115.563\%\%) = 23.113\%\%$$

The variance is obtained in the final step by adding together the product of the probabilities and the squared deviation from expected value.

$$\sigma^2 = \sum [x - E(X)]^2 P(X) = 23.113 + 11.572 + 38.503\%\% = 73.188\%\%.$$

The standard deviation or risk of not obtaining an expected return is:

$$\sigma = \sqrt{73.188\%\%} = 8.6\%$$

4.3   The normal and standard normal probability distributions

There are many types of probability distributions but for now we focus on the normal probability distribution. Many distributions of random variables, particularly those obtained from calculating sample statistics in large samples (Central Limit Theorem), are close enough to a normal distribution to be treated as one. And any normal distribution can be converted into a standard normal distribution to reveal the probabilities associated with intervals of values. Many of the concepts covered in this section apply to other distributions used later in the book.

4.3.1   *Normal probability distribution*

The normal probability distribution is presented in Figure 4.2.

The values of a normal distribution extend from minus to plus infinity. The normal distribution is constructed using a normal density function. The normal density function is:

$$f(X) = \left[\frac{1}{\sigma \sqrt{2\pi}} \right] exp \left[\frac{-(x - \mu)}{2\sigma^2} \right] \tag{4.9}$$

Note that only two parameters are required to determine the curve, the expected value (population arithmetic mean) and expected standard deviation (population standard deviation).

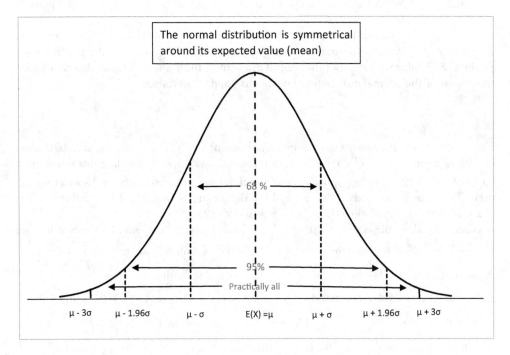

Figure 4.2 Normal probability distribution

Instead of showing the graph, it is also possible to describe a normal probability distribution using the mathematical expression:

$$X \sim N\left(\mu, \sigma^2\right) \tag{4.10}$$

This expression tells the reader that X is a normally distributed random variable with a mean μ and variance σ^2.

Although there are families of normal distributions with their own means and standard deviations, they all share common features. The first is that the distribution is symmetrical, so that the shape of the curve above and below the expected value (mean) is the same and the expected value lies in the middle of the distribution. The second is that the likelihood of achieving extreme values (values at the tails) are rare, as approximately 95% of all values lie within 1.96 standard deviations from the expected value, which implies that we can focus on dispersions around the centre of a distribution (expected value) as a measure of variation. The third feature is that 68% of values fall within one standard deviation from the expected value – there is a common likelihood associated with a standard deviation. The fourth feature is that any normal distribution can be converted to a standard normal distribution, from which probabilities have been constructed so that they can be read from a table.

4.3.2 The standard normal probability distribution

The standard normal probability distribution provides a convenient way to calculate the probability of any event occurring from a normal distribution. Any value of a variable from a normal probability distribution can be converted into a z (standardised) value and placed into the standard normal variable, Z.

Let x be a value of a random variable X which is normally distributed and characterised as $X \sim N(\mu, \sigma^2)$. A standard normal variable Z can be formed by converting the x-values into z-values using information about the central moments – those characteristics describing the centre part of the normal distribution, mean and standard deviation:

$$z = \frac{x - \mu}{\sigma} \tag{4.11}$$

This manipulation transforms any normal distribution $X \sim N(\mu, \sigma^2)$ into the standard normal distribution $Z \sim N(0,1)$. We can work through an example to validate this statement. Suppose $X \sim N(2, 5^2)$ – the population mean and the standard deviation are 2 and 5 respectively. The value of a standard deviation above the mean is 7 ($\mu + \sigma = 2 + 5$) and the value of a standard deviation below the mean is -3 ($\mu - \sigma = 2 - 5$).

When X takes the same value as its mean, x = 2, the standardised value is $z = \frac{x-\mu}{\sigma} = \frac{2-2}{5} = 0$. Since the value of a standard deviation is 5, when X takes a value one standard deviation above the mean, x = 7, the standardised value is $z = \frac{x-\mu}{\sigma} = \frac{7-2}{5} = 1$, and when it takes a value one standard deviation below the mean, x = -3, the standardised value is $z = \frac{x-\mu}{\sigma} = \frac{-3-2}{5} = -1$. This proves that any normal distribution $X \sim N(\mu, \sigma^2)$ can be converted into the standard normal distribution, $Z \sim N(0,1)$. Diagrammatically:

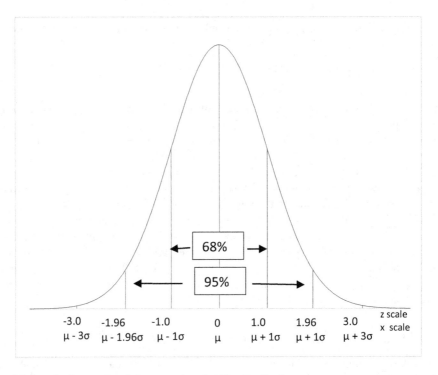

Figure 4.3 Standard normal and the normal probability distribution

This unique relationship between an interval of values from the expected value (or population arithmetic mean) implies that a table representing this relationship based on the standard normal distribution can be used to look up probabilities for any normal distribution. There are different versions of the standard normal table, but the one used in this textbook is Table A.1 in Appendix A.

This version of the standard normal table yields probabilities of an interval from the standardised value of the population mean ($z = 0$) to some positive standardised value. For example, the probability of Z taking a value between 0 and 1.96 is 0.4750, P($0 \leq Z \leq 1.96$) = 0.4750. The numbers on the edge of the table refer to the standardised value (z) to the right of the mean ($z = 0$) and the numbers inside the table the probabilities of the interval between the arithmetic mean and the standardised value to the right of it. The first column of numbers at the edge of the table represents the whole number and first decimal place of the z-value, and the first row of numbers the second decimal place.

Assume the objective is to find P($0 \leq Z \leq 1.61$). The first step is to look at the first column until we reach the whole number and first decimal place of the z-value 1.61. Then we move across this row to the column depicting a one value for the second decimal place. The intersection then yields the solution P($0 \leq Z \leq 1.61$) = 0.4463. This is illustrated in Figure 4.4.

We will now provide several examples to illustrate how probabilities can be found using the standard normal table.

z	0.0000	0.0100	0.0200	0.0300	0.0400	0.0500	0.0600	0.0700	0.0800	0.0900
1.3	0.4032	0.4049	0.4066	0.4082	0.4099	0.4115	0.4131	0.4147	0.4162	0.4177
1.4	0.4192	0.4207	0.4222	0.4236	0.4251	0.4265	0.4279	0.4292	0.4306	0.4319
1.5	0.4332	0.4345	0.4357	0.4370	0.4382	0.4394	0.4406	0.4418	0.4429	0.4441
1.6	0.4452	0.4463	0.4474	0.4484	0.4495	0.4505	0.4515	0.4525	0.4535	0.4545
1.7	0.4554	0.4564	0.4573	0.4582	0.4591	0.4599	0.4608	0.4616	0.4625	0.4633
1.8	0.4641	0.4649	0.4656	0.4664	0.4671	0.4678	0.4686	0.4693	0.4699	0.4706
1.9	0.4713	0.4719	0.4726	0.4732	0.4738	0.4744	0.4750	0.4756	0.4761	0.4767

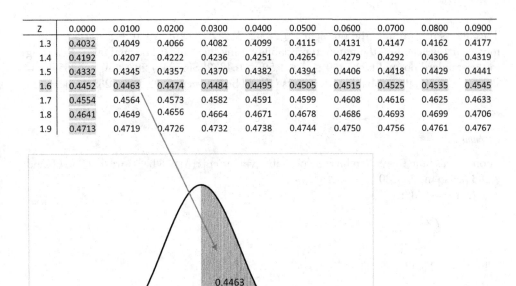

Figure 4.4 Finding probabilities from the standard normal table

4.3.3 Exploring the relationship between values and probabilities in the standard normal

Let us assume that rental values are normally distributed with mean £283 and standard deviation £1.60. In notation this is:

$$X \sim \mathcal{N}\left(\mu, \sigma^2\right)$$
$$X \sim \mathcal{N}\left(283, 1.60^2\right)$$

Example 1

Suppose we wish to find the probability that rents (X) will be between £283 and £285.40 per square metre (£ per sq m). The objective can be stated as P(£283 $\leq X \leq$ £285.40). To look up the probabilities in the standard normal table, we must convert the x-values (£ per sq m) into z (standardised) values.

When $x =$ £283:

$$z_{283} = \frac{£283 - £283}{£1.60} = 0$$

When x = £285.40:

$$z_{285.40} = \frac{£285.40 - £283}{£1.60} = 1.50$$

This allows us to restate the problem P(£283 $\leq X \leq$ £285.40) as P(0 $\leq Z \leq$ 1.50), which means that we can find the required probability using the standard normal table. Referring to the standard normal table, P(0 $\leq Z \leq$ 1.50) = 0.4332. Since P(£283 $\leq X \leq$ £285.40) = P(0 $\leq Z \leq$ 1.50), the probability that rents (X) will be between £283 and £285.40 per sq m is 0.4332.

Example 2

Suppose this time we wish to find the probability that rents (X) will be between £280.60 and £283 per sq m, P(£280.60 $\leq X \leq$ £283).

When $x =$ £283:

$$z_{283} = \frac{£283 - £283}{£1.60} = 0$$

When $x =$ £280.60:

$$z_{280.6} = \frac{£280.60 - £283}{£1.60} = -1.50$$

The restated probability is P(−1.50 $\leq Z \leq$ 0). Although not directly observed from the standard normal table, we can exploit the symmetry around the population mean characteristic of the normal distribution to find the probability of an interval below the mean. Using the symmetry argument, P(−1.50 $\leq Z \leq$ 0) = P(0 $\leq Z \leq$ 1.50). Thus, P(0 $\leq Z \leq$ 1.50) = 0.4332.

Example 3

There are multiple ways in which the symmetry characteristic may be manipulated to find probabilities for intervals. Suppose we wish to compute the probability that rents (X) will be less than £285.40 per sq m, $P(X \leq £285.40)$.

When $x = £285.40$:

$$z_{285.40} = \frac{£285.40 - £283}{£1.60} = 1.50$$

Restating the problem, $P(X \leq £285.40) = P(Z \leq 1.50)$. Figure 4.5 depicts the required probability as the shaded area under the standard normal distribution.

As the standard normal distribution is symmetrical around the mean and the cumulative probabilities in any distribution reflecting all possible values must sum to 1, it is possible to simplify by splitting the required probability calculation into two parts:

$$P(Z \leq 1.50) = P(0 \leq Z \leq 1.50) + P(Z \leq 0)$$

The centre of the standard normal distribution is its arithmetic mean ($z = 0$), implying that cumulative probability of all values below the mean has to be equal to 0.50, $P(0 \leq Z \leq 0) = 0.5$. Therefore:

$$P(Z \leq 1.50) = P(0 \leq Z \leq 1.50) + P(Z \leq 0)$$
$$= 0.4332 \text{ (from the standard normal table)} + 0.50$$
$$= 0.9332.$$

Example 4

Let us compute the probability that rents (X) will be greater than £285.40 per sq m.
When $x = £285.40$:

$$z_{285.40} = \frac{£285.40 - £283}{£1.60} = 1.50$$

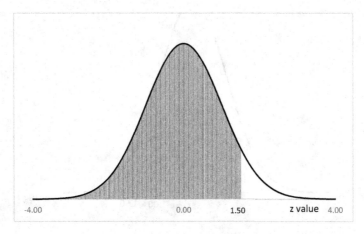

Figure 4.5 Finding probabilities from the standard normal table in example 3

The problem can be restated as: $P(X \geq £285.40) = P(Z \geq 1.50)$. Again, we make use of the symmetry argument to break the computation into two parts:

$$P(Z \geq 1.50) = P(Z \geq 0) - P(0 \leq Z \leq 1.50)$$
$$= 0.5 - 0.4332 \text{ (from the standard normal table)}$$
$$= 0.0668.$$

Example 5

It is possible to compute the probability that the rent achieved (X) will lie between £284.60 and £285.40 per sq m.
When $x = £284.60$:

$$z_{284.60} = \frac{£284.60 - £283}{£1.60} = 1.00$$

When $x = £285.40$:

$$z_{285.40} = \frac{£285.40 - £283}{£1.60} = 1.50$$

The required probability is $P(£284.60 \leq X \leq £285.40)$ and when restated is equivalent to $P(1.00 \leq Z \leq 1.50)$.

$$P(1.0 \leq Z \leq 1.50) = P(0 \leq Z \leq 1.50) - P(0 \leq Z \leq 1.00)$$
$$= 0.4332 \text{ (from the standard normal table)} - 0.3413 \text{ (from the standard normal table)}$$
$$= 0.0919$$

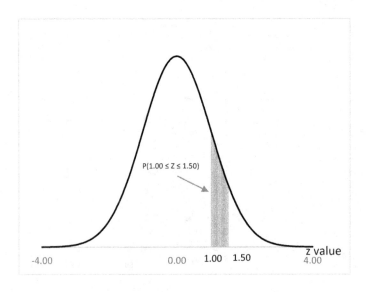

Figure 4.6 Finding probabilities from the standard normal table in example 5

4.3.4 *Finding the values for a probability using the standard normal table*

The standard normal table also can be used to find out the z-values for a given probability. Assume rental values are normally distributed with mean 283 and standard deviation 1.6, $X \sim N(283, 1.60^2)$.

Suppose the objective is to find an interval of rental values around the arithmetic population mean (expected value) which have a 95% chance of being achieved.

In this instance we have the problem: $P(x_1 \leq X \leq x_2) = 0.95$
Restating this in terms of z-values: $P(z_1 \leq Z \leq z_2) = 0.95$

Diagrammatically:

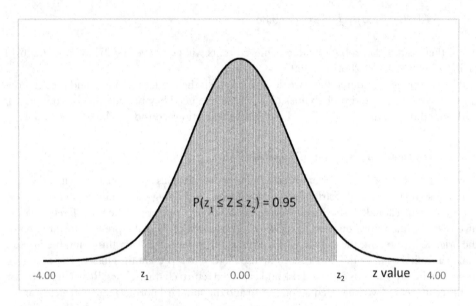

$$P(z_1 \leq Z \leq z_2) = 0.95$$

-4.00 z_1 0.00 z_2 z value   4.00

Figure 4.7 Finding the interval value for a probability

The combined shaded area under the curve from z_1 to z_2 represents the 95% probability. We can locate the value of z_2 using the standard normal table by dividing the given probability by 2:

$$P(0 \leq Z \leq z_2) = 0.95/2 = 0.475$$

Looking at the numbers representing probabilities in the standard normal table, the z-value is 1.96. More succinctly stated, $P(0 \leq Z \leq 1.96) = 0.475$.

Applying the symmetry argument, $P(-1.96 \leq Z \leq 1.96) = 0.475 + 0.475 = 0.95$.

To obtain the values denoted as £ per sq m, the z-values must be converted back into x-values. A rearrangement of the formula, $z = \dfrac{x - \mu}{\sigma}$, is required.

When $z = 1.96$:

$$z = \frac{x - £283}{£1.60} = 1.96$$

Rearranging:

$$x - \pounds283 = 1.96(\pounds1.60)$$
$$x = 1.96(\pounds1.60) + \pounds283 = \pounds286.14$$

When $z = -1.96$:

$$z = \frac{x - \pounds283}{\pounds1.60} = -1.96$$

Rearranging:

$$x - \pounds283 = 1.96(\pounds1.60)$$
$$x = -1.96(\pounds1.60) + \pounds283 = \pounds279.86$$

Thus, there is a 95% probability that achieved rents will lie between £279.86 and £286.14 or P(£279.86 ≤ X ≤ £286.14) = 0.95.

The examples considered here are designed to help the reader to understand the relationship between values and probabilities of a random value. They provide the foundations for understanding the more complicated statistical applications considered later in our book.

4.3.5   *Application – return of capital*

As well as illustrating the important relationships between values and probabilities in a distribution, the standard normal distribution can be applied to finding out the likelihood of achieving a return of capital. The return of capital is a different concept to the rate of return of an investment or the return on capital employed. The return of capital represents the amount of the original investment, the capital, which an investor can recoup. After this point, the investor makes a capital gain. In this application, we use the CV to represent the efficiency of an investment, as it incorporates its expected risk and expected return characteristics. Recall the formulae of the coefficient of variation (equation (3.34)) and the standard normal (equation (4.11)) are:

$$CV = \frac{\sigma}{\mu}$$

$$z = \frac{x - \mu}{\sigma}$$

Let the value x represent the threshold point at which an investment begins to earn capital or return more than the initial invested capital. At this point, $x = 0$. Plugging this value into the standard normal formula yields:

$$z = \frac{-\mu}{\sigma} \tag{4.12}$$

The probability of the return of capital is represented by the investment achieving standardised return values to the right of this z-value, that is:

$$P(Z > z) \tag{4.13}$$

Note that the CV is the negative reciprocal of this relationship, that is:

$$z = \frac{-\mu}{\sigma} = \frac{-1}{CV} \tag{4.14}$$

Let us assume that we are considering an investment in a commercial property. Using information reported in Table 3.18, the CV of commercial property is:

$$CV = \frac{14.3}{6.0} = 2.383$$

Substituting this into equation (4.14):

$$z = \frac{-1}{2.383} = -0.420$$

As we have shown earlier when examining the relationship between probabilities and values in the standard normal, the probability of making a return of capital from investing in commercial property is:

$$P(\mathcal{Z} > -0.420) = P(\mathcal{Z} > 0) + P(0 < \mathcal{Z} < 0.420)$$

The shaded area in Figure 4.8 depicts the probability of making a return of capital.
From the standard normal table (Table 4.6):

$$P(\mathcal{Z} > -0.420) = 0.5 + 0.1628 = 0.6628$$

Thus, the probability of making a return of capital from commercial property is 66.28%.

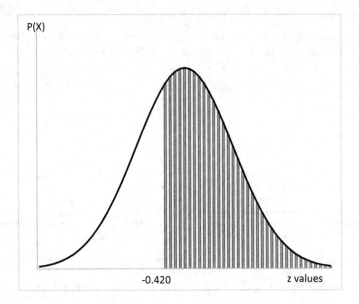

Figure 4.8 Return of capital

4.4   Measures of association: covariance and correlation

The values of random variables may be systematically related to each other from being jointly distributed. Covariance and correlation are linear measures of association (relationship) between two random variables. Covariance primarily provides a measure of the direction of their association, while correlation additionally provides an indication of the relative magnitude (strength).

4.4.1   Covariance and correlation

Covariance examines the *expected* deviations of values of each random variable from its expected value in determining their relationship.

$$Cov(X,Y) = E\big[(x - E(X))(y - E(Y))\big] \tag{4.15}$$

The shared expectations operator denotes that there is a joint (bivariate) probability distribution explaining the deviations from the expected values for random variables X and Y. A joint (bivariate) probability distribution $f(X, Y)$ reflects the probabilities associated with the deviations from the mean that occur – in other words, a systematic relationship. For example, negative deviations of asset X from its expected return could be associated with a high likelihood of positive deviations of asset Y from its expected return.

The population measure for the covariance is:

$$Cov(X,Y) = \sigma_{XY} = \frac{1}{N}\sum (x - \mu_X)(y - \mu_Y) \tag{4.16}$$

The magnitude of the covariance is influenced by the units of measurement of the data. The correlation coefficient overcomes this problem as it reports the relative magnitude of their relationship. The correlation between two random variables is the ratio of the covariance and the product of their standard deviations. The population correlation coefficient is:

$$\rho_{X,Y} = \frac{\sigma_{XY}}{\sigma_X \sigma_Y} \tag{4.17}$$

The correlation coefficient is interpreted using a scale ranging from +1 to −1 (see Figure 4.9). The random variables are perfectly positively correlated when the correlation coefficient

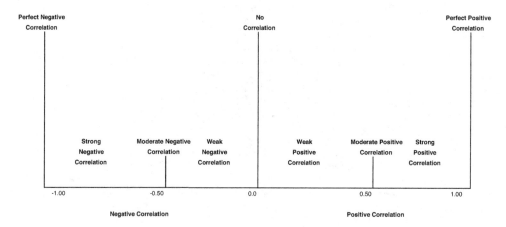

Figure 4.9 Scale for correlation coefficient

equals 1, perfectly negatively correlated when the correlation coefficient equals −1 and are not correlated when it equals 0. For magnitudes between 0 to 0.5, the correlation can be described as weakly to moderately positive; for magnitudes between 0 to −0.5 as weakly to moderately negative; for magnitudes between 0.5 to 1 as moderately to strongly positive; and moderately to strongly negative for magnitudes between −0.5 to −1.

The sample formula for the covariance is:

$$Cov(X,Y) = s_{X,Y} = \frac{1}{n-1}\sum(x-\bar{x})(y-\bar{y}) \tag{4.18}$$

Similarly, to the sample variance, there is a correction for the degrees of freedom. The sample correlation is calculated using:

$$r_{X,Y} = \frac{s_{X,Y}}{s_X s_Y} \tag{4.19}$$

where s_Y and s_X are sample standard deviations for Y and X, respectively.

There are a few caveats to note before undertaking a covariance or correlation calculation. These measures of association only identify linear relationships among random variables. A spurious correlation may be captured when there is a third common factor influencing the values of both random variables. For example, the correlation between UK house prices and the population of India is high due to both having time trends. Consequently, economic or financial theory should underpin any correlation analysis. Third, correlations may be unstable over time, especially if a sample has a relatively short time period or covers unusual events. We will now illustrate an application of covariance and correlation analysis to identify possible linear relationships which may exist among random variables.

4.4.2 Application: portfolio construction

Portfolio construction and performance analysis is based on the theorem set out by Markowitz (1952), which utilises information about the expected return and expected variance of assets and their association. The intuition behind portfolio construction is that the deviations of asset returns from their expected values vary systematically for each asset – a fall in the return from one asset may be compensated by a rise in the return from another asset. By appropriately selecting the right combination of assets, a portfolio can match the highest expected return of a particular asset for less risk or match the lowest expected risk of a particular asset for a higher expected return. Portfolios are constructed in the present in order to achieve a targeted risk-adjusted return in the future. As the future inflation rate is unknown, we will use the annual percentage nominal return data displayed in Table 3.11 for equities and bonds and in Table 3.16 for commercial property (de-smoothed). Note that assets could be included in a portfolio to act as specific instruments to hedge against future inflation.

The first step in correlation analysis is to plot the data in a scatter diagram. A scatter plot displays pairs of values of two variables on a graph. For example, one of the data points in the graph on the top left-hand corner is the percentage nominal return of commercial property (21.2%) and the percentage nominal return of equities (13.6%) in 1981. Figure 4.10 displays the scatter plots of percentage nominal returns of commercial property and equities, percentage nominal returns of commercial property and bonds and percentage nominal returns of equities and bonds. Included in the graphs are linear trend lines as an aid to view the pattern in data values.

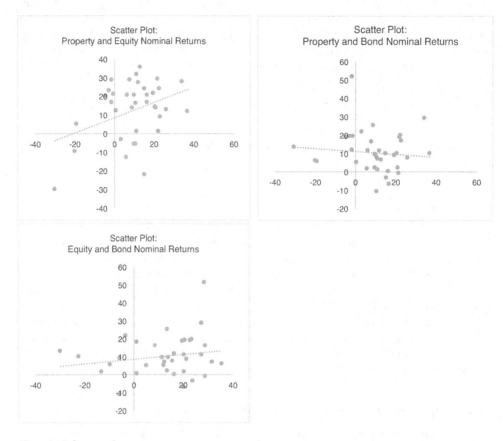

Figure 4.10 Scatter plots

The scatter plots reveal the following patterns: commercial property and equity returns appear to be positively correlated; commercial property and bond returns appear to be negatively correlated; and equity and bond returns are positively correlated. If the objective is to form a simple portfolio (two asset) with commercial property, they indicate that bonds would be the better asset to include than equities. Scatter plots are useful for a quick overview of patterns, but they often do not provide the detailed information required for an application or investigation.

The required inputs for constructing a portfolio are expected returns, risk and correlation. These inputs are then used in an algorithm to obtain the proportion of funds that should be invested in each asset, which are commonly referred to as portfolio weights, or weights. For a three-asset portfolio, the expected portfolio return and the expected portfolio risk are:

$$E\left(r_{Port}\right) = w_B E\left(r_B\right) + w_E E\left(r_E\right) + w_P E\left(r_P\right) \tag{4.20}$$

$$\sigma_{Port} = \sqrt{\sigma_{Port}^2} \tag{4.21}$$

$$\sigma_{Port}^2 = \left(w_B \sigma_B\right)^2 + 2\left(w_B \sigma_B\right)\left(w_E \sigma_E\right)\rho_{B,E} + \left(w_P \sigma_P\right)^2 + 2\left(w_B \sigma_B\right)\left(w_P \sigma_P\right)\rho_{B,P}$$
$$+ \left(w_E \sigma_E\right)^2 + 2\left(w_E \sigma_E\right)\left(w_P \sigma_P\right)\rho_{E,P} \tag{4.22}$$

where:

w represents the weights for bonds (B), equities (E) and property (P),

the expected portfolio return is a weighted average of the expected returns of each asset,

ρ represents the correlation coefficient,

the expected portfolio risk is a weighted average of the expected variance of each asset.

Equations (4.20) and (4.22) can be simplified for a two-asset portfolio by dropping the relevant terms, for example excluding equities from the portfolio requires excluding the terms $w_E E(r_E) (w_E \sigma_E)^2 \ 2(w_B \sigma_B)(w_E \sigma_E)\rho_{B,E}$ and $2(w_E \sigma_E)(w_P \sigma_P)\rho_{E,P}$ in calculations. The equations can also be expanded to accommodate calculations for portfolios comprising more than three assets

The preceding formulae reveal that there are potential diversification benefits to be gained by combining assets together to form a portfolio provided the correlation coefficient is below 1. The sample arithmetic mean and sample standard deviation are used to obtain estimates of the expected return and expected risk, respectively. The second set of required inputs are covariance or correlation measures. Table 4.3 illustrates how to calculate a measure of the sample covariance between bonds and commercial property.

The building block is the deviation from the mean, $(x-\bar{x})$ and $(y-\bar{y})$ (columns four and seven). From them, the product of the deviations from mean can be calculated $(x-\bar{x})(y-\bar{y})$ and summed together $\sum(x-\bar{x})(y-\bar{y})=-538.814\%\%$. From equation (4.18) the covariance is:

$$s_{B,P} = \frac{1}{n-1}\sum(x-\bar{x})(y-\bar{y}) = \frac{1}{36-1}(-538.814\%\%) = -15.395\%\%$$

In practice, the pre-programmed routines in Excel and statistical packages are used to calculate covariance and correlation. The estimated covariances are typically reported as a variance-covariance matrix (Table 4.4).

Numbers are normally only reported on and above the upper diagonal, as the numbers below will be mirror images of them. The numbers on the diagonal represent the variances, as a variance is a special type of covariance. This can be seen by substituting x for y in the covariance formula (equation (4.18)):

$$\frac{1}{n-1}\sum(x-\bar{x})(y-\bar{y}) = \frac{1}{n-1}\sum(x-\bar{x})(x-\bar{x}) = \frac{1}{n-1}\sum(x-\bar{x})^2$$

The numbers off the diagonal display the covariances, for example, the covariance between commercial property and bond returns is $-15.40\%\%$. In general, covariances reveal the direction of the association between random variables but do not convey much information about the strength. As the same unit of measurement are recorded for nominal returns, it reveals that the covariance between commercial property and equities is stronger than that between equities and bonds, but it does not provide an indication of how much stronger. We can convert the covariances into a correlation coefficient by dividing it with their respective standard deviations to address this. Table 4.5 reports the correlation matrix.

The correlation matrix is similar in appearance to the variance-covariance matrix except that the numbers represent correlations. They reveal that the positive association between commercial properties and equities is double that for equities and bonds. Since the correlation matrix reveals that the correlations among returns are below $+1$, there are diversification benefits to be gained from forming a portfolio. If the choice involved combining a

Table 4.3 Correlation bonds and property

Year	Commercial Property X (%)	Mean $\bar{x}$ (%)	Deviation From Mean $(x-\bar{x})$ (%)	Bonds Y (%)	Mean $\bar{y}$ (%)	Deviation From Mean $(y-\bar{y})$ (%)	Product of Deviation From Mean $(x-\bar{x})(y-\bar{y})$ (%)(%)
1981	21.2	10.0	11.2	1.8	10.3	−8.5	−95.474
1982	−1.3		−11.2	51.3		41.0	−460.432
1983	8.1		−1.9	15.9		5.6	−10.488
1984	11.3		1.3	6.8		−3.5	−4.707
1985	6.4		−3.6	11.0		0.7	−2.549
1986	12.3		2.4	11.0		0.7	1.661
1987	23.1		13.1	16.3		6.0	78.770
1988	37.0		27.0	9.4		−0.9	−24.154
1989	13.0		3.1	5.9		−4.4	−13.514
1990	−19.6		−29.6	5.6		−4.7	138.920
1991	−0.7		−10.7	18.9		8.6	−91.759
1992	−4.5		−14.5	18.4		8.1	−117.626
1993	34.4		24.4	28.8		18.5	451.990
1994	10.9		1.0	−11.3		−21.6	−20.912
1995	−2.5		−12.4	19.0		8.7	−108.195
1996	10.9		1.0	7.7		−2.6	−2.488
1997	22.8		12.8	19.4		9.1	116.592
1998	9.0		−0.9	25.0		14.7	−13.685
1999	15.5		5.6	−3.5		−13.8	−76.890
2000	9.9		−0.1	9.2		−1.1	0.073
2001	6.3		−3.7	1.3		−9.0	33.216
2002	15.3		5.3	9.8		−0.5	−2.644
2003	10.2		0.2	1.6		−8.7	−2.154
2004	26.3		16.3	7.2		−3.1	−50.476
2005	19.7		9.7	8.4		−1.9	−18.369
2006	16.7		6.7	−0.1		−10.4	−69.532
2007	−19.0		−28.9	5.2		−5.1	147.471
2008	−30.1		−40.0	12.9		2.6	−104.288
2009	21.8		11.8	−1.0		−11.3	−133.113
2010	20.8		10.8	9.4		−0.9	−9.666
2011	3.3		−6.6	21.4		11.1	−73.805
2012	1.0		−9.0	4.8		−5.5	49.269
2013	16.9		6.9	−7.2		−17.5	−120.860
2014	22.2		12.3	18.3		8.0	98.221
2015	11.4		1.4	0.5		−9.8	−13.773
2016	−1.2		−11.2	11.5		1.2	−13.446
Sum							−538.814
S	14.1			11.3			
COV(X,Y)							−15.395

Table 4.4 Variance-covariance matrix

	Commercial Property	*Equities*	*Bonds*
Commercial Property	197.938	80.73	−15.40
Equities		238.13	31.83
Bonds			127.38

Table 4.5 Correlation matrix

	Commercial Property	*Equities*	*Bonds*
Commercial Property	1.0	0.37	−0.10
Equities		1.0	0.18
Bonds			1.0

financial asset with commercial property to from a simple (two-asset) portfolio, then the correlations indicate that bonds rather than equities should be selected, as it is weakly negatively correlated with commercial property.

However, our objective is to form a three-asset portfolio. Portfolio objectives can vary. As we wish to illustrate the principle that the investment performance of a portfolio can be superior to investing in a single asset, we set as the sole objective to minimise the coefficient of variation (CV). In practice, there may be additional objectives and constraints, but we have abstracted from such considerations. Using Solver, an algorithm in Excel, the optimal solution is to invest 52% in bonds, 18% in equities and 30% in commercial property, as this leads to a portfolio CV of 0.79. The measures of expected investment performance are displayed in Table 4.6.

The optimal portfolio is expected to yield a return of 10.6%, higher than the returns expected to be achieved individually by commercial property and bonds, for a much lower risk of 8.4%. Although the optimal portfolio expected return is lower than equities, the expected risk is almost halved. The CV report the expected improvement in performance – the risk of achieving a unit of an expected return is lowest for the optimal portfolio.

Manually, we demonstrate the calculated return and risk. Our expected portfolio return is:

$$E(r_{Port}) = w_B E(r_B) + w_E E(r_E) + w_P E(r_P)$$
$$E(r_{Port}) = 0.52(10.3\%) + 0.18(12.6\%) + 0.30(10.0\%) = 10.6\%$$

Our expected portfolio variance is:

$$\sigma^2_{Port} = (w_B \sigma_B)^2 + 2(w_B \sigma_B)(w_E \sigma_E)\rho_{B,E} + (w_P \sigma_P)^2 + 2(w_B \sigma_B)(w_P \sigma_P)\rho_{B,P}$$
$$+ (w_E \sigma_E)^2 + 2(w_E \sigma_E)(w_P \sigma_P)\rho_{E,P}$$
$$\sigma^2_{Port} = (0.52*11.0\%)^2 + 2(0.52*11.0\%)(0.18*15.4\%)(0.18)$$
$$+ (0.30*14.1\%)^2 + 2(0.52*11.0\%)(0.30*14.1\%)(-0.10)$$
$$+ (0.18*15.4\%)^2 + 2(0.18*15.4\%)(0.30*14.1\%)(0.37)$$
$$\sigma^2_{Port} = 67.841\%\%$$

Table 4.6 Portfolio expected investment performance

	Commercial Property (%)	Equities (%)	Bonds (%)	Optimal Portfolio (%)
Expected Return	10.0	12.6	10.3	10.6
Expected Risk	14.1	15.4	11.0	8.4
CV	1.41	1.22	1.10	0.79

Our expected portfolio risk is:

$$\sigma_{Port} = \sqrt{67.841\%\%} = 8.2\%$$

The manual calculations are not exact, as we have used numbers corrected to one and two decimal places. The implicit assumptions in this application are that the annual percentage nominal returns of each asset are normally distributed and that the sample measures provide good estimates of the expected return and risk. We turn our attention to addressing both these issues.

4.5 Samples and sampling distributions

A topic in statistics which many students find confusing is distinguishing between the application of population and sample measures (calculations). Population measures are 'true' measures since they utilise all observations or reflect all values of a data generation process (DGP). However, information about a population is not always readily available nor are all the values of a DGP observed. For example, the census of population in the UK is undertaken once every decade as it is very time-consuming and expensive. In-between the censuses, information is still required by various government agencies about the population to draw up appropriate polices and to allocate resources such as funding for schools. Governments often commission surveys based on a sample (subset) of the population to update their records to aid in decision making. The term 'population' is also used to express a belief about a theory characterising the distribution of values. In this context the information at our disposal, our data, is a realization of this DGP and represents a value that is observed from a 'population' of possible values. In our example, applying conventional statistical methods to analyse investment performance, our belief is that percentage returns are normally distributed.

4.5.1 Population and sample measures

The symbols representing the moments of a population probability distribution are depicted by Greek letters. Population measures, also known as parameters, are always unbiased and accurate (efficient). A sample contains a subset of data drawn from the population. Statistical methods can be applied to sample data to describe the population parameters, for example, the arithmetic mean and standard deviation. The statistics calculated from a sample are estimates (estimated values) of the population parameters. There are two issues surrounding the use of estimates obtained from a sample: bias and precision (accuracy). So long as the sample values are representative of the population, the sample calculations (estimates) will be unbiased but not necessarily precise (efficient) due to the possibility of the sample including extreme values. Calculations or estimations using data from larger representative samples are less likely to be affected by extreme values and are likely to be more accurate compared to calculations undertaken from smaller samples. To take these points into consideration, sample measures are often accompanied by a report of the standard error and sample formulae contain adjustments for 'degrees of freedom', which is why they can be slightly different from the population formulae.

4.5.2   Random sampling

Random sampling describes the process in which every observation (value) from a population has a chance (non-zero probability) of being selected into the sample. A sample derived from a random sampling scheme is called a random sample. A statistic is a calculation computed from the data in a sample. Consequently, a random sample is always representative of a population and any statistic calculated is an unbiased estimate of a population value. Suppose we want to find out the typical price of a dwelling (apartments and houses) in a location. An appropriate statistical method (calculation) is the arithmetic mean. However, if the sample is designed to exclude houses and is restricted to apartments only, then the sample estimate of the typical price of a dwelling will be biased – downwards in this case, as apartments are less expensive than houses. No matter how many different samples are drawn, the exclusion of houses ensures that the sample arithmetic mean will be almost inevitably lower than the population (true) mean price.[1] Thus, random sampling and the resulting random sample will always ensure that a calculation is representative of the population ('true') value and will always provide unbiased estimates of population parameters. Unbiased estimates allow valid inferences about population parameter values to be made.

4.5.3   Random sampling distributions: the issue of bias

An inference based on a 'statistic' from a random sample utilises the concept of 'sampling', which refers to the theoretical possibility of drawing repeated random samples of the same size (n) from a population (N). Repeated random samples lead to different data values from the observations selected and different values of a 'statistic' calculated from the samples. In other words, the statistics themselves are values of a random variable and follow a probability distribution known as a sampling distribution. Treating a sample statistic as a value of a random variable is useful in addressing the twin issues of bias and precision in the estimation of population parameters. We shall illustrate these concepts using the sample arithmetic mean, but the principle applies to other statistics calculated from a sample.

Let us resume our example of finding the typical (mean) price of a dwelling in a location. We will generate artificial data to illustrate the concept of the sampling distribution of the sample means. Suppose the prices of dwellings in £000s in a location is normally distributed: $X \sim N(500, 25^2)$. Figure 4.11 displays the distribution of x-values. Although unnecessary in this case, the population arithmetic mean and standard deviation can be calculated from this

data using the formulae $\mu = \dfrac{\sum x_i}{N} = £500{,}000$ and $\sigma = \sqrt{\dfrac{\sum(x-\mu)^2}{N}} = £25{,}000$.

But let us suppose information about the population is unavailable. Instead let us generate a random sample of 500 dwellings drawn from all dwellings (the population) in the location. The sample arithmetic mean $\left(\bar{x} = \dfrac{\sum x_i}{n}\right)$, gives us an estimated value of the population arithmetic mean equal to £499,634. The sample standard deviation, $s = \sqrt{\dfrac{\sum(x-\bar{x})^2}{n-1}}$, is £25,548.

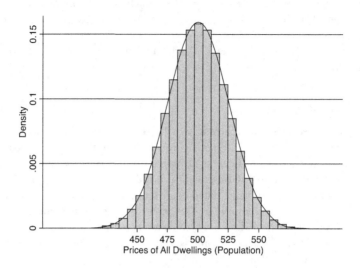

Figure 4.11 Distribution of prices of all dwellings in a location (population)

Is this a biased estimate of the population mean? No, this is an unbiased point estimate of the population mean, as every observation in the population had a chance of being selected into this sample. The difference in value between the population and sample mean is due to random sampling error. As the sample does not include all members of a population, sample statistics such as the arithmetic mean will generally differ from the characteristics of the population (population parameters). The difference arises from our example sample having a higher proportion of observations (dwellings) with lower prices compared to the population. There is always the issue of precision concerning any estimate of a population parameter.

Is this a good estimate? It is an unbiased estimate as the difference from the population arithmetic mean value occurs by chance. We can prove this statement if we repeatedly draw the same size random sample, n, from the population and calculate the sample arithmetic mean and standard deviation for each sample drawn. The sampling distribution of the sample means represents the distribution of all sample means calculated from all possible samples of a given size n drawn from a population. The distribution of the sample mean values are displayed in Figure 4.11. Their values are represented by a histogram which is overlaid by a smooth curve depicting what they would look like when normally distributed. Unsurprisingly, there is not much difference as the normal distribution was used to generate values of dwelling prices in the population.

Figure 4.12 proves that on average the estimates (sample arithmetic means) of the population parameter (population arithmetic mean) of the prices of dwellings yields the correct value as the expected value (mean) of the sample arithmetic means is equal to the population arithmetic mean. More succinctly, $E\left[\bar{X}\right] = \mu$.

Let us examine the resulting sampling distribution of sample means if houses are excluded from a sample. This is an example of non-random sampling as houses are systematically excluded from any sample drawn from the population of dwellings. We assume that the flat prices are $X_F \sim N(250, 20^2)$. The population arithmetic mean is $\mu_F = \dfrac{\sum x_i}{N} = £250,000$ and

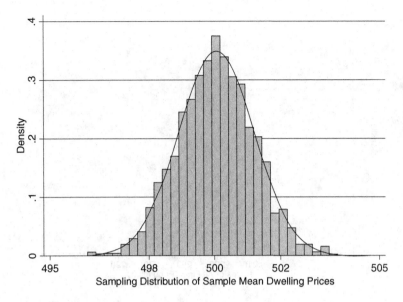

Figure 4.12 Sampling distribution of sample mean values of prices of dwellings

the population standard deviation is $\sigma_F = £20,000$ for flats. As before, the sampling distribution of sample means can be derived by repeatedly drawing same-sized samples from all flats in that location.

The lower graph in Figure 4.13 represents the sampling distribution and reveals that a sample arithmetic mean yields an unbiased estimate of the mean flat price in a location, £250,000, but does not provide an estimate of the mean dwelling price, £500,000. The estimates are biased downwards due to flats being generally less expensive than houses. Hence, the systematic exclusion of houses from sampling yields a biased estimate of the mean price of a dwelling in a location.

4.5.4   *Random sampling distributions: the issue of precision*

The dispersion around the expected value of a sampling distribution is called the standard error of the estimate. In our example it is called the standard error of the mean, $\sigma_{\bar{x}}$, as we are attempting to obtain an estimate of the population arithmetic mean. A standard error represents the loss of precision in having an estimated value rather than the actual value of the population parameter. We use a Greek letter $\sigma_{\bar{x}}$ because it is a population value (all possible sample means), and the subscript reminds us that it refers to a sampling distribution of the sample means. The standard error captures the loss in precision due to the chance selection of observations nearer the tails of the population distribution in the sample. It is affected by sample size, as extreme values and outliers have a disproportionate impact on the calculation of a statistic from a smaller sample compared to a larger sample, as the latter tend to have more values drawn from the centre of the population distribution. A standard error is always smaller than the population and sample standard deviations since the sample mean values will tend to be close to the population mean. The standard error in the sampling distribution in Figure 4.11 is $\sigma_{\bar{x}} = £1,150$.

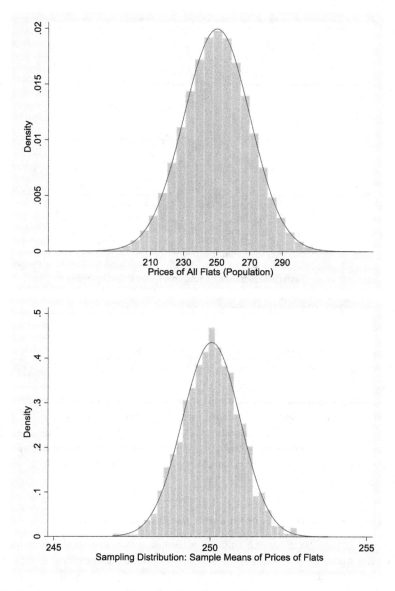

Figure 4.13 Distribution of prices of flats in a location (population) and the sampling distribution of
sample means of prices of flats

In practice, only one sample is drawn from a population and only one estimate is available
for each population parameter. Thus there is need to account for random sampling error.
The concept of a sampling distribution is helpful, as it can be used to assert that an unbiased
estimate is a reasonable measure of a population parameter and additionally to account
for the precision of the estimate. But before addressing the latter issue, let us examine the
important role of the central limit theorem, which can be applied to many types of sampling
distributions.

4.5.5 *Random sampling distributions: the Central Limit Theorem*

The Central Limit Theorem (CLT) informs us that the sampling distribution of the sample means will be approximately normal even when the population distribution of values may be non-normal, as long as the sample size is large enough (usually at least 30) and all repeated samples drawn are the same size. This is an extremely useful property, as it implies that the normal distribution may be used to address issues surrounding the precision of estimates of the population mean when the population values are not normally distributed.

Suppose the prices of dwellings in a location (population) is uniformly rather than normally distributed, with a mean $\mu = £250,000$ and standard deviation $\sigma = £25,000$. We can draw repeated random samples of 500 dwellings from this population and calculate the arithmetic mean for each sample using the formula $\bar{x} = \dfrac{\sum x_i}{n}$. The upper graph in Figure 4.13 reveals how far the population values deviate from a normal distribution, and the

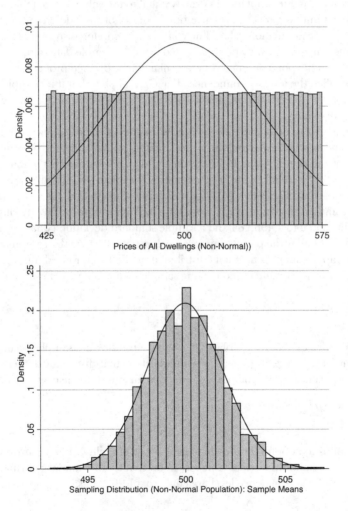

Figure 4.14 Distribution of prices of dwellings from a non-normal population and its sampling distribution of sample means

lower graph reveals that its sampling distribution of sample means is approximately normal. This example demonstrates the CLT.

4.6   Estimation

4.6.1   Estimation: point estimates

As we pointed out earlier, only one sample is usually drawn from the population. A sample arithmetic mean, a sample statistic, is a point estimate of a population arithmetic mean, a population parameter. Provided the sample is representative of the population (random sample), the concept of sampling distributions of the sample mean prove that this point estimate is unbiased.

4.6.2   Estimation: confidence intervals

However, a point estimate is not always exactly equal to the value of a population parameter due to random sampling error. A point estimate does not provide any information about how close it is to the population value. The concept of sampling distributions suggests that a confidence interval estimate may be constructed from the sample data to provide an interval of values in which the population parameter will lie for a specified probability. The specified probability is called the level of confidence. The CLT implies that it is possible to adopt the normal distribution to obtain this interval of values even when the population values are non-normally distributed.

We will initially focus on confidence interval estimation of the population arithmetic mean. We can combine its standard error with a probability to obtain a degree of confidence about the accuracy of our sample estimate. The combined term is known as the margin of error.

Let us return to our initial example of estimating the mean price of all dwellings in a location (population). Recall that the 'true' (population) mean price is £500,000 and the 'true' (population) standard deviation is £25,000. Our random sample of 500 dwellings yields an estimated value equal to £498,484 and a sample standard deviation equal to £25,548. Our point estimate lies below the population mean value £500,000. As this sampling distribution is normally or approximately normally distributed, it can be converted into a standard normal distribution. The formula for converting sample mean values, $\bar{x}$, into z-values is:

$$z = \frac{(\bar{x} - \mu)}{\sigma_{\bar{x}}} \tag{4.23}$$

Notice that $\bar{x}$ replaces x and $\sigma_{\bar{x}}$ replaces σ when we transform a sample mean from a sampling distribution into a z-value. As it is a standard normal distribution, there is a specific relationship between the probabilities and intervals surrounding the central value:

$$P\left(z_1 \leq \frac{(\bar{x} - \mu)}{\sigma_{\bar{x}}} \leq z_2\right) \tag{4.24}$$

The convention is to adopt a 95% confidence level (probability). From our previous discussion (and from the standard normal table), we know that for the standard normal distribution the z-values for a 95% confidence level are equal to ± 1.96. Thus:

$$P\left(-1.96 \leq \frac{(\bar{x} - \mu)}{\sigma_{\bar{x}}} \leq 1.96\right) = 0.95 \tag{4.25}$$

Equation (4.33) states the well-known fact that 95% of the z-values in a standard normal will lie within ±1.96 of the central value 0. However, the margin of error is usually reported as a value measured in the units of measurement of the sample mean value, $\bar{x}$, which in our example is £. By a bit of manipulation, the terms inside the brackets can be rearranged to obtain:

$$P\left(\bar{x}-1.96\sigma_{\bar{x}}\leq\mu\leq\bar{x}+1.96\sigma_{\bar{x}}\right)=0.95 \tag{4.26}$$

The probability can now be reinterpreted as representing a 95% degree of confidence that the population mean will be larger than $\bar{x}-1.96\sigma_{\bar{x}}$ but smaller than $\bar{x}+1.96\sigma_{\bar{x}}$. In other words, the margin of error associated with a point estimate is $\pm1.96\sigma_{\bar{x}}$ for a 95% confidence level.

Returning to our example, the point estimate is £499,634 and the standard error of the sampling distribution is £1,150. The margin of error associated with the point estimate is $\pm1.96\sigma_{\bar{x}} = \pm1.96(£1,150) = \pm£2254$. Hence we can be 95% confident that the mean dwelling price in the location (population) lies between £498,484($= £499,634-£1,150$) and £500,784($= £499,634+£1,150$). Recall that the mean dwelling price in the location is £500,000.

The generic formula for constructing a confidence interval (without specifying the probability level) is:

$$\bar{x}+z\sigma_{\bar{x}} \tag{4.27}$$

The second component, $z\sigma_{\bar{x}}$, represents the margin of error. If a 95% confidence level is specified, then $z = \pm1.96$, whereas the z-value for a 90% confidence level is ±1.645. The appropriate z-value can be obtained from the standard normal table. There are however additional considerations in accounting for accuracy in estimation.

Issue 1: the standard error is unobserved

There is a problem in that a sampling distribution is unobserved (ours is artificially created) and there is no information available about the standard error $\sigma_{\bar{x}}$. Fortunately, the CLT reveals that there is a relationship between the standard error and population standard deviation:

$$\sigma_{\bar{x}} = \frac{\sigma}{\sqrt{n}} \tag{4.28}$$

With information about the population standard deviation, the standard error and subsequently the confidence interval can be computed. In our example, the population standard deviation is £25,000. The standard error is: $\sigma_{\bar{x}} = \dfrac{£25,000}{\sqrt{500}} = £1,118.03$. Note how close this calculated value is to the standard error derived from our sampling distribution.

Using this information, the margin of error associated with our estimate is $\pm1.96\sigma_{\bar{x}} = \pm1.96(£1,118.03) = \pm£2,191.34$. We can be 95% confident that the mean dwelling price in the location lies between £497,442.66($= £499,634-£2,191.34$) and £501,825.34($= £499,634+£2,191.34$).

Issue 2: the population standard deviation is unobserved

Very often, the population standard deviation (σ) is unknown. But we know that the best estimate of the population standard deviation is the sample standard deviation(s). The standard
error can be calculated using:

$$\sigma_{\bar{X}} = \frac{s}{\sqrt{n}}$$

(4.29)

The standard error is:

$$\sigma_{\bar{X}} = \frac{£25,548}{\sqrt{500}} = £1,142.54$$

The margin of error is $\pm 1.96\sigma_{\bar{X}} = \pm 1.96(£1,142.54) = \pm£2,239.38$. Consequently, we
can be 95% confident that the mean dwelling price in the location lies between £497,394.62
(= £499,634 − £2,239.38 and £501,873.98 (= £499,634 + £2,239.38).

Issue 3: small samples

When the population standard deviation is unknown and replaced by a sample estimate of
it (the sample standard deviation), the continued use of the standard normal (z) value is an
approximation which is only suitable for relatively large samples, deemed in statistics to be
when the sample size is 30 or more ($n \geq 30$) as it relies on the CLT. A more accurate confidence interval can be constructed using the Student's t distribution.

The Student's t distribution is a continuous probability distribution, with many similar
characteristics to the standard normal distribution. The t distribution addresses the discrepancy between a sample standard deviation(s) and the population standard deviation (σ) when
s is calculated from a very small sample ($n < 30$). The t distribution (with 10 degrees of freedom) and the standard normal distribution are shown graphically in Figure 4.15. Note that

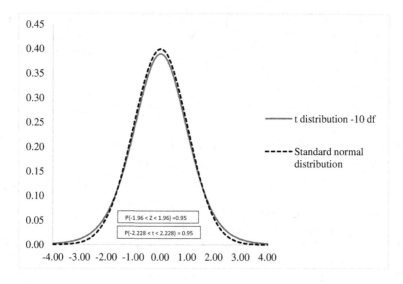

Figure 4.15 Standard normal and Student's t distribution

the t distribution is flatter, more spread out, than the standard normal distribution because the standard deviation of the t distribution is larger than the standard normal distribution.

The main properties of the t distribution are:

- It is a continuous bell-shaped and symmetrical distribution.
- Unlike the standard normal distribution, there is not one t distribution but a family of t distributions.
- All t distributions have a mean value equal to 0, but their standard deviations differ according to the available degrees of freedom, which partly depend on sample size, n. There is a separate t distribution for each degree of freedom.
- The standard deviation of a t distribution with 10 degrees of freedom is wider than for a t distribution with 20 degree of freedom.
- The t distribution has a wider spread and a flatter peak at the centre than the standard normal distribution. As the sample size and subsequently the available degrees of freedom increases (in the limit), the t distribution converges to the standard normal distribution as the random sampling error associated with using the sample standard deviation as an estimate of the population standard deviation decreases.

The value of t for a confidence level is larger in magnitude than the corresponding z-value because the t distribution has a greater spread than the z distribution. For example, the z- and t-values for a 95% level of confidence for the arithmetic mean when the sample size $n = 11$ are ±1.96 and ±2.228, respectively. Families of t distribution are determined by degrees of freedom (df). It is difficult to explain this concept formally to readers without a degree in mathematics or statistics. We adopt an approximate and practical explanation. The degrees of freedom, typically represented by the Greek letter v or Latin letters df, describe the number of values in the final calculation of a statistic that are 'free to vary'. It measures the amount of information from the sample data that has been used up previously in an intermediate calculation. Every time a statistic is calculated from a sample and used to replace a population parameter, one degree of freedom is used up ('lost'). In constructing a confidence interval, the population standard deviation (σ) is replaced by the sample standard deviation(s) in calculating the standard error $(\sigma_{\bar{x}})$, and therefore one degree of freedom is lost. A confidence interval constructed for a point estimate of a population mean from a sample of 11 observations has $v = 11 - 1 = 10$ degrees of freedom.

The t table can be found in Appendix A (Table A2). This table of the t distribution can be used in several different applications. It may be used to undertake a one- or two-tailed hypothesis test and to construct a confidence interval. It can be applied with different probabilities. The column df represents the degrees of freedom available for the relevant t-value. We will just focus on the application of constructing a confidence interval. For a point estimate computed from a sample of 11 observations, the available degrees of freedom is $v = 11 - 1 = 10$. As we are applying a 95% confidence level, we focus on the column 95% under the title Confidence Intervals and extract the number in that column intersecting the row representing 10 degrees of freedom, giving us a t-value of 2.228. Notice that the table only gives us consecutive t-values for small samples ($n < 30$), which have low degrees of freedom.

Table 4.7 provides a guide on when to use the t or z distribution in constructing a confidence interval. The first column requires the researcher to consider whether the population values are normally or non-normally distributed, the second column the sample size, the third and fourth columns whether the population standard deviation is known or a sample standard deviation has to be used instead respectively for computing the standard error. In

Table 4.7 Guide to using the standard normal or *t* distribution to construct confidence intervals

Population	Sample Size	Population standard deviation known	Population standard deviation unknown
Normally Distributed			
	Large (n ≥ 30)	z value	t value
			or
			z value
			(Central Limit Theorem)
	Small (n < 30)	z value	t value
Population	Sample Size	Population standard deviation known	Population standard deviation unknown
Non-normally Distributed			
	Large (n ≥ 30)	z value	t value
			or
			z value
			(Central Limit Theorem)
	Small (n < 30)	non-parametric statistic values	

general, the *t*-value should be used to construct a confidence interval when the population standard deviation in unknown, but the *z*-value can be used as an approximation in the case of a large sample ($n \geq 30$). When the population values are non-normally distributed and the sample size is small, non-parametric statistics should be employed. We do not consider them as it is beyond the scope of our book.

We return to our example of addressing the issue of precision of the sample mean as an estimate of the mean dwelling price in a location but this time using *t* instead of a standard normal value *z*. The generic formula for constructing a confidence interval using a *t*-value is:

$$\bar{x} + t\sigma_{\bar{x}} \tag{4.30}$$

The standard error remains the same: $\sigma_{\bar{x}} = \dfrac{£25,000}{\sqrt{500}} = £1,118.03$. The sample size is 500.

The *t*-value for $v = n - 1 = 500 - 1 = 499$ degrees of freedom can be found from a software package, for example, typing in Excel T.INV.2T(5%,499) yields 1.96, the same value as using the standard normal distribution because the sample is large. In cases when the sample is smaller, the *t*-value will be larger than the *z*-value for a specified confidence level.

4.6.3   *Application: future investment performance analysis I*

To illustrate the concepts outlined earlier, we undertake an analysis of the future investment performance of the three assets considered earlier. As we are interested in the future rather than past investment performance, the analysis uses nominal percentage returns.

Conventional investment performance analysis adopts the arithmetic mean and standard deviation as measures of expected return and expected risk, respectively, but their validity as estimates depend upon the assumption of asset percentage nominal returns being normally distributed. Let us for the moment consider this assumption to be valid. Since nominal percentage returns can be treated as random variables, there exists an expected value which can be expressed generically as:

$$E[X] = \mu \qquad (4.31)$$

Provided the percentage return data are derived from information representative of each asset market and are observed over enough time periods to capture market cycles, they can be treated as a sample to obtain an estimate of their expected return. The concept of the sampling distribution of sample means asserts that the sample arithmetic mean is an unbiased point estimate of the population arithmetic mean, in this case the expected percentage return, since the mean of the sample means is equal to the population mean:

$$E[\bar{X}] = \mu \qquad (4.32)$$

Table 4.8 is generated using the Data Analysis/Descriptive Statistics function in Excel on the nominal percentage return data. For commercial property, the calculations are undertaken on the de-smoothed nominal percentage returns. We only report the relevant parts of the full table for this discussion.

The sample arithmetic means provide estimates of the expected return and are reported in the first row. The estimates of the expected risk are the sample standard deviations displayed in the third row. Given the characteristics of the cash-flows of each asset considered in Chapter 3, the point estimates of the expected return and risk indicate an anomaly, as the expected return from bonds is marginally higher than commercial property even though the latter's expected risk as measured by the standard deviation is higher.

Any unbiased estimate of a population parameter is likely to be subject to random sampling error. Although the arithmetic mean may be an unbiased point estimate of the expected return when returns are normally distributed, this estimate could be affected by random sampling error. We can apply a 95% confidence interval to establish the margin of error associated with our estimates of the expected return ($\bar{x} \pm t\sigma_{\bar{x}}$).

Excel and many statistical packages routinely report the standard error of the sample mean and its margin of error. However, it is instructive to calculate both manually. Focusing on commercial property, Excel computed the standard error using $\sigma_{\bar{x}} = \dfrac{s}{\sqrt{n}} = \dfrac{14.1\%}{\sqrt{36}} = 2.35\%$

Table 4.8 Descriptive statistics percentage nominal returns

Statistical Measures	Commercial Property	Equities	Bonds
Mean $(\bar{x})$	10.0	12.6	10.3
Standard Error $\left(\sigma_{\bar{x}} = \dfrac{s}{\sqrt{n}}\right)$	2.3	2.6	1.9
Standard Deviation (s)	14.1	15.4	11.3
Count (n)	36	36	36
Confidence Level (95.0%) $(\sigma_{\bar{x}}t)$	4.76	5.22	3.82

(again our manual calculations are subject to rounding error). The confidence level reported in the last row of the table is the margin of error, $to_{\bar{x}}$. Since a confidence interval is being constructed, there is a loss of one degree of freedom. The *t*-value of a 95% confidence level at $v = 35$ (36 − 1) degrees of freedom is 2.03; this can be obtained in Excel by typing T.INV.2T(5%,35). The margin of error is therefore $to_{\bar{x}} = 2.03(2.35\%) = 4.77\%$ (rounding error).

The table reveals that we can be 95% confident that the expected return of commercial property, equities and bonds will lie at $(10.0 \pm 4.76)\%$, $(12.6 \pm 5.22)\%$ and $(10.3 \pm 3.82)\%$, respectively. The margin of error reveals that the point estimates of the expected return of equities and bonds are the least and most reliable respectively. The interval estimates provide a possible explanation of the anomaly observed between the investment performance estimates reported for commercial property and bonds. They indicate that it is possible for the expected return of commercial property to be higher than bonds.

4.6.4   Estimation: constructing confidence intervals using the chi-square (χ_v^2) distribution

Confidence interval construction is not limited to z and t distributions or for obtaining interval estimates of the sample mean. It is possible to construct a confidence interval for a standard deviation. In our application of analysing investment performance, this would be to address the margin of error of the estimate of the expected risk. But as variances and standard deviations are a very different type of measure compared to an arithmetic mean, there are major differences in its construction. For example, the variance and resulting standard deviation is non-negative. Assuming returns are normally distributed, the relevant sampling distribution is a chi-square (χ_v^2) Figure 4.16 displays examples of chi-square distributions having two (chi-sq2), five (chi-sq5) and ten (chi-sq10) degrees of freedom.

The chi-square distribution is used directly or indirectly in many tests of significance, typically when the test statistic involves the summing of squared normally distributed variables. There are multiple formulae for the calculation of a chi-square value. In the context of

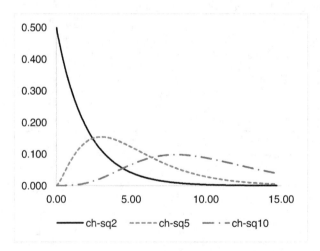

Figure 4.16 Chi-square distributions with various degrees of freedom

our discussion, a basic chi-square distribution concerns the sampling distribution of z-score squared:

$$z^2 = \frac{(X - \mu)^2}{\sigma^2} \tag{4.33}$$

The centre and shape of distribution of the chi-square (or its PDF) depends on the available degrees of freedom. As the degrees of freedom increase, the curve becomes less skewed and eventually converges in the limit to a normal distribution. A summary of the key properties of the chi-square distribution are:

- It takes only positive values ranging from zero to infinity
- It is a skewed distribution but becomes more symmetrical as the degrees of freedom increase
- The expected value of the chi-square distribution is its degrees of freedom
- The expected variance of the chi-square distribution is twice the number of the degrees of freedom.

The general formula for a confidence interval for the standard deviation is:

$$P\left(\frac{(n-1)s^2}{\chi^2_{(\alpha/2),n-1}} < \sigma^2 < \frac{(n-1)s^2}{\chi^2_{(1-\alpha/2),n-1}} \right) = 1 - \alpha \tag{4.34}$$

or

$$P\left(\sqrt{\frac{(n-1)s^2}{\chi^2_{(\alpha/2),n-1}}} < \sigma < \sqrt{\frac{(n-1)s^2}{\chi^2_{(1-\alpha/2),n-1}}} \right) = 1 - \alpha \tag{4.35}$$

where $v = n - 1$ degrees of freedom.

α represents the probability remaining after specifying the confidence level, $1 - 0.95 = \alpha$. The significance level α is considered in greater detail in our discussion of hypothesis testing. For a 95% confidence level, $\alpha = 0.05$. The lower limit confidence interval for a standard deviation is:

$$s\sqrt{\frac{(n-1)}{\chi^2_{(\alpha/2),35}}} \tag{4.36a}$$

The upper limit confidence interval for a standard deviation is:

$$s\sqrt{\frac{(n-1)}{\chi^2_{(1-\alpha/2),35}}} \tag{4.36b}$$

The chi-squared values can be found from a comprehensive chi-square distribution table or obtained in Excel by typing CHIINV((alpha/2), n−1)) and CHIINV(1−(alpha/2), n−1)) for the lower and upper limits, respectively. The chi-squared values are:

$$\chi^2_{(0.05/2),35} = 53.20$$

$$\chi^2_{(1-0.05/2),35} = 20.57$$

For commercial property, the lower limit confidence interval for a standard deviation is:

$$s\sqrt{\frac{(n-1)}{\chi^2_{\alpha/2}}} = 14.1\%\sqrt{\frac{(36-1)}{53.20}} = 11.4\%$$

and the upper limit confidence interval for a standard deviation is:

$$s\sqrt{\frac{(n-1)}{\chi^2_{1-\alpha/2}}} = 14.1\%\sqrt{\frac{(36-1)}{20.57}} = 18.4\%$$

Unlike the arithmetic mean, the confidence intervals for the standard deviation are not symmetrical. As the standard deviation is always a positive number, the lower confidence limit cannot be less than zero, which means that the upper confidence interval usually extends further above the sample standard deviation and the lower limit extends below it. This asymmetry is particularly noticeable in small samples.

Applying equations (4.36a) and (4.36b), we can be 95% confident that the risk is expected to lie between:

11.4% and 18.4% for commercial property
12.5% and 20.1% for equities
9.2% and 14.7% for bonds

The interval estimates of expected risk confirm that commercial property is likely to be riskier than bonds, which is line with their cash-flow properties. The caveats are that the standard deviations are assumed to be an appropriate measure of risk, and the application of the chi-square distribution in the construction of confidence intervals assumes that the estimates of the standard deviations are normally or approximately normally distributed.

4.7   Hypothesis testing

Rather than estimating an interval of values within which the population parameter is expected to lie, hypothesis testing sets out a procedure to test the validity of a statement about a population parameter. A hypothesis is a statement about a population or the DGP. In an application, a hypothesis is stated, the required data (sample) collected and then used to test the assertion. The problem with using a sample information is that it is prone to random sampling error. A hypothesis test allows us to distinguish between an implausible hypothesis statement and random sampling error in deciding whether to reject or not to reject the proposition. It is a procedure based on sample evidence and probability theory to determine whether the hypothesis is a reasonable statement. There are three related methods in hypothesis testing: the confidence interval approach, the critical value approach and the *p*-value approach. The focus in this book is on the critical value and *p*-value approaches as these are the conventional procedures applied in real estate data analysis.

4.7.1   Stylised depiction of a hypothesis test

A stylised depiction of a hypothesis testing procedure is shown in Figure 4.17.

Figure 4.17 lists the sequential considerations of executing a hypothesis test. In the research undertaken by most real estate researchers and analysts, a conventional testing procedure has

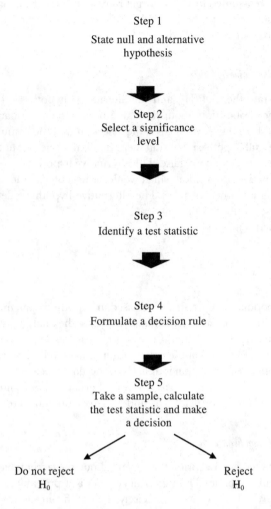

Figure 4.17 Stylised hypothesis testing procedure

already been formulated. The researcher or analyst is concerned with the execution of the hypothesis test and the hypothesis statements do not need to be formulated from scratch. For example, they do not have to be concerned about designing a hypothesis statement to act as 'straw man' to take advantage of the fact that it is easier to reject a false hypothesis than a true hypothesis. Our book focuses on explanations concerning the execution of testing of conventional hypotheses rather than in the design of them.

4.7.2   *Principles underlying a hypothesis test: two-tailed test*

We will explain the underlying concepts of hypothesis testing and the steps listed in Figure 4.16 by testing an argument. A risk premium denotes that an asset return must be above the return available from a riskless asset to compensate for the additional risk of investing in it. The conventional assumption is that commercial property has 2% risk premium. This is the risk premium often assumed in setting target rates of return involving property. Does the data support this assumption (proposition)? We will use de-smoothed commercial property percentage returns to test this hypothesis.

Step 1: state null and alternative hypothesis

The first step is to state the null (H$_0$) and the alternative hypothesis (H$_a$ or H$_1$). The null hypothesis (H$_0$) is the proposition about the value of a population parameter – it reflects the purpose of the test. In our example, it is the assertion that the commercial property risk premium is 2%. The null hypothesis is a statement that is not rejected unless our sample data provide convincing evidence that it is false. The alternative hypothesis can be denoted by (H$_a$) or (H$_1$). It is the alternative proposition which would be accepted if the sample data provides evidence that the null hypothesis is false. The alternative hypothesis determines whether a one- or two-tailed test is undertaken.

The statement is written as:

$$H_0: \mu = 2\%$$
$$H_a: \mu \neq 2\% \quad (4.37)$$

Representations of population parameters are used in hypothesis statements as they reflect a belief that this is the 'truth'. In stating the null hypothesis, the analyst is describing a hypothesised distribution and the value stated is the hypothesised expected value. The alternative statement reflects the possibility that the risk premium may not be 2%. It could be higher or lower. Since the alternative proposition is unlikely, it should be located at the tail of a hypothesised distribution. In other words, there are two alternative possibilities, hence the term two-tailed test. We explain the principles of undertaking a one-tailed test in a later example.

Step 2: select the level of significance

The level of significance is an a priori statement about the acceptable likelihood that an incorrect decision will be made. The significance level is a probability and conventionally represented by the Greek letter α. More precisely, the significance level represents the probability of rejecting the null hypothesis when it is true. The lower the significance level, the less chance of incorrectly rejecting the null hypothesis. The choice of the significance level determines the likelihood of making a type 1 or type 2 error. Table 4.9 outlines the relationship between them.

The decision column represents the decision to either reject or not reject a null hypothesis. The true situation is unknown. The second column reflects the situation when the null hypothesis is true and the third when it is false. The table reveals that an analyst may reject

Table 4.9 Type 1 and type 2 errors

Decision	True Situation	
	Null Hypothesis true	*Null Hypothesis false*
Reject null hypothesis	Type 1 error	correct
Do not reject the null hypothesis	correct	Type 2 error

the null hypothesis when it is in fact true, thereby making a type 1 error. However, if the null hypothesis is false, then the analyst has made the correct decision. It is possible for an analyst to fail to reject a false hypothesis, thereby making a type 2 error. Why do decision errors occur even when a sample is representative of the population? The answer is random sampling error. By chance, it is possible that the sample includes extreme values which would affect the calculated test statistic. Stating the significance level is designed to address this issue. Making a type 1 error is considered to be more serious than making a type 2 error. The conventional approach in most applications is to adopt a 5% ($\alpha = 0.05$) level of significance, which means that there is a possibility of making a mistake in rejecting the null hypothesis 5 times out of 100. It reflects the concept of repeated sampling in testing the proposition. But other levels of significance may also be used, such as 10% or 1%. A reader may wonder why we do not set a significance level to eliminate or almost eliminate making a type 1 error. The problem is that there exists a trade-off between making a type 1 and type 2 error. When you reduce the likelihood of making a type 1 error, you automatically increase the likelihood of making a type 2 error. Thus, you can never be certain that you have made a correct decision in hypothesis testing.

The significance level can be expressed as a probability as shown or converted into a critical value. The method of conversion to a critical value depends upon the test statistic applied in the test. In our examples, the conventional 5% level of significance is adopted.

Step 3: identify the sample test statistic

The sample test statistic reflects the sampling distribution chosen to undertake the hypothesis test and incorporates the sample information to determine whether the null hypothesis should be rejected. Examples of test statistics used in this textbook are the standard normal (z), the t-statistic, the F-statistic and the chi-squared statistic. We use the z-statistic in this example. The z-statistic can be calculated from our data using:

$$z = \frac{\bar{x} - \mu}{\sigma_{\bar{x}}} x \qquad (4.38)$$

$\bar{x}$ is the sample mean or sample estimate of the typical risk premium, μ is the hypothesised value of the risk premium and $\sigma_{\bar{x}}$ is the standard error of the mean.

Step 4: formulate a decision rule

A decision rule states the specific conditions under which the null hypothesis is rejected and not rejected. The region or area of rejection defines the location of the values that are large or small enough to make the probability of their occurrence under the true null hypothesis highly unlikely. Figure 4.18 illustrates this graphically.

The distribution at the centre of Figure 4.18 is the hypothesised distribution and reflects the values that would be realised when the null hypothesis asserting the typical risk premium is 2% is valid. The alternative hypothesised distributions represented by the dotted lines characterise the typical risk premium being greater or less than 2%. The diagram also shows the risk premium as a z-value. If the sample supports the proposition, then the sample mean should be also 2%, and hence have a z-value equal to zero ($\bar{x} - \mu = 2\% - 2\% = 0$). The alternative distributions are depicted as overlapping the tails of the null hypothesised distribution. The point at where they overlap is called the critical value; when the test statistic value exceeds the critical value, it suggests that the null hypothesis is implausible.

We will initially apply the critical value approach to hypothesis testing. A two-tailed test indicates that there are two possible regions (areas) of rejection: the right (upper) and the left (lower) tail of the curve. The significance level therefore must be divided equally into two to capture both possibilities. Thus, the probability of taking a value in each tail is 0.05/2 = 0.025 or 2.5%. The critical values of the hypothesised sampling distribution separate the area of non-rejection from the area of rejection. Since we have already identified the appropriate test statistic to apply in this case, we can use the standard normal tables to find the critical values. We may also use the t table, as the t distribution turns into the standard normal when there is an infinite number of degrees of freedom. The t table in (Table A.2 in Appendix A) informs us that the critical z-value for a two-tailed test at a 5% significance level is ±1.96. If our z-test statistic falls outside ±1.96, it is unlikely that the assertion that the risk premium in commercial property is 2% is plausible. Figure 4.19 presents a graphical

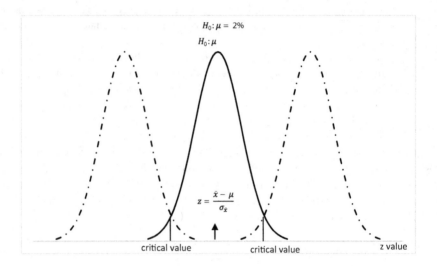

Figure 4.18 The null and alternative hypothesised distributions in a two-tailed z-test

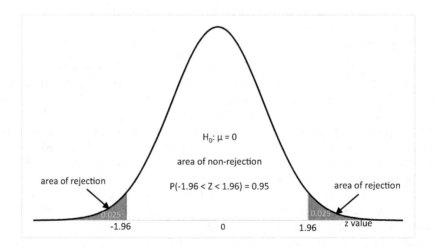

Figure 4.19 Decision rule: significance level and critical values

depiction of the decision rule in our example. The shaded area represents the region of rejection of the null hypothesis.

Step 5: take a sample, calculate the test statistic and make a decision

In hypothesis testing we either reject or do not reject the null hypothesis. We use the words 'do not reject' or 'fail to reject' instead of 'accepting' the null hypothesis because even if we fail to reject the null hypothesis, it does not mean the null hypothesis is true. A hypothesis test does not determine which hypothesis is true or most likely. It only assesses whether the available evidence exists to reject the null hypothesis.

The standard formula for extracting the risk premium from commercial property percentage nominal returns (de-smoothed) is:

$$risk\ premium_t = \frac{\left(1 + r_t^{prop}\right)}{\left(1 + r_t^{T\ Bill}\right)} - 1 \quad (4.39)$$

Table 4.10 displays the sample data in percentages and their arithmetic means, standard deviations and standard errors. The sample arithmetic mean of the risk premium represents the typical risk premium. The standard error of the risk premium is obtained using equation (4.40).

$$\sigma_{\bar{x}} = \frac{s}{\sqrt{n}} = \frac{14.3\%}{\sqrt{36}} = 2.4\%$$

The z-statistic used in tests involving the sample arithmetic mean is:

$$z = \frac{\bar{x} - \mu}{\sigma_{\bar{x}}} = \frac{\bar{x} - \mu}{s / \sqrt{n}} \quad (4.40)$$

$$z = \frac{3.5\% - 2.0\%}{14.3\% / \sqrt{36}} = 0.63$$

Since the z-test statistic value of 0.63 lies within the critical values ± 1.96 (area of non-rejection of the null hypothesis), a decision should be taken not to reject the null hypothesis at a 5% level of significance (see Figure 4.19). We conclude that there is sample evidence to support the assertion of a 2% risk premium in investing in commercial property.

4.7.3 Principles underlying a hypothesis test: one-tailed test

Suppose it is argued that the typical risk premium is less than 2% as there is a new economic paradigm (this was an argument some commentators used in the years leading up to the global financial crisis). Is this a reasonable proposition? The null (H_0) and alternative hypothesis (H_a) statements are:

H_0: $\mu \geq 2\%$
H_a: $\mu < 2\%$

The null hypothesis statement should always include an equals sign. The null hypothesis is the statement being tested, and a specific value has to be included in the calculations. The

Table 4.10 Sample data

Year	Commercial Property De-Smoothed Nominal Returns	T Bill Nominal Returns	Risk Premium
	(%)	(%)	(%)
1981	21.2	13.8	6.5
1982	−1.3	12.4	−12.2
1983	8.1	10.1	−1.8
1984	11.3	9.6	1.6
1985	6.4	11.9	−5.0
1986	12.3	10.9	1.3
1987	23.1	9.6	12.3
1988	37.0	11.0	23.4
1989	13.0	14.6	−1.4
1990	−19.6	15.9	−30.6
1991	−0.7	11.6	−11.0
1992	−4.5	9.5	−12.8
1993	34.4	5.9	26.9
1994	10.9	5.4	5.3
1995	−2.5	6.7	−8.6
1996	10.9	6.2	4.5
1997	22.8	6.9	14.9
1998	9.0	7.9	1.1
1999	15.5	5.5	9.5
2000	9.9	6.2	3.5
2001	6.3	5.5	0.7
2002	15.3	4.1	10.8
2003	10.2	3.8	6.2
2004	26.3	4.6	20.7
2005	19.7	5.0	14.0
2006	16.7	4.9	11.2
2007	−19.0	5.9	−23.5
2008	−30.1	5.2	−33.5
2009	21.8	0.7	20.9
2010	20.8	0.5	20.2
2011	3.3	0.5	2.8
2012	1.0	0.3	0.7
2013	16.9	0.3	16.5
2014	22.2	0.4	21.8
2015	11.4	0.4	10.9
2016	−1.2	0.4	−1.6
Mean ($\bar{x}$)	10.0	6.5	3.5
Std Dev (s)	14.1	4.5	14.3
Count (n)	36	36	36
Std Error ($\sigma_{\bar{x}}$)	2.3	0.7	2.4

inequality sign in the alternate hypothesis identifies the region of rejection, in this case the left tail of the hypothesised distribution. The alternative hypothesis statement puts forward explicitly the argument that the risk premium is less than 2%.

The conventional significance level, 0.05, is adopted but this time this probability does not have to be split between two tails. Unlike a two-tailed test, there is only one critical value.

The z-test statistic value remains unchanged as the same information in the two-tailed test is applied to test the new proposition.

$$z = \frac{\bar{x} - \mu}{\sigma_{\bar{x}}} = \frac{3.5\% - 2\%}{2.4\%} = 0.63$$

Since the test statistic is a standard normal value, the critical value is found using the standard normal table. As before, the t table can be used for convenience as the distribution of a t-value with infinite degrees of freedom is equivalent to a standard normal distribution. The t table reveals that the critical z-value for a 5% significance level is −1.645. See Figure 4.20.

The shaded area in the diagram represents the region (area) of rejection, where the probability is equal to 0.05 and the critical value (−1.645) defines the start of its boundary on the horizontal axis.

Comparing the test statistic against the critical value is equivalent to identifying which region the test statistic value lies on the horizontal axis in the graph. In this case, it is in the area of non-rejection of the null hypothesis.

As the test statistic value, $z = 0.63$, lies above the critical z-value −1.645, we cannot reject the null hypothesis at a 5% level of significance and conclude that the risk premium has not fallen below 2%.

It is also possible to undertake a right-sided one-tailed test of a hypothesis, using the same principles described earlier.

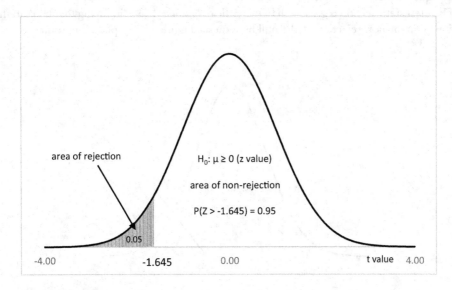

Figure 4.20 One-tailed test: testing for a reduction in the risk premium

4.7.4   *Hypothesis testing for small samples: an example using the t distribution*

The standard normal is appropriate for testing a hypothesis using large samples ($n \geq 30$). In our risk premium example, there are 36 observations. A conservative approach could be adopted by using a test statistic designed for smaller samples: the *t*-test statistic. We will now employ the *t*-test statistic to examine whether the risk premium is less than 2%. The null (H_0) and alternative hypothesis (H_a) statements are:

$H_0: \mu \geq 2\%$
$H_a: \mu < 2\%$

The significance level remains at 5%. The formula for the *t*-test statistic involving tests of the sample mean is:

$$t = \frac{\bar{x} - \mu}{\sigma_{\bar{x}}} = \frac{\bar{x} - \mu}{s / \sqrt{n}} \tag{4.41}$$

$$t = \frac{\bar{x} - \mu}{s / \sqrt{n}} = \frac{3.5\% - 2.0\%}{14.3\% / \sqrt{36}} = 0.63$$

Finding the critical value for a *t* sampling distribution is trickier compared to the standard normal as it is necessary to account for the available degrees of freedom. Since the population standard deviation is replaced by the sample standard deviation in the calculation of the standard error, there is a loss of one degree of freedom, resulting in $v = 36 - 1 = 35$ available degrees of freedom. The critical *t*-value for 35 degrees of freedom is not reported in the *t* table but can be found in Excel by typing =T.INV(0.05,35), which yields a critical value equal to -1.690.

The *t*-test statistic 0.63 is greater than the critical *t*-value -1.690, implying that it lies in the region (area) of non-rejection of the null hypothesis. Figure 4.21 displays the result.

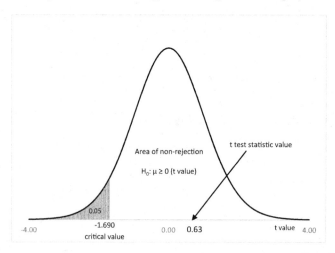

Figure 4.21 A diagrammatical representation of the result testing the risk premium using a one-tailed *t*-test

Table 4.11 Critical values of z and t sampling distributions

Significance Level	Type of Test	
	One-Tailed Critical Values	Two-Tailed Critical Values
Standard normal 5%	+1.645 (−1.645)	+1.96 and −1.96
t distribution 5% (35 df)	+1.690(−1.690)	+2.030 and −2.030

We conclude that the sample evidence supports the proposition that the risk premium of commercial property is equal to or greater than 2%.

The main implication of using the t distribution rather than the standard normal concerns the critical values. For a specified level of significance, the t critical values are larger to reflect the possibility of greater random sampling error due to a smaller sample size. Table 4.11 illustrates the differences between the critical values of the z and t sampling distributions when the sample size is 36.

4.7.5   The p-value approach to hypothesis testing

Many software packages often report a *p*-value as a standard output in statistical analysis involving hypothesis testing, avoiding the need to refer to statistical tables for critical values. The *p*-value stands for probability value. It is the probability of obtaining a value of the sample statistic that is at least as extreme as the value observed if the null hypothesis were true. It represents the lowest significance level at which the null hypothesis can be rejected when determined by the sample test statistic. The *p*-value approach to hypothesis testing is similar to the significance level approach except that it is the probabilities associated with the test statistic value and the critical value which are compared. Let us demonstrate using our previous example. Recall that the null (H_0) and alternative hypothesis (H_a) statements are:

$H_0: \mu \geq 2\%$
$H_a: \mu < 2\%$

The significance level remains at 5%. The *t*-test statistic value is:

$$t = \frac{\bar{x} - \mu}{s / \sqrt{n}} = 0.63$$

The *p*-value approach requires the analyst to calculate probability of the test statistic taking a value less than or equal to 0.63 in the sampling distribution when the null hypothesis is true. The *p*-value can be obtained in Excel by typing T.DIST(0.63,35,1), which reports a value equal to 0.734. Figure 4.21 illustrates the *p*-value approach.

The *p*-value provides an indication of the probability of making a type 1 error, represented by the shaded area in the diagram. The *p*-value approach requires the investigator to compare the *p*-value against the significance level. The null hypothesis should be rejected when the *p*-value is less than or equal to the significance level and not rejected when the *p*-value is larger than the significance level. A *p*-value of 0.734 reveals that rejecting the null

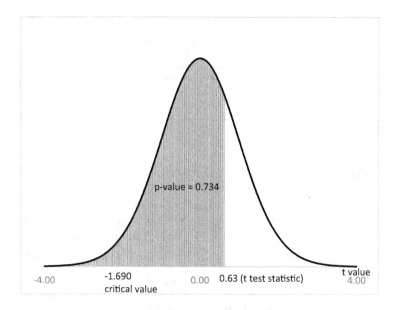

Figure 4.22 The *p*-value approach to testing the risk premium using a one-tailed test

hypothesis in our example could lead to it being rejected incorrectly 73 times out of 100 – a far larger likelihood than is deemed acceptable as stated in setting the significance level. Consequently, the null hypothesis should not be rejected.

4.7.6   Chi-square hypothesis test: future investment performance analysis II

Hypothesis testing may involve other types of sampling distributions such as the chi-square distribution. We resume our analysis of future investment performance among three assets. Table 4.12 is created using the Data Analysis/Descriptive Statistics function in Excel on the nominal percentage return data. Again, we only report the relevant parts of the full table for this discussion.

The validity of adopting the sample arithmetic mean and sample standard deviation as estimates of expected return and risk depend upon the assumption of asset percentage nominal returns being normally distributed. Otherwise, the formulae will yield biased estimates of the expected return and risk. The assumption of normality may not hold because returns may be asymmetrically distributed and/or have fat or thin tails. Excel actually reports an estimate of excess kurtosis even though it refers to it as kurtosis. We highlight this discrepancy in brackets in the table. Row 3 reports that all asset percentage nominal returns are leptokurtic, implying that asset returns may be prone to tail risk, a risk not captured by the standard deviation measure. The estimates presented on the fourth row imply that commercial property and equity returns are negatively skewed while bond returns are positively skewed. As these are estimates, they may be prone to random sampling error. A normality hypothesis test can be used to assess whether the asset returns are significantly affected by excess kurtosis or skewness. The Jarque-Bera (JB) test is a goodness-of-fit test of whether the sample data have

Table 4.12 Descriptive statistics percentage nominal returns

Statistical Measures	Commercial Property	Equities	Bonds
Mean	10.0	12.6	10.3
Standard Deviation	14.1	15.4	11.3
Kurtosis (*Excess Kurtosis*)	1.28	0.62	3.92
Skewness	−0.80	−1.01	1.25
Count	36	36	36

the skewness and kurtosis matching a normal distribution. The null and alternative hypothesis statements for our application are:

H$_1$: asset percentage return values are normally distributed
H$_0$: asset percentage return values are not normally distributed

The significance level is 0.05. The JB test statistic follows a chi-squared distribution with 2 degrees of freedom. The chi-squared distribution is asymmetrical. The JB test statistic is calculated using the formula:

$$JB = \frac{n}{6}\left(SK^2 + \frac{(K-3)^2}{4}\right)$$

(4.42)

where:
SK = skewness
$(K-3)$ = excess kurtosis
n = sample size

It should be obvious that $JB = 0$ if the normality assumption holds. The JB statistic for:

commercial property: $JB = \dfrac{36}{6}\left((-0.80)^2 + \dfrac{(1.28)^2}{4}\right) = 6.26$

equities: $JB = \dfrac{36}{6}\left((-1.01)^2 + \dfrac{(0.62)^2}{4}\right) = 6.66$

bonds: $JB = \dfrac{36}{6}\left((1.25)^2 + \dfrac{(3.92)^2}{4}\right) = 32.32$

The critical value for this normality test is always $\chi^2_{0.05,2} = 5.99$, which can be obtained from the chi-square table or typing '=CHISQ.INV.RT(0.05,2)' in Excel. Figure 4.23 displays the areas of non-rejection (unshaded) and rejection (shaded area) of the null hypothesis.

Since the JB test statistic values are larger than the critical values and lie in the area of rejection, the assumption of normality can be rejected at a 5% significance level. The arithmetic mean and standard deviation formulae provide incorrect estimates of the expected return and risk respectively for all asset returns considered. We may obtain p-values to assess the likelihood of making a type 1 error. Typing '=CHISQ.DIST.RT(6.26,2)' in Excel yields the *p*-value of the JB test statistic for commercial property. The *p*-values for equities and bonds can be similarly obtained. The *p*-value for commercial property is 0.044, implying that there is a 4.4% chance that we are rejecting the assumption of normality incorrectly. The *p*-value for equities is 0.036 and bonds is 0.000. It appears that bond percentage nominal returns are almost certainly not normally distributed. These results indicate that other measures should be used to obtain estimates of the expected return and risk.

We investigated further and found that a potential explanation could involve the Great Financial Crash in 2007/2008 (GFC) and subsequent policy interventions in its aftermath. Table 4.13 compares the descriptive statistics and components of the normality test for the full sample period (1981 to 2016) and a sample restricted to the period before the GFC (1981 to 2006).

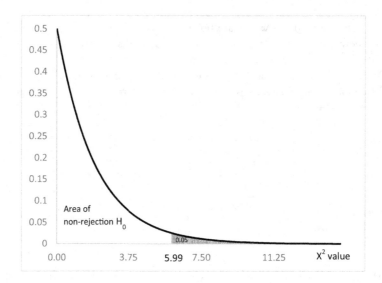

Figure 4.23 JB test for normality

Table 4.13 Investment performance measures

Statistical Measures	Full Sample: 1981–2016			Pre-GFC: 1981–2006		
	Commercial Property	Equities	Bonds	Commercial Property	Equities	Bonds
Mean (%)	10.0	12.6	10.3	12.0	14.9	11.3
Standard Deviation (%)	14.1	15.4	11.3	12.1	14.8	12.1
Excess Kurtosis	1.28	0.62	3.92	1.1	0.5	3.9
Skewness	−0.80	−1.01	1.25	−0.2	−1.1	1.3
Count	36	36	36	26	26	26
JB Test Statistic	6.26	6.66	32.32	1.51	5.41	24.03
Critical Value	5.99	5.99	5.99	5.99	5.99	5.99
p-value	0.04	0.04	0.00	0.47	0.07	0.00

The normality tests indicate that commercial property and equity percentage nominal returns are normally distributed in the period prior to the GFC. Conversely, bond returns are not normally distributed, which implies that the arithmetic mean and the standard deviation for it may not be appropriate as measures of expected return and risk for this asset.

4.7.7   Testing the significance of the correlation coefficient

Hypothesis testing can be applied to testing the significance of relationships between random variables. Let us apply it to testing for the significance of a correlation. The statistical significance of a correlation can be examined with the *t*-test. The sample correlation coefficient, *r*, provides an estimate of the 'true' correlation but is affected by random sampling error. The null hypothesis is represented by *p*, the population correlation coefficient. The hypothesis statements for this test are:

$$H_0 : \rho = 0$$

$$H_1 : \rho \neq 0 \tag{4.43}$$

The general correlation test of statistical significance is a two-tailed test. The conventional significance level adopted is 0.05.

The *t*-test statistic is:

$$t = \frac{r\sqrt{n-2}}{\sqrt{1-r^2}} \tag{4.44}$$

We shall now apply this test to investigate the correlation between the annual percentage nominal returns of commercial property and equities. The correlation is a useful indicator of the potential diversification benefits of forming an investment portfolio, but this relies on returns being normally distributed. Table 4.13 indicates that commercial property and equity percentage nominal returns are normally distributed in the period 1981 to 2006 (26 years). Using equations (4.18) and (4.19), the estimate of their correlation is 0.1367, positive and weak. We would like to know if this correlation estimate is significant as it is going to be used to make an investment decision. The degrees of freedom available are $\nu = n - 2 = 26 - 2 = 24$, the loss of 2 degrees of freedom resulting from replacing two sets of population parameters with sample estimates in (4.18) and (4.19). The critical *t*-values for a two-tailed test at a 5% significance level when there are 24 available degrees of freedom is 2.064.

Applying equation (4.44), the *t*-test statistic value is:

$$t = \frac{0.1367\sqrt{26-2}}{\sqrt{1-(0.1367)^2}} = 0.68$$

Since the *t*-test statistic value lies in-between the critical values, we cannot reject the null hypothesis ($H_0 : \rho = 0$) at a 5% level of significance and conclude that there is no correlation between commercial property and equity returns. We should be wary about basing the composition of a portfolio on the sample correlation coefficient. This is also an example of a contemporaneous correlation test. There are other applications of correlation tests, such as testing the significance of autocorrelations which will be covered in later chapters.

4.8 Concluding remarks

The analytical framework underlying applied statistical analysis is the random variable and the extensions to joint and conditional probability distributions characterising the relationships among random variables. The conditional probability distribution is explained in Chapter 5. Understanding the concept of a random variable is important, as it explains the reasons for employing summary measures characterising distributions to explain not only past events and outcomes but also the future. It further provides a rationale for adopting sample measures in order to obtain estimates of population parameters (true values). The relationships between a population, sampling distribution and a sample provide the foundation for undertaking the more advanced statistical analysis covered in later chapters. The branch of statistics focusing on issues surrounding estimation and hypothesis testing is known as inferential statistics. In this chapter we explained the concept of bias and precision in estimation using the sampling distribution of sample means and showed how a point estimate may be adjusted to allow for a margin of error through the construction of a confidence interval. Three important considerations in the construction of confidence intervals include the type of distribution of the population values, the population parameter of interest and the sample size. The same concepts, albeit applied in a slightly different way, are employed in hypothesis testing. Hypothesis testing provides a framework for testing propositions about the population or DGP. It controls for the fact that an estimate obtained from a sample in testing may be measured with random error in the evaluation of whether the proposition is plausible or implausible. The hypothesis testing framework is flexible and can be extended to test more complicated propositions such as relationships between populations or differences between populations using information contained in variables as sample data. All of these are important concepts underlying the application of statistics in data analysis which are often implicit in calculations and their interpretation, particularly in investigations attempting to explain and draw conclusions about phenomena observed in a market, markets or in more general statistical parlance, populations. The real estate profession is increasingly embracing statistical methods which embody these concepts in analysis as data become more readily available.

Chapter 4 online resource

In the online resource supporting this chapter, a brief guidance document and a dataset are provided so that you can replicate the calculations using Excel.

- Chapter 4 accompanying notes (covering autocorrelation, autocorrelation patterns – acf and pacf – and significance of autocorrelations)
- Excel file: "ch4_excel"

Note

1 There is the small possibility that the apartment sample will contain exceptionally expensive apartments so that the resulting sample mean will be close to the population mean.

5   Simple regression analysis

5.1   Introduction

The focus in this and the remaining chapters is on regression analysis. Regression analysis is an important econometric technique in quantitative analysis. It has been the dominant method in empirical research in real estate and a popular technique in the analysis of time-series and cross-section data. It provides a rigorous method for the real estate student and analyst to test theories and investigate relationships between real estate and the broader economic and investment environment. Empirical results from regression models are easy to interpret and communicate, which makes regression analysis quite appealing in practice. Forecasting from regression models and basic scenario analysis are also straightforward and further increases its attractiveness among practitioners. The growing availability and sophistication of real estate data allow the researcher to deploy more complex econometric methods. A good understanding of regression analysis is a prerequisite for more advanced work and certainly it is the starting point not only in real estate but in other fields too.

In this chapter we present the fundamentals of regression analysis by studying the single-equation linear regression model (also referred to as univariate or bivariate regression models). The theoretical notions are accompanied by illustrative examples so that the reader achieves a strong grasp of the notions and applications of regression to real estate.

The chapter begins by pointing out the difference between correlation and regression analysis, as past experience has revealed that many students are unable to differentiate between the two concepts. We then explain the concepts underlying the theoretical regression model, the population regression function, and its representation in estimation, the sample regression function, and the ordinary least squares estimation. We examine issues relating to bias and efficiency in estimation and briefly consider the assumptions required for the ordinary least squares to be the Best Linear Unbiased Estimator (BLUE) to ensure that is the appropriate formula to apply to the data.

Having explained the key concepts, we take the reader through a practical example of modelling commercial prices in the US covering manual calculations involving ordinary least squares (OLS) and interpretation of simple regression results. Given the prominence of regression models in real estate, another example of how to construct, estimate and evaluate both a simple regression model (studied in this chapter) and a multiple or multivariate regression model (the focus of the next chapter) is given in the online resource of this chapter. The online resource also illustrates regression analysis in EViews. Before presenting the theory and applications, it is important to understand the model building process.

* *Online note #5.1: Model building and applied econometric analysis*

5.2 Regression versus correlation – the difference

Correlation or covariance analysis allows us to evaluate the direction and the significance of the linear relationship between two variables but does not assign any direction of causality, the latter referring to one variable determining the value of another. Recall the concept underlying correlation is:

$$Cov(X,Y) = E\big[(x - E(X))(y - E(Y))\big] \tag{5.1}$$

The expectations operator denotes that there is a joint (bivariate) probability distribution explaining the deviations from the mean values of random variables X and Y. A joint (bivariate) probability distribution $f(X,Y)$ reflects the probabilities associated with the deviations from the mean that occur simultaneously – a systematic relationship where negative deviations of X from its expected value $E(X)$ are associated with a likelihood of positive (or negative) deviations of Y from its expected value $E(Y)$ when the correlation is positive (negative). The main point is that there is no statement about the direction of causality. If we were to apply correlation to analysing office rents for example, we would not be able to explicitly state that a particular variable (factor) is a driver of rent values. Moreover, correlation analysis is restricted to examining a non-causal relationship between two variables.

Regression analysis on the other hand forces the analyst to explicitly state the direction of causality by categorising variables into dependent and explanatory variables. A dependent variable is the variable of interest in that the analyst is concerned about how its value is determined and an explanatory variable is the variable whose values determine the value of the dependent variable. It is clear that the dependent variable, represented conventionally by the letter Y, depends upon the value of the explanatory variable, represented conventionally by the letter X.

Table 5.1 lists the alternative ways of describing the variables in regression analysis. More precisely, the main focus in applying regression as a tool in analysis is to explain the conditional mean of the dependent variable: $E(Y|X)$.

Notice the difference between the conceptual expressions of covariance (or correlation) and regression. The perpendicular line separating Y from X denotes that the values of Y are conditional upon the values of X. The expectations operator, $E(.)$, informs us that we are looking at the expected value which the dependent variable Y takes for the given value of an explanatory variable X.

Table 5.1 Terminology for variables in a regression model

Y	$X(X_s)$
Dependent variable	Independent variable(s)
Predicted variable	Regressor(s)
Explained variable	Explanatory variable(s)
Regressand	Causal variable(s)
Left-hand side (LHS) variable	Right-hand side (RHS) variable(s)
	Determinant(s)
	Control variable(s)

5.3 Population regression function (PRF): key concepts

The population regression function is a representation of a relationship which we believe to be true, usually derived from theory. For example, the life-cycle housing demand and supply theory underpins the regression modelling of real house prices, which informs us that real house prices are positively influenced by the level of real household incomes and population but negatively influenced by the housing user cost (special discount rate) and the housing stock.

In a simple (bivariate) regression, the value of the dependent variable is assumed to be explained by one explanatory variable. We will adopt the subscript i to represent the observations relevant to the variables. For example, i could refer to particular real estate sector, market, building, locations or households. In a time-series context, the subscript is usually replaced by subscript t. The expected value of a linear causal relationship can be written as:

$$E(Y_i) = \alpha + \beta X_i + E(u_i) \tag{5.2}$$

where:
Y_i = dependent variable
X_i = explanatory variable
u_i = disturbance error
α = intercept or constant
β = population parameter (slope coefficient)
i = particular observation

It is convention to use Latin letters and the Greek alphabet to represent variables and parameters respectively in describing a population regression function. There are no observation subscripts for the parameter β, as it is assumed that the effects of a change in the value of an explanatory variable remain the same across observations, an assumption that can be relaxed using advanced regression techniques not considered in this book. To save on notation, the preceding equation is often expressed as:

$$E(Y_i) = \alpha + \beta X_i + u_i \tag{5.3}$$

The population parameter (β) attached to the explanatory variable indicates the direction and the magnitude by which a marginal change in the value of the explanatory variable X_i affects the value of the dependent variable Y_i. Marginal in regression analysis refers to a one-unit change. The unit can be in terms of a currency (dollars, pounds, euros, other), square metres, an index unit, percentage points or percentages depending on how the explanatory variable is measured and the functional form of the model. The latter is discussed in the book's online chapter I. The disturbance error, u_i, is a term designed to capture all other non-important influences that may affect (disturb) the values of the dependent variable – influences that by chance (non-systematically) play a role. Equation (5.3) is a univariate or bivariate linear regression model as there is only one independent variable or regressor (X).

There is no expectations operator attached to an explanatory variable as it is assumed that its value is fixed (given) and therefore is not associated with a distribution in modelling. On the other hand, the preceding population regression equation tells us that marginal changes in the value of the explanatory variable affect the expected value of the dependent variable, which refers to the most likely value that it will take if the explanatory variable assumes a particular value. Figure 5.1 graphically illustrates the regression concept.

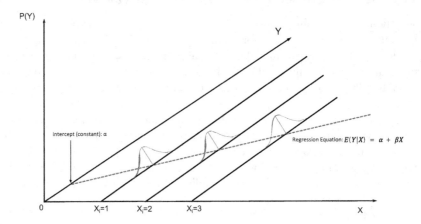

Figure 5.1 The concept of the population regression function (PRF)

The diagram introduces the important concepts underlying a linear regression by decomposing the unconditional mean into a conditional mean and disturbance error term. $P(Y)$ refers to the probability of Y taking certain values given a specific value of X. The alternative to drawing the diagram is to state:

$$E(Y_i) = E(Y_i|X_i) + E(u_i) \tag{5.4}$$

where:
$E(Y_i) =$ unconditional mean of the dependent variable
$E(Y_i|X_i) =$ conditional mean of the dependent variable
$E(u_i) =$ conditional mean of the disturbance error term

The diagram highlights how each value of an explanatory variable X_i leads to an interval of possible values of the dependent variable Y_i, the conditional distribution, depicted by the curves on the diagonal axes. Each value of an explanatory variable, represented by $X_{i=1}$, $X_{i=2}$, $X_{i=3}$, generates a different distribution of values of the dependent variable. The purpose of regression is therefore to identify the conditional mean, $E(Y_i|X_i)$, which reveals the most likely value that the dependent variable will take for given values of the explanatory variable. The most likely value that the dependent variable will take is the value associated with the highest probability (the peak of the conditional distribution). The conditional mean is represented by the dot on the dotted line intersecting the possible values the dependent variable could take for each value of the explanatory variable.

The conditional mean represents the linear regression model. Rather than drawing the diagram, the conditional mean can be written as:

$$E(Y_i|X_i) = \alpha + \beta X_i \tag{5.5}$$

Various formulae can be used to extract the conditional mean from the unconditional mean, but the formula we discuss later in this chapter is the formula applied in standard regression analysis which is ordinary least squares (OLS).

In our explanation of regression in this chapter, we are assuming that a single explanatory variable X_i is the sole determinant of the values of our dependent variable (bivariate regression). What we actually mean by this is that it is the only variable which systematically influences the values of the dependent variable. However, as the diagram shows, the conditional distributions of our dependent variable can take values other than the conditional mean value (i.e. the probability that it takes a conditional mean value is not equal to 1). The reason is that there may be other influences which unsystematically affect the values of the dependent variable. Such influences are irregular, occasional and therefore inconsequential. The regression modelling accounts for them by applying the concept of the disturbance error term, represented by u_i. The disturbance error can be expressed as:

$$u_i = Y_i - X_i \text{ or } u_i = E(Y_i) - E(Y_i|X_i) \tag{5.6}$$

All the factors unsystematically affecting the values of the dependent variable are subsumed into the disturbance error. The disturbance error term also follows a conditional distribution as it depends on the conditional mean of the dependent variable. As there is no pattern in the influence of an explanatory variable, the conditional mean of the disturbance error term is expected to be zero:

$$E(u_i|X_i) = 0 \tag{5.7}$$

For convenience, the term $E(u_i|X_i)$ is often written as $E(u_i)$ with the implicit understanding that the disturbance error term is derived from a conditional distribution.

Figure 5.2 is a two-dimensional diagram of the PRF summarising the key elements.

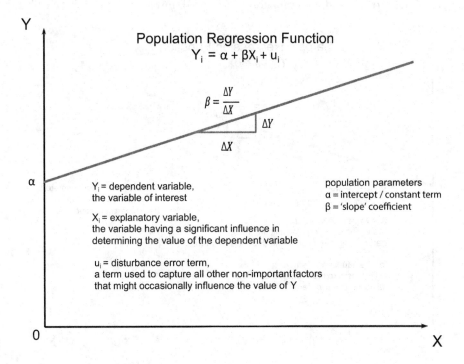

Figure 5.2 The population regression function

The influence an explanatory variable (X) has on the value of the dependent variable (Y) is represented by the population slope parameter β. The direction and magnitude of the effect is measured by how much a unit increase in the value of an explanatory variable, X, raises the value of the dependent variable, Y. In Figure 5.2 we have assumed that the explanatory variable has a positive effect as the PRF is upward sloping. For a negative effect, the PRF line would be downward sloping.

5.4   The sample regression function (SRF): key concepts

The sample regression function (SRF) represents the depiction of the calculation of estimates of the population regression function. The sample regression function can be written as:

$$Y_i = a + bX_i + e_i \tag{5.8}$$

where:
a = estimator (formula used to obtain estimate) of the intercept
b = estimator of (formula used to obtain estimate) of the slope parameter
e = residual error term

Unlike the PRF, the convention in specifying the SRF is to adopt lowercase Latin letters to represent the constant (intercept) and slope coefficient. Note that they provide a general representation of the 'estimator' or formula that will be used to derive estimates (values) from the data. Figure 5.3 provides a graphical illustration of the SRF.

The SRF contains 'estimators' (a, b) which provide a point estimate ($\hat{\alpha}$, $\hat{\beta}$) for the population parameters (α, β) and an 'estimator' (e) for the disturbance error term (u) called the

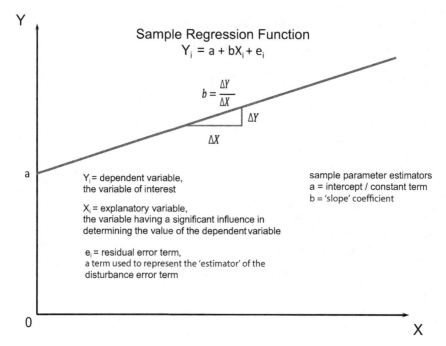

Figure 5.3 The sample regression function

residual error term. Because the components of the SRF are 'estimators', they follow a probability distribution, which implies that hypotheses tests can be undertaken, a useful property
as it means we can test theoretical propositions and examine the robustness (validity) of the
estimated regression model.

Note that the expression $b = \dfrac{\Delta Y}{\Delta X}$ in Figure 5.3 represents the decomposition of b (how to
interpret a value derived for b) and not the formula for its calculation.

An alternative describing the SRF is to use predicted values, that is, the value the regression model predicts that the dependent variable Y will take for the chosen estimator and
values of the explanatory variable:

$$Y_i = a + bX_i + e_i$$

$$Y_i = \hat{Y}_i + e_i \tag{5.9}$$

where:

$$\hat{Y}_i = a + bX_i \text{ or } \hat{Y}_i = \hat{\alpha} + \hat{\beta}X_i \tag{5.10}$$

The predicted value (also commonly referred to as Y-hat) represents values of the dependent
variable on the dotted line in Figure 5.3.

A stylised diagram of the relationship between the PRF and the SRF is shown in Figure 5.4.
The objective in regression analysis is to ensure that the SRF is a good approximation of the
PRF. Graphically, it refers to the line representing the SRF being as close as possible to the

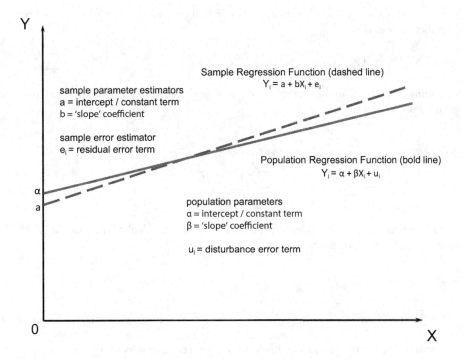

Figure 5.4 The population and sample regression function

line representing the PRF. How do we know when a sample SRF is a good approximation of the PRF? Truly we never know, but we are able to ascertain with a degree of confidence that it may be a good approximation by assessing the plausibility of the estimates obtained and diagnostic testing to ensure that the a priori properties of the 'estimator' are not violated.

The estimator employed in conventional regression analysis is ordinary least squares (OLS), as it is the best linear unbiased estimator (BLUE). In this case, the SRF is likely to be a good approximation of the PRF when the estimates are plausible and the assumptions underlying the classical linear regression model (CLRM) are satisfied. The desirable properties of the estimator depend upon these assumptions being met. We discuss these terms in detail in the next sections.

5.5   The ordinary least squares estimator (OLSE)

The ordinary least squares (OLS) estimator refers to a specific method in deriving formulae used to calculate estimated values of the population parameters. To understand the formulae, we can re-express the simple linear sample regression function as the 'residual errors' made (prediction mistakes) after fitting a regression line.

Re-arranging equation (5.10), we get:

$$e_i = Y_i - a - bX_i \tag{5.11}$$

or

$$e_i = Y_i - \hat{Y}_i \tag{5.12}$$

The method of OLS involves fitting a straight line that minimises the sum of squared residuals (prediction mistakes):

$$\min \sum e_i^2 = \sum (Y_i - \hat{Y}_i)^2 \tag{5.13}$$

This minimisation leads to the formulae for a simple regression for calculating the slope and the intercept/constant:

5.5.1   Slope

$$b = \frac{\sum (X_i - \bar{X})(Y_i - \bar{Y})}{\sum (X_i - \bar{X})^2} \quad \text{or } b = \frac{\sum X_i Y_i - n\bar{X}\bar{Y}}{\sum X_i^2 - n\bar{X}^2} \tag{5.14}$$

These are different ways of expressing the same formula to calculate an estimate of β. We may also denote the slope b as $\hat{\beta}$.

5.5.2   Intercept/constant

$$a = \bar{Y} - b\bar{X} \tag{5.15}$$

The derivation of the preceding formulae can be found in Appendix 6A of this chapter. Note that they are only valid for a simple regression. A multiple regression has different formulae.

The formulae reveal that the estimate, $\hat{\beta}$, the value derived from applying the estimator

$$b = \frac{\sum(X_i - \bar{X})(Y_i - \bar{Y})}{\sum(X_i - \bar{X})^2}$$ is not a correlation coefficient, and that the regression line has to

pass through the unconditional mean values of Y and X in order to identify the location of the intercept, $a = \bar{Y} - b\bar{X}$.

Figure 5.5 illustrates the concept of applying OLS in the SRF.

The black dots represent data values and the grey dot represents the special coordinate on the regression line, with the latter having to pass through the arithmetic mean values of Y and X, $(\bar{x}, \bar{y})$. The difference between the regression line (or predicted value) and the data value is known as the residual value (e_i) and reflects the deviation of the actual value of the dependent variable from its predicted value. The method of OLS yields the best possible fitted straight line to the data values as it ensures that the sum of squared residuals (prediction mistakes from fitting the line) is at a minimum. There is no other linear estimator which can improve on this fit. However, the validity of applying OLS in regression analysis depends upon the plausibility of the estimates and meeting the assumptions of the classical linear regression model. We will explain both concepts later on in this chapter.

In the sample regression function model, the sample estimates of the coefficients are themselves random variables due to the concept of sampling distributions. Recall the concept of sampling distribution of an estimate in the previous chapter. They will vary in value from sample to sample but will on average converge to the 'true value' of the relationship (α, β). The coefficient estimates are useful if accompanied by a measure of their variability or dispersion around their 'true value'.

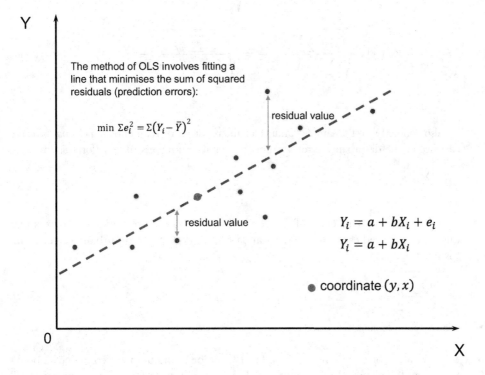

Figure 5.5 Sample regression function: the ordinary least squares estimator

5.6 Sampling variability of OLS estimators

In the empirical regression model, the sample estimates for the coefficients are average estimates. The regression coefficients themselves are random variables since they depend on the extent of the deviation of the linear combinations of the explanatory variables from the stochastic dependent variable. The value of the coefficients will vary from sample to sample. The coefficient estimates are useful if accompanied by a measure of variability or dispersion around their mean values. The standard error of an estimate is a similar concept to the standard error of the estimate of the arithmetic mean which we explained in Chapter 4.

The focus is on the slope coefficient. What is the dispersion of the random variable $(\hat{\beta})$? A smaller dispersion from its central location or expected value is preferred. Suppose the empirical estimation of a model results in a slope coefficient $(\hat{\beta})$ of 5. At a confidence level of 90% or 95%, does $\hat{\beta}$ in our model vary between, say, 4 and 6 or between 2 and 8? In the first case, we would have greater confidence in the coefficient estimate as it yields a more confined interval for the value of the true (population) value of β. It is therefore important to provide a measure of coefficient dispersion to assess the reliability of the OLS estimators $\hat{\alpha}$ and $\hat{\beta}$. The dispersion measures of the coefficients are based on the variance and its square root or the standard error of the estimate. The standard error is a concept comparable to the standard deviation, but there is a different adjustment made to the degrees of freedom.

In the bivariate model it can be shown that the standard errors for the coefficients (based on sample estimations) are given by the following formulae:

$$SE\left(\hat{\beta}\right)=\sqrt{Var\left(\hat{\beta}\right)}=\sqrt{\frac{\hat{\sigma}^2}{\sum_{i=1}^{n}X_i^2-n\bar{X}^2}}=s\sqrt{\frac{1}{\sum_{i=1}^{n}X_i^2-n\bar{X}^2}} \tag{5.16}$$

$$SE\left(\hat{\alpha}\right)=\sqrt{Var\left(\hat{\alpha}\right)}=\sqrt{\frac{\hat{\sigma}^2\sum_{i=1}^{n}X_i^2}{n\sum_{i=1}^{n}(X_i-\bar{X})^2}}=s\sqrt{\frac{\sum_{i=1}^{n}X_i^2}{n\sum_{i=1}^{n}X_i^2-n^2\bar{X}^2}} \tag{5.17}$$

where:

- $\hat{\sigma}^2$ (also denoted s^2) is the sample variance of the residuals. $\hat{\sigma}^2$ is an unbiased estimator of the variance of the disturbance error term of the population regression. The formula for $\hat{\sigma}^2$ is:

$$\hat{\sigma}^2=\frac{\sum_{i=1}^{n}\hat{u}_i^2}{n-2} \tag{5.18}$$

- s (or $\hat{\sigma}$) is the standard error of the residuals of the sample regression. It follows that this is an unbiased estimator of the standard deviation of the disturbance error term. s is given by:

$$s=\sqrt{\frac{\sum_{i=1}^{n}\hat{u}_i^2}{n-2}} \tag{5.19}$$

- n is the sample size.

The full derivation of equations (5.16) and (5.17) can be found in most econometrics books (see Brooks and Tsolacos, 2010, chapter 4). Fortunately, Excel and statistical software will compute these statistics automatically.

- ### Online note #5.2: Deriving standard errors

The variance of the slope (β) is of course of greater interest in empirical analysis. The variance and standard error formulae presented earlier for the slope coefficients reveal the following:

(i) The larger the variance of the disturbance errors, the larger the variance of β, which implies that a larger variance of the residuals $(\hat{\sigma}^2)$ results in a larger variance of the OLS estimate $\hat{\beta}$. This will make it more difficult to infer accurate values for the population parameter β, the topic of the next section. Large variability in the errors reflects the impact of other influences, both random and non-random. More variation in the unobserved influences on Y makes it harder to determine the precise impact of X on Y.

(ii) The variance of X appears in the denominator of equation (5.16). Hence the larger the variance of X the smaller the variance of β and the more precise the OLS estimate $\hat{\beta}$. There is more information to reveal the 'true' relationship between Y and X.

(iii) A larger sample is concomitant with a larger variability of X. With more observations and greater variability of X, a more precise estimate for β can be obtained. The size of the sample n appears in the quantity $\sum_{i=1}^{n}(X_i - \bar{X})^2$. That is, the larger the sample variance of an explanatory variable, the smaller the variance of the OLS estimate.

5.7 Significance of regression coefficients

In the previous sections we introduced the formulae for calculating the coefficient of an explanatory variable X and its variability. In this section we explain how these estimates can be used to examine whether X has a significant influence on determining the values of Y. We will use the estimates to perform hypothesis tests and construct confidence intervals using the procedures outlined in Chapter 4. There are two important questions that need to be answered when assessing an estimate of the population regression parameter.

(i) Does the explanatory variable have an impact on the dependent variable? Or alternatively, could the slope coefficient take the value of zero? This is an important question and hypothesis to test since OLS only provides an estimate for the unobserved population slope coefficient. If the estimate $\hat{\beta}$ is not statistically different from zero, the explanatory variable X has no influence on determining the values of Y. The hypothesis test involves testing the assertion that the population regression parameter is zero, $\beta = 0$. We can conclude that the explanatory variable X has a statistically significant or systematic influence on Y when it is rejected.

(ii) How reliable is the estimate? What is the range of values the population coefficient(s) could take at chosen levels of confidence, say 90%, 95% or 99%? We touched on this point at the beginning of the previous section.

There are three methods to perform hypothesis testing for the significance of OLS regressors: the test of significance, the p-value and the confidence interval approaches.

5.7.1 Test of significance approach

The formula for the test statistic used in examining the significance of the slope coefficient is:

$$\frac{\hat{\beta} - \beta^*}{SE(\hat{\beta})} \sim t_{n-k} \qquad (5.20)$$

where β^* is the value of β we would like to test, or the value of β under the null hypothesis.

This test statistic follows a t distribution with $n - k$ degrees of freedom, where k is the number of parameters in the equation including the intercept. In the simple (bivariate) regression there are two degrees of freedom lost due to the model containing an intercept and one slope coefficient. See online chapter for background to the derivation of the test statistic (5.20).

Our main interest is the significance of the estimated slope coefficient $\hat{\beta}$. If the estimated coefficient is zero, the population regression model becomes:

$$\hat{Y}_i = \hat{\alpha} + 0X_i \Rightarrow Y_i = \alpha + u_i$$

Hence if the coefficient $\hat{\beta}$ is 0, X has no influence on determining the values of Y. To test this hypothesis, we substitute $\beta^* = 0$ in equation (5.20) and calculate the t-statistic:

$$\frac{\hat{\beta} - 0}{SE(\hat{\beta})} \sim t_{n-k} \Rightarrow \frac{\hat{\beta}}{SE(\hat{\beta})} \sim t_{n-k} \qquad (5.21)$$

The hypotheses are:

Null $H_0 : \beta = 0$
Alternative $H_1 : \beta \neq 0$

The decision rules are:

(i) If the computed test statistic (equation (5.20)) is higher than the critical t value at the chosen level of significance (α) (usually 5%) and $n - k$ degrees of freedom, we reject the null assertion that $\beta = 0$. There is sufficient evidence from the sample that X has a statistically significant impact on Y.
(ii) Alternatively, if the p-value is less than the significance level (e.g. 5%), we can reject the null hypothesis.

With statistical software calculating the p-value, it is straightforward to make inferences about the statistical significance of coefficients.

- **Online note #5.3:Background to t-test statistics**

5.7.2 Confidence interval approach

A related test to make inferences about the significance of coefficients and produce intervals within which coefficients lie in is the confidence interval approach.

This approach will give us the interval within which the slope coefficient or indeed any other coefficient can vary with a certain level of confidence. Put differently, we have 95% confidence that the interval contains the true population value of the parameter (in our case β). Mostly confidence interval tests are two sided. The general form of confidence intervals is given by expression (5.22) – see online chapter regarding the derivation of expression (5.22).

$$\hat{\beta} - t_{crit} \times SE(\hat{\beta}) \leq \beta^* \leq \hat{\beta} + t_{crit} \times SE(\hat{\beta}) \qquad (5.22)$$

or

$$\hat{\beta} - t_{\frac{\alpha}{2}, n-k} \times SE\left(\hat{\beta}\right) \leq \beta^* \leq \hat{\beta} + t_{\frac{\alpha}{2}, n-k} \times SE\left(\hat{\beta}\right) \tag{5.23}$$

In the case of the bivariate model $k = 2$.

Assume we test the hypothesis $\beta^* = 0$. The hypotheses are:

$$H_0 : \beta = 0$$
$$H_1 : \beta \neq 0$$

We reject the null if the hypothesised value 0 lies outside the confidence interval at the chosen level of significance (usually 5%). Alternatively, we do not reject the null if 0 falls within the confidence interval. The process is similar if we test for any other value for the unknown population parameter beta.

- **Online note #5.4: Constructing confidence intervals**

5.8 Analysis of variance (ANOVA)

The generic representation of the sample regression function (SRF) is:

$$Y_i = a + bX_i + e_i \text{ (we have also written the SRF as } Y_i = \hat{a} + \hat{\beta}X_i + \hat{u}_i \text{).}$$

The SRF can be expressed using terms describing the ANOVA, which leads to:

$$TSS = ESS + RSS \tag{5.24}$$

where:

Total sum of squares: $TSS = \sum\left(Y_i - \bar{Y}\right)^2$ (5.25)

Explained (or regression) sum of squares: $ESS = \sum\left(\hat{Y}_i - \bar{Y}\right)^2$ (5.26)

Residual sum of squares: $RSS = \sum e_i^2$ or $RSS = \sum\left(Y_i - \hat{Y}_i\right)^2$ (5.27)

TSS is the sum of the squared differences between the observed *dependent variable* and its mean and represents the dispersion of the observed variables around the mean. It is a measure of the total variability of the dataset.

ESS is the sum of the squared differences between the *predicted* value and the mean of the *dependent variable*. It is a measure that describes how well the regression line fits the data.

RSS is the sum of the squared differences between the *observed* value and the *predicted* value (predicted by the regression). The smaller the error, the better the estimation power of the regression. The RSS is also known as the sum of squares error (SSE).

5.9 Overall performance of the model – goodness of fit

5.9.1 Overall performance of the model

The overall performance of the SRF concerns how well the model explains or 'fits' the values of the dependent variable. For the SRF to fit the data well, the explanatory variable in the simple regression model or at least one of the explanatory variables in the multiple regression model we study in the next chapter should have an influential role in determining the value of the dependent variable. For this assessment we compute a measure of the *goodness of fit* known as the *coefficient of determination* (R^2) and assess whether it is statistically significant. The latter involves testing restrictions involving the proposition that the model does not explain the values of the dependent variable.

5.9.2   Coefficient of determination (R-squared)

To understand the *R*-squared (R^2), it is convenient to express the SRF in analysis of variance (ANOVA) terms. From previous sections:

$$TSS = ESS + RSS \Rightarrow \sum_{i=1}^{n}(Y_i - \bar{Y})^2 = \sum_{i=1}^{n}(\hat{Y}_i - \bar{Y})^2 + \sum_{i=1}^{n}\hat{u}_i^2$$

Dividing throughout by *TSS*:

$$\frac{TSS}{TSS} = \frac{ESS}{TSS} + \frac{RSS}{TSS} \Rightarrow 1 = \frac{ESS}{TSS} + \frac{RSS}{TSS} \Rightarrow \frac{ESS}{TSS} = 1 - \frac{RSS}{TSS} \tag{5.28}$$

The R^2 is the ratio:

$$\frac{ESS}{TSS} = \frac{\sum(Y_i - \bar{Y})^2}{\sum(\hat{Y}_i - \bar{Y})^2} \tag{5.29}$$

The coefficient of determination provides a measure of how well the sample regression model explains the variation in the values of the dependent variable. In principle, $0 \le R^2 \le 1$, as *RSS* cannot be larger than *TSS*. Any R^2 value greater than zero means that the regression analysis did better than just using a horizontal line through the mean value. However, you can get a negative R^2 value. If this happens, you should probably revise your specification of the PRF.

Figure 5.6 presents two extreme or rather theoretical cases. The dots represent the actual data. The dotted lines are the fitted values from a model. In panel (a) the model does not explain any of the variation of *Y*, that is, ESS is zero. In this case $R^2 = 0$.

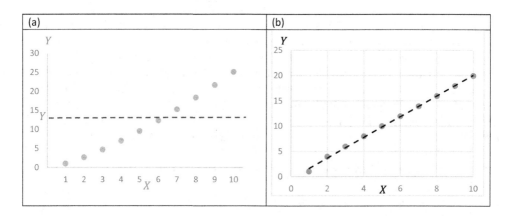

Figure 5.6 Extreme *R*-squared values

$$R^2 = \frac{ESS}{TSS} = \frac{0}{TSS} = 0, \text{ hence } TSS = 0 + RSS \Rightarrow TSS = RSS$$

We can illustrate the consequences by running a regression on a constant-only model. The horizontal fitted line is the mean of the dependent variable. The model in panel (a) is $y = \bar{y} + 0x$. The aim of building a regression model is to fit a line that would do better than the horizontal line.

If the model makes no errors $RSS = 0 \Rightarrow TSS = ESS$ and hence $R^2 = \frac{ESS}{TSS} = 1$. That is shown in panel (b). The model predicts (fits) exactly Y. The estimated line goes through all observed data points.

In practice, such extreme cases are not encountered. We will endeavour to build a model to achieve a higher explanatory power. A higher R^2 implies a better fit to the observed data values. We should note though that the maximisation of the R^2 (or the adjusted R^2 we present in the next chapter) should not be the fundamental objective in regression analysis as it only provides an indication of how 'close' the predicted values are to the actual values of the dependent variable and does not inform us whether the model makes sense. Further, the model will have to satisfy other properties too (presented in Chapters 7 and 8). The R^2 is the square of the correlation coefficient of the actual (Y) and predicted $(\hat{Y})$ values.

The R^2 can be used to compare the fit between various estimated models to aid selection, but it can only be used in this way when the sample size is similar and when the dependent variable in various models has the same unit of measurement.

5.9.3 Standard error of the regression as a goodness-of-fit measure

The standard error of the regression, also known as the standard error of the estimate (*SEE*), is an alternative presentation of a measure of goodness of fit. The *SEE* is an absolute measure of the typical distance that the data points fall from the SRF. It is measured in the units of the dependent variable. Since the *SEE* indicates how far on average the data points are from the regression line, lower values signify a better fit as the distances between the data points and the fitted values are smaller.

5.10 Assumptions of the classical linear regression model (CLRM)

The validity of applying least squares estimation in obtaining estimates of the PRF parameters (α, β) depends on certain 'classical' assumptions being met for it to be BLUE. The classical linear regression model (CLRM) makes the following assumptions:

Assumption 1: E(u₍ᵢ₎) = 0

The errors or disturbance error term has a mean (average value) of zero. In mathematical notation and using the expectations operator: $E(u_i) = 0$. Subscript i denotes time-series or cross-section observations. Given the values of the explanatory variables, the expected value of the disturbance term or its conditional mean is zero:

$$E(u_i \mid X_i) = 0.$$

The assumption of a zero mean error term is plausible since we expect the errors in the model to represent random influences on the dependent variable. This assumption tends to be violated when an important variable is omitted from the model or when an inappropriate functional form is used to model the causal relationship. The latter is related to assumption 2.

If this assumption is violated, then OLS will yield a biased estimate of the 'true' relationship.

Assumption 2: linear in parameters

The regression model is linear in the unknown parameters or coefficients (α and β or βs if we have more than one explanatory variable). That is, there is a linear relationship between the dependent variable Y and the regressor or regressors (X or Xs). More precisely, linearity refers to linear in parameters. If this assumption is violated, then OLS will yield a biased estimate of the 'true' relationship β.

Assumption 3: non-stochastic (non-random) regressors

The values of the independent variable X are assumed to be fixed in repeated samples. This assumption imposes a direction on causality of values emanating from the explanatory variable(s) X(s) to the dependent variable Y. A more precise description is that the explanatory variables are non-stochastic, meaning that they are non-random, exogenous or non-endogenous.

It is represented by the assumption that the conditional distribution of values of the dependent variable Y are statistically independent:

$$E\left(u_i \mid X_i\right) = 0$$

since

$$cov\left(u_i, X_i\right) = E\left[\left(u_i - E\left(u_i\right)\right)\left(X_i - E\left(X_i\right)\right)\right] = E\left[u_i\left(X_i - E\left(X_i\right)\right)\right] = E\left(u_i, X_i\right) - E(u_i)E\left(X_i\right)$$
$$= E\left(u_i, X_i\right) = 0$$

The strict exogeneity assumption asserts that there is no correlation between the disturbance error term and the explanatory variable among all observations $i \neq j$, i.e. $cov\left(u_i, X_j\right) = 0$. This assertion is usually expressed in a time-series context, where i represents the current period and j represents all past, current and future periods.

If this assumption is violated, then OLS will yield a biased estimate of the 'true' relationship.

However, it is possible to invoke a weaker assumption to satisfy the direction of causality criterion. The explanatory variables are required to be predetermined or weakly exogenous. The explanatory variable, X, is predetermined when it is uncorrelated with the error term, u, for particular observations.

$$corr\left(u_i, X_i\right) = 0$$

If this assumption holds, then OLS will yield consistent estimates of the 'true' relationship even though the estimates are biased.

The lack of correlation between the independent variables and the disturbance error refers to the orthogonality of the error terms with the explanatory variables.

Assumption 4: homoscedasticity

Homoscedasticity means that the disturbance errors have a constant variance. This assumption requires that the variance of the errors is constant for each given value of X. The alternative to homoscedasticity is heteroscedasticity, which refers to disturbance errors not having a constant variance. Using mathematical notation this assumption is expressed as:

$$var\left(u_i \mid X_i\right) = E[u_i - E\left(u_i \mid X_i\right)]^2 = E\left(u_i - 0\right)^2 = E\left(u_i^2\right) = \sigma^2 \text{ for all } i .$$

where:

var stands for variance

σ^2 denotes the variance is a constant number not depending on X_i.

The mathematical notation to depict heteroscedasticity is:

$$E\left(u_i^2\right) = \sigma_i^2$$

Thus, the variance attributed to observation i is different from the variance attributed to observation j: $\sigma_i^2 \neq \sigma_j^2$ for $i \neq j$.

Natural heteroscedasticity can be a feature of cross-sectional data. If the assumption of homoscedasticity is violated, then OLS will yield incorrect standard errors of the estimates. Any inferences made using them during hypothesis testing will be invalid.

Assumption 5: the disturbance error terms are not serially correlated (no autocorrelation)

This assumption necessitates the independence of errors through time, that is, the errors are stochastically independent conditional on the values of the explanatory variable(s).

For any two values in our sample X_i and X_j the correlation between u_i and u_j is zero. Hence the expected value of the correlation between u_i and u_j is zero:

$$E\left(u_i, u_j\right) = 0 \text{ for } i \neq j .$$

Or more formally the conditional covariance of u_i and u_j is zero.

$$cov(u_i, u_j \mid X_i X_j) = E\left[\left(u_i - E\left(u_i\right)\right)\left(u_j - E\left(u_j\right)\right)\right] = E\left[\left(u_i u_j\right)\right] = 0$$

$$\text{since } E\left(u_i\right) = E\left(u_j\right) = 0$$

Natural autocorrelation/serial correlation can be a feature of time-series data. When present, the standard errors calculated by OLS are no longer correct, and t- and F-tests based on their values are invalid. The solution is to obtain correct standard errors.

Provided the CLRM assumptions 1 to 5 are met, the Gauss Markov theorem states that OLS is the best linear unbiased estimator (BLUE). The next assumption is sufficient but not necessary for OLS to be BLUE. It can, however, be an important consideration in small samples ($n < 30$), as it degrades the efficiency of the estimator, implying that the test statistics such as t and F used to produce confidence intervals and in hypothesis testing no longer follow their assumed distributions. Inferences using regression results thus become unreliable.

In presence of non-normal errors, OLS is the only most efficient estimator in class of linear unbiased estimators (Wooldridge, 2012) rather than the most efficient of all unbiased estimators.

Assumption 6: the disturbance error terms are normally distributed

The disturbance error term follows the normal distribution with zero mean and constant variance:

$$u_i \sim \mathcal{N}\left(0, \sigma^2\right)$$

Recall that the disturbance term is the random error in the relationship between the explanatory variables and the dependent variable in a regression model. It represents the deviation from the conditional mean at each value of the explanatory variable. Conceptually with repeated random sampling, each sample produces different estimated values of the disturbance error term (known as residuals). It is the distribution of this disturbance term or "noise" for all cases in a sample that should be normally distributed.

   The normality assumption can be overlooked if a regression model satisfies all the CLRM assumptions as it does not affect the bias of the estimator or its efficiency in large samples. It is only important for the calculation of p-values for significance testing after specifying a SRF when the sample size is very small. The normality assumption is not needed at all when the sample size is large as the Central Limit Theorem (CLT) ensures that the distribution of disturbance term will approximate normality. When the distribution of the disturbance term deviates from normality, a feasible solution is to adopt a more conservative p-value (0.01 rather than 0.05) for conducting significance tests and constructing confidence intervals.

5.11   The issue of bias and efficiency – properties of CLRM

As we stated earlier, the objective in regression analysis in modelling causal relationships is to ensure that the SRF is a good approximation of the PRF. We can ascertain with a degree of confidence that the SRF may be a good approximation by assessing the plausibility of the estimates obtained and diagnostic testing to ensure that the a priori properties of the 'estimator' are not violated. The estimator applied in conventional regression analysis of a linear causal relationship is ordinary least squares (OLS) as it is the best linear unbiased estimator (BLUE). In this case, the SRF is likely to be a good approximation of the PRF when the estimates are plausible and the assumptions underlying the classical linear regression model (CLRM) are satisfied. Figure 5.7 illustrates the concepts of unbiasedness and efficiency in regression estimation.

   The diagram shows the theoretical sampling distributions of estimates of β, the unknown true value of the effect an explanatory variable, obtained from three different estimators (formulae). The distributions characterise estimated values obtained from all possible representative samples of data (repeated random sampling) about the causal relationship. The black line represents the distribution of the unbiased and efficient estimates, the dashed line the distribution of the unbiased but inefficient estimates, and the grey line the distribution of the biased estimates.

   Instead of drawing a diagram, the property of an unbiased estimator can be represented mathematically by $E(\hat{\beta}) = \beta$. This states that the expected value of the estimates obtained

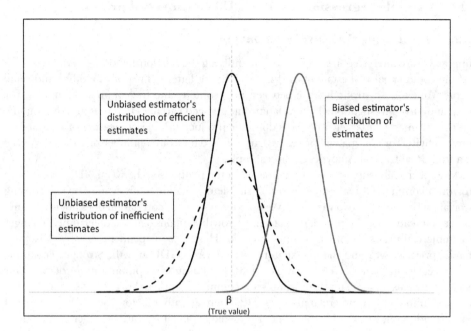

Figure 5.7 Bias and efficiency in estimation

from applying the OLS estimator on repeated random sampling of a fixed size is equal to the 'true' value of the population parameter.

The grey line in Figure 5.7 highlights the first estimation issue, bias. An inappropriate estimator applied to representative data is almost certainly never going to provide an estimate of the true value of the causal relationship, β.

The dashed line highlights the second estimation issue which concerns the precision or efficiency of an estimate. This estimator provides an unbiased estimate, but it is less reliable since the standard error reflected by the width of the distribution is relatively large (not the smallest possible). The black line represents the distribution of estimates of β obtained by applying OLS. It illustrates description of OLS as the best linear unbiased estimator (BLUE).

Furthermore, the large sample or asymptotic properties also mean that OLS yields consistent estimates provided that it is an unbiased estimator. Mathematically, this is represented by $plim\hat{\beta} = \beta$. *plim* stands for probability limit, which relates to the theoretical result that as the sample size increases without limit, the sampling distribution of the estimate, $\hat{\beta}$, becomes more concentrated around the true value β. In the limit, the sampling distribution collapses on a single point equal to the 'true' value of β.

However, it is also possible for OLS to be consistent even when OLS yields biased estimates due to not meeting some of the stringent CLRM assumptions. For example, OLS is biased but consistent as an estimator when the explanatory variable is predetermined (weakly exogenous) rather than strictly exogenous.

• ***Online note #5.5:Illustrating the OLS principle with a simple example***

5.12 A simple regression model of US commercial prices

5.12.1 *Illustrating the causal relationship*

Suppose we are interested in studying and quantifying the relationship between the economy and the commercial real estate market in the US, the latter defined as the office and retail sectors. We expect a strong linkage between the economy and the commercial market. For the quantitative investigation of this relationship we need variables to represent the economy and the commercial market. The gross domestic product (GDP) is a measure of overall economic activity. For commercial prices we can obtain data from a number of sources. We use data from Real Capital Analytics in this example.

More specifically, suppose we are interested in the relationship between GDP growth and growth in commercial real estate prices. A simple correlation analysis may reveal a strong association between the two variables. We noted that a strong correlation does not imply anything about causality. We postulate that GDP growth explains (causes, determines) changes in commercial prices (commercial price growth). Hence our dependent variable Y is commercial price growth and the explanatory variable X is GDP growth. Stronger economic activity results in higher levels of employment and business output and more demand for space that will exert some upward pressure on commercial prices. We are not saying that the only determinant of commercial price is GDP. There are other factors too, such as supply of space. In addition, we could employ better targeted explanatory variables such as output or employment in services, as GDP may be too general. The selection of the most appropriate variable is discussed in detail in the next chapter and further in Chapter 7 where we outline criteria, processes and tests to specify the most appropriate model.

Regression analysis will help us evaluate the relationship between GDP growth and commercial price growth and quantify the impact. We seek answers to a number of practical questions:

- Is GDP growth a significant driver of changes in commercial real estate prices in the US?
- If so, what is the sensitivity of these two sectors to GDP growth changes?
- How well does it explain commercial prices? We know there are unaccounted factors; how does this affect the simple model?
- What is the expected impact on commercial prices if the US economy moves into a recession and GDP growth is -2%? Would commercial prices decline too?

Regression analysis can help to answer these questions and enable us to study this relationship more comprehensively. For example, has the relationship between commercial price growth and GDP growth become stronger or weaker post global financial crisis (2008/2009)?

5.12.2 *Specifying and estimating the empirical model*

The theoretical relationship between growth in commercial estate prices and growth in GDP at the national level in the US is given by the following bivariate model:

$$comg_t = \alpha + \beta\, gdpg_t + u_t \tag{5.30}$$

where *comg* is the percentage growth in real commercial prices, *gdpg* is the percentage growth in US GDP and t denotes time.

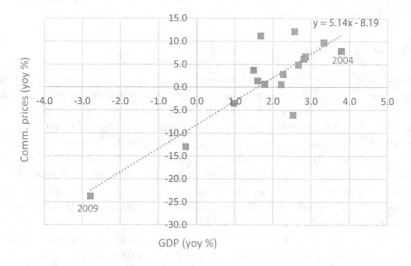

Figure 5.8 Regression line for equation (5.31)

The data for commercial prices is provided by Real Capital Analytics and are deflated and converted to growth rates. Hence in equation (5.30) we test the simple argument that overall economic activity in the US determines (causes) real price growth in commercial real estate. Equation (5.30) also assumes a contemporaneous relationship, which is denoted by the subscript 't'. The data are given in the Excel file for this chapter. We have 18 observations in levels and 17 observations in growth rates. Is this a sample too small for regression analysis? Yes, it is. The objective is to illustrate the regression statistics presented in the earlier sections of this chapter, and in order to keep the analysis compact and manageable we use 17 observations.

Figure 5.8 presents the scatter plot of the two series of growth rates. On the *y* axis we measure growth in commercial prices and on the *x* axis GDP growth. Each dot represents a year and shows the combination of the observed commercial price and GDP growth rates in that particular year. The observation at the far left is the combination for the year 2009, when in the aftermath of the global financial crisis commercial prices fell by 23.8% and GDP contracted by 2.8%. At the other end of the spectrum, when GDP growth was the strongest at 3.8% in our sample in 2004, commercial prices increased by 7.9% in real terms.

We should also note that stronger GDP growth is not associated with higher increase in commercial prices in a proportionate manner. In 2012 GDP growth was 2.2% with commercial prices barely rising at 0.6% growth. The following year, GDP growth was 1.7% and growth in commercial prices was 11.2%. In 2013, other factors not reflected in GDP growth could have pushed commercial price growth to 11.2% whereas the economy delivered modest growth by US standards. It is apparent that GDP does not fully explain commercial real estate price growth (as argued in the previous section).

A visual inspection of the pairs of the data values suggests a positive relationship – stronger (weaker) growth in GDP is associated with stronger (weaker) growth in real commercial prices. The contemporaneous correlation of the two series is strong. The correlation coefficient is +0.86, indicating a relatively strong positive association between the two variables. The upward sloping estimated regression line (also shown in Figure 5.8) confirms the expected positive relationship. The regression model is given by equation (5.31).

$$\widehat{comg}_t = \hat{\alpha} + \hat{\beta} gdp_t \Rightarrow \widehat{comg}_t = -8.19 + 5.14\, gdp_t \tag{5.31}$$

The regression line crosses the vertical axis at point -8.19. The slope is $+5.15$. Hence if GDP growth is 0% (the US economy stagnates), real commercial prices will tend to fall on average by 8.16 percentage points (8.2 percentage points). The slope of 5.15 (or 5.2) gives the sensitivity of real commercial prices to the general economy – the responsiveness of commercial real estate price growth to changes in GDP growth. Therefore, a 1 percentage point increase in real GDP (e.g. GDP growth accelerates from 2% to 3%) leads to an increase of commercial price growth by 5.2 percentage points. This increase is an average expectation based on the specific sample used in the estimation and gives us a fairly good idea of the responsiveness of commercial real estate prices to changes in real GDP. The preceding point applies in the opposite direction: if GDP growth slows by 1 percentage point (say from -0.5% to -1.5% or from 0.5% to -0.5%), commercial price growth is expected to decelerate by -5.2 percentage points. It is apparent that any 1 percentage point change in GDP growth, for example GDP growth picking up from 1% to 2% or from 2.2% to 3.2%, the expected impact on commercial growth remains constant at 5.2 percentage points. This is because of the linearity assumption.

Obtaining estimates of the slope coefficient in regression analysis has practical benefits. Suppose we replicate the analysis using data from another country and discover that the estimated impact is 3.6 or 7.1. This is useful information to have. An analyst would like to know why the impact is greater or lower in different countries and markets. Is it a data issue, whether the data are collected or measured differently or whether GDP is the most appropriate series for representing activity in the commercial real estate market, or whether there are market or institutional factors limiting the response of commercial prices to changes in GDP growth?

5.12.3   *Calculation of OLS coefficients*

In Table 5.2, we demonstrate how the values of OLS estimates of the parameters $\hat{\alpha}$ $(= -8.19)$ and $\hat{\beta}\,(= 5.14)$ estimated in equation (5.31) are derived.

We apply formulae (5.14) and (5.15) to calculate the coefficients. As this is time-series data, the subscript 'i' is replaced with 't'.

$$\hat{\beta} = \frac{\sum_{t=1}^{T} x_t y_t - T\, \overline{yx}}{\sum_{t=1}^{T} x_t^2 + T \overline{x}^2} = \frac{229.80 - 17 \times 1.28 \times 1.84}{94.50 - 17 \times 3.39} = \frac{189.64}{36.87} = 5.14$$

$$\hat{a} = \overline{y} - \hat{\beta}\overline{x} = 1.28 - 5.14 \times 1.84 = -8.18$$

These calculations are corrected to two decimal points. Expect some minor differences if you perform these calculations using the data in this chapter's datafile.

In sections 5.3 and 5.4, a distinction was made between the SRF and the PRF. The dataset for our example is a subset of the unknown DGP of real commercial price growth and GDP growth. There are 17 observations. We attempt to infer the 'true relationship' or population parameters α and β from our sample.

Table 5.2 Calculation of OLS coefficients for equation (5.30)

Year	Commercial prices (%)	GDP (%)	$Y = X$	X^2
	$Comg = Y$	$gdpg = X$		
2001	−3.54	0.98	−3.45	0.95
2002	0.64	1.79	1.14	3.19
2003	6.15	2.81	17.26	7.88
2004	7.85	3.79	29.72	14.33
2005	9.71	3.34	32.47	11.19
2006	4.86	2.67	12.96	7.11
2007	0.56	1.78	1.00	3.16
2008	−12.98	−0.29	3.78	0.08
2009	−23.77	−2.78	65.99	7.71
2010	−6.12	2.53	−15.50	6.41
2011	1.37	1.60	2.19	2.57
2012	0.59	2.22	1.32	4.94
2013	11.15	1.68	18.71	2.81
2014	12.12	2.57	31.15	6.60
2015	6.66	2.86	19.05	8.19
2016	3.69	1.49	5.49	2.21
2017	2.86	2.27	6.51	5.17
$T = 17$	$\bar{Y} = 1.28$	$\bar{X} = 1.84$ $\bar{X}^2 = 3.39$	$\sum_{1}^{17} YX = 229.80$	$\sum_{1}^{17} X^2 = 94.50$

5.12.4 Estimation of standard errors

Equations (5.16) through to (5.18) are employed to calculate the standard error of the regression and standard errors of the estimates. Table 5.3 contains the data and calculations of the constituent parts of these formulae.

SE of regression: $SE(\hat{u}_t)\ or\ s = \sqrt{\dfrac{338.14}{17-2}} = 4.75$

SE of slope coefficient: $SE(\hat{\beta}) = 4.75\sqrt{\dfrac{1}{36.89}} = 0.78$

SE of intercept: $SE(\hat{\alpha}) = 4.75\sqrt{\dfrac{94.56}{17 \times 36.89}} = 1.84$

5.12.5 Statistical significance

The measures of variability (standard errors) are used to formally examine the statistical significance of *gdpg* in explaining the growth in commercial prices. The following calculations are based on the discussion in section 5.7.

Table 5.3 Calculation of standard errors

(i)	(ii)	(iii)	(iv)	(v)	(vi)
Year	*gdpg (X)*	*Residual*	*Squares of residuals*	*(gdpg−1.84)²*	*GDPG²*
2001	0.98	−0.36	0.13	0.74	0.96
2002	1.79	−0.36	0.13	0.00	3.20
2003	2.81	−0.09	0.01	0.94	7.90
2004	3.79	−3.42	11.68	3.80	14.36
2005	3.34	0.71	0.50	2.24	11.16
2006	2.67	−0.66	0.43	0.69	7.13
2007	1.78	−0.39	0.15	0.00	3.17
2008	−0.29	−3.30	10.87	4.54	0.08
2009	−2.78	−1.31	1.72	21.36	7.73
2010	2.53	−10.95	119.85	0.47	6.40
2011	1.6	1.32	1.75	0.06	2.56
2012	2.22	−2.65	7.01	0.14	4.93
2013	1.68	10.72	114.90	0.03	2.82
2014	2.57	7.11	50.54	0.53	6.60
2015	2.86	0.14	0.02	1.04	8.18
2016	1.49	4.25	18.05	0.12	2.22
2017	2.27	−0.63	0.40	0.18	5.15
Average	1.84				
Sum			338.14	36.89	94.56
$T = 17$					

(i) Test of significance approach – application

We examine whether the coefficient on *gdpg* in the population regression is zero ($\beta^* = 0$). The *t*-ratio (expression (5.21)) is:

$$t\text{-ratio} = \frac{5.14}{0.78} = 6.58$$

We compare the *t*-ratio value with the *t* critical value from the *t* tables. The chosen level of significance α (commonly taken to be 10%, 5% or 1%) is divided in half since we have a two-tailed test (hence 5% or 0.05, 2.5% or 0.025, and 0.5% or 0.005 in each tail). Let us for the purpose of our example focus on the conventional significance level (α) of 5%. The critical values can be found from the *t* table.

Our sample has 17 observations (*n* or $T = 17$), leaving us with $17 - 2 = 15$ degrees of freedom (*df*). From Table 5.4 (another version of the *t*-table containing a subset of critical values) we get:

$$t_{\frac{\alpha}{2},T-2} = t_{0.025,15} = 2.131 \quad t_{\frac{\alpha}{2},T-2} = t_{0.025,15} = 2.131$$

The estimated *t*-statistic value is greater than the critical value at the chosen level of significance (5%). We reject the null hypothesis and conclude with 95% confidence that the

Table 5.4 Finding the *t*-critical values

df/α	(df: degrees of freedom, α: level of significance)					
	0.15	0.10	0.05	0.025	0.01	0.005
1	1.963	3.078	6.314	12.706	31.821	63.656
...	...	...	...	...	...	...
14	1.076	1.345	1.761	2.145	2.624	2.977
15	1.074	1.341	1.753	2.131	2.602	2.947
16	1.071	1.337	1.746	2.12	2.583	2.921
17	1.069	1.333	1.74	2.11	2.567	2.898
...	...	...	...	...	...	...

coefficient on *gdpg* is statistically significant. Statistical software will provide the *t*-statistic routinely, but not necessarily the critical values. What is also given as standard output is the associated probability (*p*-value) with the computed *t*-ratio. In our model this probability is 0.00, which is lower than 0.05 and 0.01, confirming the significance of the *gdpg* in explaining *comg*.

We can perform tests for additional hypotheses, for example one-sided hypotheses. We will apply the same inferential process. Suppose we would like to test the one-sided hypothesis of whether the slope coefficient can take the value of 8 (β^*), denoting a possible much higher impact than the average impact of 6.14. For example, the slope coefficient of 8 could have been the finding in another market/country and we want to test it in the context of the US. We apply the same process.

$$H_0 : \beta = 8$$
$$H_1 : \beta < 8$$

The test statistic is:

$$\frac{\hat{\beta} - \beta^*}{SE(\hat{\beta})} \sim t_{T-2} \Rightarrow \frac{5.14 - 8}{0.78} = -3.67$$

For a one-sided test, the column is 0.05 (5% level of significance) and the degrees of freedom is 15. The critical value is 1.753. Since the distribution of the test statistic is symmetrical and we are undertaking a one-tailed test, the critical value is −1.753. Again, the estimated test statistic is more negative than the critical value and the null hypothesis that the slope coefficient could be 8 is rejected. This test tells us that with 95% confidence the population coefficient is less than 8. A more straightforward process is to make comparisons using the absolute value of the *t*-statistic. The absolute value of the test statistic −3.67 is 3.67, which is compared to the critical value of 1.753. Since it is larger, we can reject the null hypothesis.

(ii)   Confidence interval approach – application

We can also adopt the confidence interval approach to examine both the significance of the coefficient on *gdpg* and calculate boundaries for its value as per discussion in section 5.7.2.

Table 5.5 Construction of confidence intervals

Null hypothesis: $\beta^* = 0$

$T = 17$, $df = 17-2 = 15$

Significance Level	Confidence level	t-critical (2-sided test)	Interval estimation based on equation (5.22)	Interval
5%	95%	2.131	$5.14 - 2.131 \times 0.78, 5.14 + 2.131 \times 0.78$	3.5–6.8
1%	99%	2.947	$5.14 - 2.947 \times 0.78, 5.14 + 2.947 \times 0.78$	2.8–7.4
10%	90%	1.753	$5.14 - 1.753 \times 0.78, 5.14 + 1.753 \times 0.78$	3.8–6.5

Applying equation (5.22), we get the confidence interval at different levels of significance. Table 5.5 summarises the calculations.

As expected, the interval for the value of population coefficient gets wider if the confidence level increases. In order to be more confident that the range contains the true (population) β, the boundaries are wider. The hypothesised value of 0 falls outside the confidence interval at the 5% level of significance, hence we reject the null hypothesis that real GDP growth has no effect on growth in real commercial prices. The estimated coefficient can be between 3.5 and 6.8 with 95% level of confidence. Earlier we also tested the assumption of $\hat{\beta} = 8$. The null hypothesis is rejected since 8 lies outside the confidence interval at any significance level. Both the significance test and confidence interval approaches give us the same result.

5.12.6   *Fitted values and residuals*

The model for commercial price growth (equation (5.31)) will give us the fitted (predicted) values for commercial real price growth (*comg*). The difference between the actual and fitted values is the estimated residuals ($\hat{u}_t$):

$$\hat{u}_t = comg_t - \widehat{comg_t}$$

Hence the fitted value for 2012 is:

$$\widehat{comg_t} = -8.19 + 5.14\, gdp_{2012} = -8.19 + 5.14 \times 2.22\% = 3.24\%$$

and the error is:

$$\hat{u}_{2012} = comg_{2012} - \widehat{comg}_{2012} = 0.59\% - 3.24\% = -2.65\%$$

In 2012 the model indicates a stronger growth (3.24%) than what was actually achieved (0.59%), resulting in a negative error of 2.65%.

In Table 5.6 we calculate the fitted values and errors for each year. Figure 5.9 plots the actual and fitted values along with the errors.

In some years we observe large errors, whereas in other years the fitted values are pretty close to the actual values. In 2015, the actual growth in real commercial prices was 6.66% and the model predicted 6.52%. Similarly, the big decline in 2009 by a massive 23.77% in the aftermath of the global financial crisis is replicated by this simple model, which predicts

Table 5.6 Fitted values and errors (equation (5.31))

Year	Comm. prices (%)	GDP (%)	Fitted values (equation (5.31))	Error (equation (5.31))
2001	−3.54	0.98	−3.18	−0.36
2002	0.64	1.79	0.99	−0.36
2003	6.15	2.81	6.24	−0.09
2004	7.85	3.79	11.27	−3.42
2005	9.71	3.34	9.00	0.71
2006	4.86	2.67	5.52	−0.66
2007	0.56	1.78	0.95	−0.39
2008	−12.98	−0.29	−9.69	−3.30
2009	−23.77	−2.78	−22.46	−1.31
2010	−6.12	2.53	4.83	−10.95
2011	1.37	1.60	0.04	1.32
2012	0.59	2.22	3.24	−2.65
2013	11.15	1.68	0.43	10.72
2014	12.12	2.57	5.02	7.11
2015	6.66	2.86	6.52	0.14
2016	3.69	1.49	−0.55	4.25
2017	2.86	2.27	3.49	−0.63

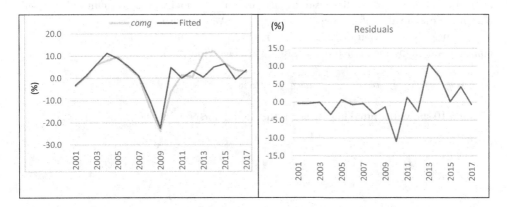

Figure 5.9 Actual, fitted and residual values of equation (5.31)

a fall of 22.46%. Hence GDP growth in the US which was −2.78% took its toll on real commercial prices and the economy was primarily responsible for the fall in commercial prices that year. The following year, 2010, the model makes the largest prediction error in our sample. Real commercial prices fell by 6.12% whereas the US economy had recovered with GDP growth of 2.53%. The model prediction (fitted value) for 2010 is 4.83%.

A large error occurred in 2013, when the US economy grew by a mere 0.43% with the model indicating growth of 1.68%. The actual growth was much stronger, at 11.15%. Other factors beyond GDP growth influenced the market that year and supported price growth.

This is why market knowledge is important. Market knowledge can offer explanations for large errors that represent influences other than GDP growth. These influences, unlike GDP growth, can be unsystematic or random, affecting prices in specific years. GDP growth is a systematic influence on commercial price growth, although this will formally be examined in the next section where we will perform the coefficient significance tests.

The graphic representation of actual values, fitted values and the residuals portray some interesting patterns. Our model fits the growth in real commercial prices well in the first part of our sample as the errors are small. In the second half of the sample and after the global financial crisis, we observe large errors in some periods and greater error volatility. This is not a welcome feature of the residuals. Do the residuals indicate that something is wrong with our model? In Chapter 8, we will explain how to perform a series of diagnostic tests on theresiduals to determine whether OLS is BLUE.

5.12.7 Properties of the intercept (constant)

What is the role of the constant or intercept in the regression model? In general, the intercept has no interpretation. Only in certain models will it have a theoretical interpretation. Yet the constant is always included in empirical regression analysis. It is the point at which the regression line crosses the y-axis, in this case at -8.19%. If real GDP growth is zero, the model predicts that on average real commercial price growth will be -8.19%. If another explanatory variable is used instead of real GDP growth, the constant will take a different value.

The presence of the intercept in the model ensures that residuals add up to zero, which is one of the properties of the residuals. By running a model with a constant only, we will obtain the mean value of the dependent variable. We run our model (equation (5.30)) with a constant only. Equation (5.32) gives the result of this estimation. The coefficient on the constant will be the mean of the dependent variable. That is:

$$comg_t = \alpha + u_t \Rightarrow \widehat{comg}_t = \hat{\alpha} \Rightarrow \widehat{comg}_t = 1.28\% \tag{5.32}$$

Figure 5.10 shows the fitted line, which is the dotted line crossing the y-axis at 1.28%.

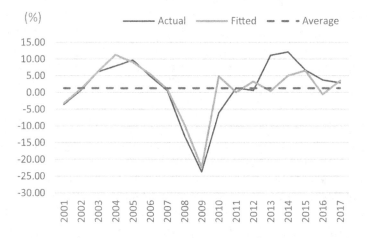

Figure 5.10 Actual and fitted values in relation to the average of the dependent variable

The constant in a model estimated by OLS replicates the mean value of the dependent variable. By including independent variables such as GDP growth the model explains the variation around this mean value.

5.12.8 Goodness-of-fit statistics

In equation (5.31) we have a highly significant explanatory variable for real commercial price growth. The next piece of information we would like is to know is how much of the variation of the dependent variable, *comg*, is explained by equation (5.31). This is revealed by the R^2 and its underlying components: *TSS* (total sum of squares), *ESS* (explained sum of squares) and *RSS* (residual sum of squares) as per sections 5.8 and 5.9. Table 5.7 shows the calculations in detail as the preceding quantities are fundamental in regression analysis.

In Table 5.7 column (iv), we take the deviations of *comg* from its mean of 1.28 and square them to obtain the total sum of squared residuals, or total sum of squares (*TSS*). Hence the value of *TSS* in 2001 is $(-3.54 - 1.28)^2 = 23.24$. The fitted line illustrates how much of this variation is explained by the model. The sum of the squared deviations of the fitted line from the mean is the sum of explained squared residuals (*ESS*). The value for *ESS* of 19.84

Table 5.7 Estimation of *TSS*, *ESS* and *RSS*

(i)	(ii)	(iii)	(iv)	(v)	(vi)	(vii)
Year	comg	Fitted $(\widehat{comg})$	$(comg - 1.28)\hat{}2$ [required to estimate TSS)	$((\widehat{comg}) - 1.28)\hat{}2$ (required to estimate ESS)	Residuals	Squares of residuals (RSS)
2001	−3.54	−3.18	23.24	19.84	−0.36	0.13
2002	0.64	0.99	0.42	0.08	−0.36	0.13
2003	6.15	6.24	23.68	24.65	−0.09	0.01
2004	7.85	11.27	43.13	100.02	−3.42	11.68
2005	9.71	9.00	70.99	59.81	0.71	0.50
2006	4.86	5.52	12.80	18.03	−0.66	0.43
2007	0.56	0.95	0.52	0.10	−0.39	0.15
2008	−12.98	−9.69	203.52	120.30	−3.30	10.87
2009	−23.77	−22.46	627.75	564.03	−1.31	1.72
2010	−6.12	4.83	54.84	12.63	−10.95	119.85
2011	1.37	0.04	0.01	1.52	1.32	1.75
2012	0.59	3.24	0.48	3.87	−2.65	7.01
2013	11.15	0.43	97.40	0.71	10.72	114.90
2014	12.12	5.02	117.55	14.02	7.11	50.54
2015	6.66	6.52	28.91	27.53	0.14	0.02
2016	3.69	−0.55	5.81	3.35	4.25	18.05
2017	2.86	3.49	2.50	4.94	−0.63	0.40
Average	1.28					
Sum			TSS = 1313.54	ESS = 975.40		RSS = 338.14
				(RSS = 1313.54 − 975.40 = 338.14)		

in 2001 is computed as $(-3.18 - 1.28)^2$. The difference between *TSS* and *ESS* is the sum of squared residuals (*RSS*).

If an alternative regression model, for example a model using employment growth instead of GDP growth, explains more (less) of the variation of *comg*, it will have a higher (lower) *ESS* and lower (higher) *RSS* than equation (5.31).

Using these metrics, we can calculate the R^2 of equation (5.31). As we noted in section 5.9, this statistic gives an indication of the explanatory power of a regression model in a more meaningful way and facilitates comparisons of models with the same dependent variable.

$$R^2 = \frac{ESS}{TSS} = \frac{975.40}{1313.54} = 0.74 \text{ or}$$

$$R^2 = \frac{ESS}{TSS} = \frac{TSS - RSS}{TSS} = 1 - \frac{RSS}{TSS} = 1 - \frac{338.14}{1313.54} = 0.74$$

Equation (5.31) explains 74% of the variation of *comg* (or the variation around its mean). This indicates a good to strong explanatory power.

- **Online note #5.6: Simple regression analysis in EViews**

5.13   Forecasting

We use our simple model (equation (5.31)) to forecast real commercial price growth in the US. Forecasts for US GDP growth from a third source can be fed into our model to obtain a forecast for the years after our last observation. We use GDP forecasts from the OECD for the US economy available to the authors at the time of writing. GDP growth is forecast to be 2.9% in 2018 and 2.8% in 2019. Hence,

$$\widehat{comg}_{2018} = -8.19 + 5.14\,gdp_{2018} = -8.19 + 5.14 \times 2.9\% = 6.7\% \tag{5.33}$$

$$\widehat{comg}_{2018} = -8.19 + 5.14\,gdp_{2019} = -8.19 + 5.14 \times 2.18\% = 6.2\% \tag{5.34}$$

Given steady economic growth of just under 3%, real commercial prices are expected to post growth of 6.7% and 6.2%, respectively. We know of course that other factors will affect *comg* in 2018 and 2019, perhaps variables we have omitted and unpredictable events. We therefore get an expected value for *comg* in 2018 and 2019.

Using equations (5.33) and (5.34), one can conduct a 'what if' or scenario analysis. What would be, for example the impact on commercial price growth if GDP growth was merely 1% due to a shock in 2018? We do not know the nature of the shock. At the time of writing there was much talk of a global trade war. Many economists would expect that such a war will affect global economic growth. If this view is credible, we would like to know what the downside risks are to commercial prices in the US. A number of organisations provide scenario analyses for the economy. Suppose that according to one scenario US GDP growth will slow down to 1%. In such a case commercial price growth would be negative (−3%). Hence scenario analysis is straightforward, transparent and quite illustrative using regression analysis.

We can also make use of the confidence intervals constructed for the sample regression slope coefficients. The 90% confidence interval limits are 3.8 and 6.5 (see Table 5.8). A confidence interval can also be constructed for the constant. Suppose the constant remains

Table 5.8 Forecast interval for *comg*

$\hat{\beta}$		*gdpg for 2018*	*comg for 2018*
5.14	(central value)	2.9%	6.7%
3.8	(lower 90% value)	2.9%	2.8%
6.5	(upper 90% value)	2.9%	10.7%

at −8.19, then we can provide a range of forecasts for commercial price growth utilising the confidence interval for $\hat{\beta}$.

We expect the growth in real commercial prices (*comg*) to be between 2.8% and 10.7% in 2018 if real GDP growth is 2.9%. This is the 90% forecast interval for *comg*. Due to unforeseen factors, there is a 10% chance that *comg* could be less than 2.8% or over 10.7%. It should be noted that shocks in the past are incorporated in the estimated response of *comg* to changes in *gdpg* in the model (as reflected in the value of the estimate ($\hat{\beta} = 5.14$) and the standard error of $\hat{\beta}$ and the range of the confidence interval).

This section touched on the basics of regression forecasting. Forecasting and forecast evaluation is the topic of Chapter 9. Before using a regression model for forecasting, we need to ensure that the model is fit for forecasting, that is the model provides unbiased or consistent and efficient estimates. This is the subject of Chapter 7.

5.14 Concluding remarks

This chapter introduces the reader to regression analysis, a fundamental method in applied quantitative work in real estate. Regression analysis enables us to determine the nature of relationships and test various propositions in practice. The chapter contains a discussion of simple regression analysis, covering several statistics that have to be interpreted. The objective is twofold. First, we attempt to make the discussion less technical but at the same time shows that the output of statistics from running a regression is not a black box and can be replicated. Econometric software automatically calculates a large number of statistics. We attempt to explain some of these statistics. In the next three chapters, we will interpret and analyse more regression statistics.

The chapter provides an introduction to the usefulness of regression analysis in real estate, highlighting several practical tasks involved in its execution:

- The application of the method of regression should be underpinned by a theory about the nature of the relationships between the dependant and explanatory variables, represented by stating the PRF. It is up to the researcher or analyst to determine the appropriate theory to be used in an application. The theory serves as a guide for the selection of explanatory variables to be included in the regression model. The SRF is used to test various propositions. The standard formulae to estimate the SRF is OLS.
- Determine from the regression output the effect of changes in the explanatory (predictor) variable on the dependent variable (target), which could include examining the range of values the predictor can vary.

- Determine from the regression output whether our selected predictor (explanatory variable) is statistically significant and, therefore, has a systematic effect on determining the values of the dependent variable.
- Assess the explanatory power of the model. The discussion introduces the R^2, a goodness-of-fit statistic.

In the present chapter we introduce the Classical Linear Regression Model (CLRM) assumptions, the properties a regression model should satisfy. In Chapter 7 these properties are treated in detail both theoretically and with examples.

This chapter contains illustrative examples and the estimation of a bivariate regression model using US data. In online chapter 1, we present a further example of a simple (bivariate) regression model. Guidance on running the models in EViews along with interpretation of regression outputs enhances the understanding of regression analysis. The regression output is standard across different software packages. The explanation contained in the online chapter extends the discussion to interpreting the coefficients of models having different functional forms.

Finally, a key application of regression models is forecasting. We derive forecasts from the estimated model and provide guidance for scenario forecasting. In the real estate profession, both point forecasts and scenario forecasting are employed. We further show how to construct naïve forecast intervals utilising the spread of forecasts that can be available for the independent variables (such as GDP growth in our example). We should note that the estimation of intervals for forecasts is a much more complicated exercise than the naïve calculation presented in Table 5.8.

As we are about to embark on the examination of multiple or multivariate regression in the next chapter, a regression model having more than one explanatory variable, a note about the single regression model is appropriate. The fact that in the univariate model the dependent variable is explained by reference to a single independent variable may sound restrictive. But it has provided us with an opportunity to acquaint ourselves with the fundamentals of regression analysis, a helpful starting point.

Chapter 5 online resource

- Chapter 5 accompanying notes
- Excel file: "ch5_excel"
- EViews file: "ch5_eviews"

6 Multiple regression

6.1 Introduction

In this chapter we generalise the univariate regression model outlined previously to multiple or multivariate regression, a statistical tool that allows us to examine how a number of explanatory variables can be used to explain causal relationships. It is rare to have a theory which would suggest that the dependent variable reacts to a single explanatory variable. Multiple regression enables us to examine more general and complex relationships between the dependent variables and its causal factors. The majority of regression analyses undertaken in real estate and other fields are based on multiple regression. A number of tests and metrics presented both in this and the next chapter will determine the number of variables to include in multivariate regression models.

Estimating a multivariate regression model follows similar principles to the univariate model. We will still apply the OLS procedure to estimate the parameters and their standard errors. But multiple regression involves further considerations. This chapter introduces additional important concepts associated with estimating a multiple regression model. We need to utilise model selection criteria to help select significant variables to be included in the final model. There is also an opportunity to include lagged effects – to reflect slow adjustments in the real estate market.

Within the framework of multiple regression, multiple hypothesis testing is commonly undertaken. We introduce the F distribution, which is appropriate for testing multiple restrictions such as the joint significance of regressors, the significance of a block (list) of variables in a model and parameter equality restrictions. In addition, the F-test is used in the diagnostics analysis which is presented in detail in the next chapter.

Using a general framework of office rent determination, we build and estimate a model of office rents in Hong Kong to illustrate the concepts associated with multiple regression analysis. The online note 6.8 contains another example of the application of multiple regression analysis, which examines US apartment price growth. The online chapter I which takes you through extending a simple regression model to a multiple regression model. This online chapter also illustrates how to apply the general-to-specific methodology to specifying a regression model.

6.2 Multiple regression model: an overview

The multiple or multivariate regression model is a generalisation of the univariate model. We noted earlier that real estate market theories often indicate that there is more than one explanatory variable, which can explain the variation in the dependent variable. We augment the univariate model to include more regressors to allow for the influence of additional causal factors and discuss the additional concepts that should be considered in conducting multiple regression analysis.

6.2.1   Conditional mean

As in a simple regression, the objective is to explain the conditional mean of the dependent variable, except this time it is conditioned upon more than one explanatory variable. Using an expectations operator, a model containing four explanatory variables $(X_{1i}, X_{2i}, X_{3i}, X_{4i})$ can be described by:

$$E(Y_i) = E(Y_i | X_{1i}, X_{2i}, X_{3i}, X_{4i}) + u_i \qquad (6.1)$$

where:
$E(.)$ = is the expectations operator
i = particular observation

The conditional mean is modelled as a linear relationship between the dependent and explanatory variables:

$$E(Y_i | X_{1i}, X_{2i}, X_{3i}, X_{4i}) = \alpha + \beta_1 X_{1i} + \beta_2 X_{2i} + \beta_3 X_{3i} + \beta_4 X_{4i} \qquad (6.2)$$

It is standard practice to attach a subscript to a parameter (in this case 1, 2, 3, 4) to match it to the relevant variable. There are no observation subscripts for the parameters β as it is assumed that the effects of a change in the value of an explanatory variable remain the same across observations.

6.2.2   The population regression function

The population regression function for a model containing four explanatory variables can be written as an equation excluding the expectations operators, though they remain implicit in the equation:

$$Y_i = \alpha + \beta_1 X_{1i} + \beta_2 X_{2i} + \beta_3 X_{3i} + \beta_4 X_{4i} + u_i \qquad (6.3)$$

As in a simple regression, the generic representation does not assign signs to the parameters even though theory informs us for example that X_1 and X_4 should be negative and X_2 and X_3 should be positive. We can make separate statements to clarify the impacts:

$$\beta_1 < 0;\ \beta_2 > 0;\ \beta_3 > 0 \text{ and } \beta_4 < 0$$

In formulating the population regression function, we do not make a statement that a population parameter is zero, $\beta_i = 0$, as this would mean that variable X_i has no influence on determining the (expected) value of the dependent variable – it is redundant as a determinant of the values of the dependent variable.

6.2.3 Multiple regression: general form

The general form of the multiple regression model including *k* regressors can be written as:

$$Y_i = \alpha + \beta_1 X_{1i} + \beta_2 X_{2i} + \dots \beta_K X_{Ki} + u_i \qquad u_i \sim N\left(0, \sigma^2\right) \qquad (6.4)$$

K is the number of regressors and the sample size is *n* with *i* = 1, 2 . . ., *n*. The sample could consist of time-series observations or cross-sectional data. α is the constant or intercept and $\beta_k s$ (*k* = 1, 2, . . ., *K*) are the slope coefficients. The total number of parameters we estimate including the constant is *K* + 1. If we have three regressors, *K* = 3, we will estimate 3 + 1 = 4 parameters inclusive of the constant. The disturbance error terms are required to be normally distributed with zero mean and constant variance and satisfy the same properties as in the single regression model. In this chapter we will work with time-series data. The principles apply to models estimated using cross-sectional data. We estimate a model using cross-sectional data in the online chapter II.

- *Online comment #6.1: Alternative expressions for the general form of multiple regressions*

6.2.4 The ordinary least squares (OLS)

Although derived from the same concept as a simple regression, minimising the sum of squared residuals, the OLS formulae for a multiple regression has to take into consideration the number of explanatory variables. Let us illustrate using a two-variable model. The PRF is:

$$Y_i = \alpha + \beta_1 X_{1i} + \beta_2 X_{2i} + u_i \qquad (6.5)$$

The method of OLS involves fitting a straight line that minimises the sum of squared residuals (prediction mistakes):

$$\min \sum e_i^2 = \sum \left(Y_i - \hat{Y}_i\right)^2 = \sum \left(Y_i - \hat{\alpha} - \hat{\beta}_1 X_{1i} - \hat{\beta}_2 X_{2i}\right)^2 \qquad (6.6)$$

6.2.5 Coefficient estimation

This minimisation of (6.6) leads to the formulae for calculating the:

- slope estimate $\hat{\beta}_1$

$$b_1 = \frac{\left(\sum (X_{1i} - \bar{X}_1)(Y_i - \bar{Y})\right)\left(\sum (X_{2i} - \bar{X}_2)^2\right)}{\left(\sum (X_{1i} - \bar{X}_1)^2\right)\left(\sum (X_{2i} - \bar{X}_2)^2\right)} \\ \frac{-\left(\sum (X_{2i} - \bar{X}_2)(Y_i - \bar{Y})\left(\sum (X_{1i} - \bar{X}_1)(X_{2i} - \bar{X}_2)\right)\right)}{-\left(\sum (X_{1i} - \bar{X}_1)(X_{2i} - \bar{X}_2)\right)^2} \qquad (6.7a)$$

- slope estimate $\hat{\beta}_2$

$$b_2 = \frac{\left(\sum(X_{2i}-\bar{X}_2)(Y_i-\bar{Y})\right)\left(\sum(X_{1i}-\bar{X}_1)^2\right)}{\left(\sum(X_{2i}-\bar{X}_2)^2\right)\left(\sum(X_{1i}-\bar{X}_1)^2\right)}$$
$$\frac{-\left(\sum(X_{1i}-\bar{X}_1)(Y_i-\bar{Y})(\sum(X_{2i}-\bar{X}_2)(X_{1i}-\bar{X}_1))\right)}{-\left(\sum(X_{2i}-\bar{X}_2)(X_{1i}-\bar{X}_1)\right)^2}$$

(6.7b)

- intercept/constant estimate $\hat{\alpha}$

$$a = \bar{Y} - b_1\bar{X}_1 - b_2\bar{X}_2$$

(6.8)

6.2.6   Standard error and variance estimation

The formulae for the variances of the estimate of the slope coefficients $(\hat{\beta}_1, \hat{\beta}_2)$ and the intercept $(\hat{\alpha})$ in a simple regression are:

$$Var(b_1) = \frac{\sigma^2}{\sum(X_{1i}-\bar{X}_1)^2\left(1-r_{1,2}^2\right)}$$

(6.9a)

$$Var(b_2) = \frac{\sigma^2}{\sum(X_{2i}-\bar{X}_2)^2\left(1-r_{1,2}^2\right)}$$

(6.9b)

$$Var(a) = \left[\frac{1}{n} + \frac{\bar{X}_1\sum(X_{2i}-\bar{X}_2)^2 + \bar{X}_2\sum(X_{1i}-\bar{X}_1)^2 - 2\bar{X}_1\bar{X}_2\sum(X_{1i}-\bar{X}_1)(X_{2i}-\bar{X}_2)}{\sum(X_{1i}-\bar{X}_1)^2\sum(X_{2i}-\bar{X}_2)^2 - \left(\sum(X_{1i}-\bar{X}_1)(X_{2i}-\bar{X}_2)\right)^2}\right]\sigma^2$$

(6.9c)

where:
σ^2 = the variance of the error term.
$r_{1,2}^2$ = square of the sample correlation coefficient between variables X_{1i} and X_{2i}.

The standard errors are obtained by taking the square root of the variances. An unbiased estimator of the variance of the error term is:

$$\hat{\sigma}^2 = \frac{\Sigma\hat{u}_i^2}{n-K-1}$$

(6.10)

As there are two explanatory variables and a constant in the model (equation (6.5)), the denominator is $n - K - 1 = n - 3$ in (6.10).

Excel and statistical software programmes will perform the calculations for us.

6.2.7   Generalisation to more variables

For regression models containing three or more variables, it is more convenient to state the formulae using vector and matrix notation; see online note #6.1.

6.3   Coefficient interpretation in multiple regression

In a multiple regression we study the significance and impact of an explanatory variable in the presence of other explanatory variables in the model. A powerful feature of multiple regression analysis is that it allows us to examine the unique effect of an independent variable holding the values of all other variables constant. In the regression model (equation (6.4)) we control for the influences of X_2, X_3, X_4 when we examine the impact of X_1 on Y.

More generally, when several variables are found to determine the values of the dependent variable, it is useful to examine the impact of a particular independent variable whilst controlling for the influences of the other variables in the model. Suppose we are examining the returns on real estate investment trusts (REITs) and the impact the underlying (private) market has on them. The literature suggests that REIT returns at least in the short run could be influenced by the general stock market (stock market returns). A multiple regression model containing both variables (stock and private real estate market returns) would allow us to quantify the impact of the private market on REIT returns whilst controlling for general stock market influences (keeping the latter's influence constant). In a cross-sectional hedonic model of house prices, economic theory indicates both age and size of the house should be included as explanatory variables. In a multiple regression model, we examine and quantify the impact of age on house prices controlling for size.

6.4   Coefficient of determination: the adjusted *R*-squared

When more variables are added in a regression model, we tend to explain incrementally more of the variation in the dependent variable but at the cost of having more parameters to estimate. Is it worth having a more complex model if the contributions made by the extra variables are really small? A problem with the R^2 metric is that it will always increase in value when we include extra variables in the model, even if they only yield marginal (very little) information. A more appropriate metric of the explanatory power of the regression is the adjusted *R*-squared or *R*-bar squared ($\bar{R}^2$). This metric imposes a penalty when we add more variables. It is given by:

$$\bar{R}^2 = 1 - \left[\frac{n-1}{n-K-1} \left(1 - R^2 \right) \right] \tag{6.11}$$

where:
n = the number of observations
K = the number of parameters to estimate excluding the constant

The inclusion of more explanatory variables raises both K and R^2. The $\bar{R}^2$ will only increase if the rise in R^2 is more than offset by the rise in K. A higher $\bar{R}^2$ is indicative of a model which explains the variance of the dependent variable well. We reiterate that the maximisation of the $\bar{R}^2$ should not be the fundamental objective in regression analysis. Maximising the value of $\bar{R}^2$ is considered a soft rule since we may still end up with a large model and coefficients which make almost meaningless (marginal) contributions. The adjusted *R*-squared is also useful in helping us to compare competing regression models estimated for a particular dependent variable, that is, when the dependent variable in various models has the same unit of measurement.

6.5   The *F*-test of multiple restrictions in the model

When assessing the overall performance of a simple regression model, we examined the statistical significance of a single slope coefficient, a hypothesis test involving a single restriction that a slope coefficient could take. We used the *t*-statistic test in the single restriction test. We tested whether a coefficient in the regression would take the value of zero or be different from zero and hence statistically significant. However, a multiple regression contains more than one slope coefficient and to assess its overall significance requires the testing of significance of all slope coefficient simultaneously. We now generalise the topic of hypothesis testing in regression and consider testing multiple and slightly more complicated restrictions.

Consider the following multiple regression time-series model:

$$Y_t = \alpha + \beta_1 X_{1t} + \beta_2 X_{2t} + \beta_3 X_{3t} + u_t \tag{6.12}$$

The overall significance of the model tests the hypothesis:

$$H_0 : \beta_1 = \beta_2 = \beta_3 = 0$$

$$H_1 : at\ least\ one\ \beta_i \neq 0$$

H_0 represents the statement that all the coefficients are simultaneously equal to zero, in other words all explanatory variables are not relevant to explaining the dependent variable, while H_1 represents the alternative statement that at least one slope coefficient is non-zero.

The *F*-test statistic of imposing restrictions in the form of zero values on the parameters of the slope coefficients can be computed as:

$$F = \left(\frac{R^2}{1-R^2}\right)\frac{N-K-1}{m} \tag{6.13}$$

where:
N = number of observations
K = number of regressors in unrestricted regression excluding the constant
m = number of restrictions

In this hypothesis test there are three restrictions: $\beta_1 = \beta_2 = \beta_3 = 0$. The term m represents the number of restrictions being tested.

There is an alternative way in which this *F*-test statistic can be expressed:

$$F = \frac{ESS / df}{RSS / df} = \frac{ESS / m}{RSS / (N-K-1)} \tag{6.14}$$

where:
ESS = explained sum of squares
RSS = residual sum of squares

The preceding two methods can be calculated from the results reported by running a single multiple regression model.

There is also a third way, using a more general formula based on the sum of squared residuals of the unrestricted and restricted regression equations:

$$F = \frac{(RRSS - URSS)/m}{URSS/(N-K-1)} \tag{6.15}$$

where:
$URSS$ = residual sum of squares (RSS) from unrestricted regression
$RRSS$ = RSS from restricted regression
m: number of restrictions
N: number of observations
K: number of regressors in unrestricted regression excluding the constant

In the econometric literature, it is not uncommon to define K as the number of regressors in unrestricted regression including the constant. In this case, equation (6.15) becomes:

$$F = \frac{(RRSS - URSS)/m}{URSS/(N-K)} = \frac{RRSS - URSS}{URSS} \times \frac{N-K}{m}$$

The third method may be used to test alternative multiple restrictions but requires two regressions to be run, the regression representing the general model (no restrictions imposed) which is also known as the unrestricted model, and the restricted model with the restrictions imposed. In the case of testing the overall significance of the model, the restricted model would be a constant-only model.

The RSS from the restricted model should be higher than the unrestricted (more general) model since we would expect a lower proportion of the variance of the dependent variable to be explained with a model containing fewer explanatory variables. Thus, the F-test statistic should be greater than zero since $RRSS \geq URSS$.

For the restrictions to be valid, the F-test statistic has to be small in order to reflect the small differences between $RRSS$ and $URSS$. A small F-ratio signifies little loss of fit as a result of the restrictions. However, this is an arbitrary statement. To define what is small or negligible we need to know the distribution of this test statistic.

If the disturbance errors in the regression equation are distributed normally, it can be shown that under the null hypothesis (if the null is true) the F-statistic is distributed as an F distribution with degrees of freedom m, $N - K - 1$. The number of restrictions m are the degrees of freedom of the numerator. $N - K - 1$ are the degrees of freedom of the denominator. Since the test statistic is positive ($RRSS \geq URSS$), there is only one critical value which will depend on the degrees of freedom and the significance level chosen. The F distribution is plotted in Figure 6.1.

Unlike the standard normal and t distributions, the F distribution has a positive skew. The degree of skewness is determined by the degrees of freedom of the numerator and the denominator. The critical value divides the distribution into an area of rejection and non-rejection of the null hypothesis, where the null hypothesis represents the proposition that the restrictions are valid.

The critical value is described by the term $F_{a,(m,N-K-1)}$. Note that subscript α is the standard statistical symbol for stating the significance level. If the significance level is 5% or 0.05, then the critical F-value can be written as $F_{0.05}(m, N - K - 1)$. The decision to reject the null hypothesis is made when the F-test statistic exceeds the critical F-value.

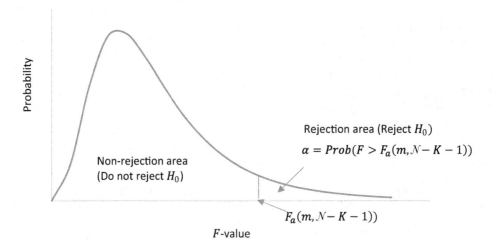

Figure 6.1 The F distribution

Table 6.1 Hypothesis testing using the F-test

	Hypothesis	*Number of restrictions*
(i)	$\beta_2 = \beta_3 = 0$	2
(ii)	$\beta_1 = \beta_2 = 1$	2
(iii)	$\beta_1 = 1$ and $\beta_2 = -1$	2
(iv)	$\beta_2 + \beta_3 = 1$	1
(v)	$\beta_1 \times \beta_2 = 2$	Cannot be tested
(vi)	$\beta_1^2 = 1$	Cannot be tested

In Appendix 6A, there is a diagram displaying F distributions with different degrees of freedom. Appendix 6B shows the critical values of the F distribution for a 5% significance level. The shape and position of the F distribution changes at different significance levels (1% or 10%) – see Appendix 6A. The changes require different F tables with critical values to be used. The critical values at a given significance level and degrees of freedom (df_1 and df_2) for the numerator (m) and denominator ($N - K - 1$) can be obtained in Excel using the function: $= FINV(\alpha, df_1, df_2)$.

We can impose other sets of restrictions in a multiple regression model. At the same time there are restrictions that cannot be tested including non-linear or multiplicative restrictions using OLS. Table 6.1 shows examples of restrictions.

As an example, consider the restriction $\beta_2 = \beta_3 = 0$ in equation (6.16).

The unrestricted and restricted PRFs are:

$$Y_t = \alpha + \beta_1 X_{1t} + \beta_2 X_{2t} + \beta_3 X_{3t} + u_t \tag{6.16}$$

$$Y_t = \alpha + \beta_1 X_{1t} + u_t \tag{6.17}$$

Equation (6.16) is the *unrestricted* model. When we impose the restrictions, $\beta_2 = 0$ and $\beta_3 = 0$, we obtain the *restricted* model (6.17) where the terms $\beta_2 X_2$ and $\beta_3 X_3$ are not present.

In effect, this test is investigating the loss in the fit of the equation – the loss of power to explain Y when we impose restrictions. Once we exclude variables X_2 and X_3, even if they only have a small impact on the dependent variable, the sum of squared residuals (*RSS*) should be inflated. But if their exclusion from the equation does not have a notable impact (as reflected in *RSS*), it is better to have a more parsimonious model by leaving these variables out, as it allows for the marginal effects to be more precisely estimated. It follows that the sum of squared residuals of the restricted model (*RRSS*) will always be higher than the unrestricted sum of squared residuals or unrestricted residual sum of squares (*URSS*): *RRSS* ≥ *URSS*. Statistical tests will tell us whether we should reject or not reject the null hypothesis (hence whether we should accept the restrictions). In general, these tests will tell us whether imposing the restriction results in a restricted model which is statistically different from the unrestricted model.

It is apparent that one of the tasks in multiple hypothesis testing is to specify the restricted equation. When the restriction involves the coefficients to be zero (such as $\beta_2 = \beta_3 = 0$), it is easy to work out the restricted equation. The explanatory variables X_1 and X_2 are simply excluded and the model re-estimated.

For the second set of restrictions in the table, $\beta_1 = \beta_2 = 1$, some algebra is needed. Imposing the restrictions implies the equation:

$$Y_t = \alpha + X_{1t} + X_{2t} + \beta_3 X_{3t} + u_t \tag{6.18}$$

We now rearrange so that the right-hand side has terms in βs only:

$$Y_t - X_{1t} - X_{2t} = \alpha + \beta_3 X_{3t} + u_t \tag{6.19}$$

The left-hand side is a new variable created from the original data. Let us name it Q.
　　The restricted model is:

$$Q_t = \alpha + \beta_3 X_{3t} + u_t \tag{6.20}$$

where:

$$Q = Y_t - X_{1t} - X_{2t}$$

In a similar way we can obtain the restricted equation for the other two restrictions. Restrictions (iii) and (iv) require a bit more algebra.

6.5.1　Steps in conducting an F-test

The steps to carry out the F-test are:

(i)　Specify the null and alternative hypotheses to reflect the multiple restrictions.
(ii)　Estimate the unrestricted and restricted equations and obtain the sum of squared residuals, hence the values for *URSS* and *RRSS*.
(iii)　Calculate the F-test statistic.
(iv)　Determine the level of significance α for rejecting the null, e.g. 5%. Find the critical value for F at that significance level (e.g. 5% level of significance) with m and $N - K - 1$

degrees of freedom (m are the degrees of freedom in the numerator and $N - K - 1$ of the denominator).

(v)   Apply the decision rule: If computed F > critical F at the given level of significance, we reject the null hypothesis. That is:

Reject H_0 at level of significance of α if: $F_{estimated} > F_{\alpha}(m, T - K - 1)$

If we apply the Excel function to get the critical value, Excel will return the critical value above which we will reject the null hypothesis.

(vi)   Econometric software will compute the F-value and associated p-value. Another decision rule is to reject the null hypothesis if the computed p-value is less than the chosen significance level α (say 5%).

We should note that the regression F-test applies to models with both time-series or cross-section data. It can also be used for testing the significance of individual coefficients. We will obtain the same inference results as the t-ratio. In practice, the t-ratio rather than the F-statistic is used to test the significance of individual coefficients due to its simplicity.

6.6   Model specification – dynamics and lags in the real estate market

We turn our attention to another consideration in multiple regression analysis which is associated with modelling time-series relationships. The simple regression model considered in the last chapter is referred to as a static model, since it is the current value of an explanatory variable, GDP, which determines the likely value of the dependent variable rental growth.

In many markets such as real estate markets, there exist frictions which makes 'price' adjustment to 'shocks' relatively slow. The implication is that previous values of variables, or lagged variables, may also be influential in determining the current conditional mean of the dependent variable. The specification of the type and length of lags in modelling is referred to as 'dynamics'. The following are some examples:

(i)   A prime example is modelling new building construction orders (development starts). Decisions to deliver more space are based on the assessment of current and future economic and market trends. Feasibility analysis takes time to conclude, and it will take some time before actual development begins. Even when the decision is made, the process to initiate construction takes time. Plans are drawn, planning permissions are sought and finance should be arranged. These are some factors that can introduce lags in the relationship between building construction and business conditions. Hence, the study of the volume of building starts at time t should include, among a number of causal factors, both current and past economic conditions, in particular if these are extrapolated into the future.

(ii)   Another example is models of take up or absorption, metrics for real estate demand as we discussed in Chapter 2. Firms will seek more space when they expect better economic conditions and when they have decided to increase employment levels. Suppose firms hire more workers, but they cannot expand immediately as they are not able to find space. The search time may be long, especially if availability is low. If firms take up space after, say, a year due to a newly refurbished building in the market, take up is the result of past assessments. Therefore, explaining leasing activity at time t could reflect past employment and output growth.

(iii) Economic time-series variables used in models display a degree of autocorrelation (see Chapter 4 online notes) due to smoothness and measurement methods. In this case the impact of an economic variable on the dependent variable may span over a few past periods and requires the inclusion of lags of the explanatory variables in the model. This is more likely when higher frequency data are used, that is quarterly or monthly.

(iv) When we model rents and prices, lagged values of the dependent variable could appear as regressors. Smoothed valuations data (see Chapter 2) could be an issue, particularly in markets with thin transactions. Lack of comparative information can lead to anchoring values to previous estimates and reduce responsiveness to changing market conditions. Smoothness calls for the use of lags.

(v) Lagged values of the dependent variable can be used to capture momentum effects which are particularly relevant in the determination of investment performance variables. To the degree that the market believes that recent performance is a predictor of near future performance, then these types of lagged effects are important. Periods of persistent over- or under-reaction in the market will cause autocorrelation in the dependent variable, which would require its lagged values to be included in the model.

Dynamics may be incorporated into a regression model by including lagged explanatory variables, lagged dependent variables or a combination of both. *Distributed lag models* include lags of the explanatory variable(s).

For example:

$$Y_t = \alpha + \beta_1 X_{1t} + \beta_2 X_{2t} + \beta_3 X_{2t-1} + u_t \tag{6.21}$$

Distributed lag models permit us to test whether the effect of X_2 on Y occurs over two periods, t and $t-1$. Of course, it is possible to test for longer lags (e.g. X_{2t-2}). One of the significance tests that can be used is the t-test, for example, this test can be used to examine the statistical significance of the lag of X_2 in equation (6.21). We should note that theory is often not helpful for deciding on the maximum number of lags (maximum lag length) and which lags to include in a model. In practice, annual time series data is unlikely to require more than two lags of each independent variable. More lags are usually required in modelling when using quarterly and monthly data. On the other hand, the signs of the lags in the model should be consistent with theory.

Models can also contain lagged dependent variables – these models are called *autoregressive models*. *Autoregressive distributed lag models* (ARDL) comprise of lagged effects for both the dependent and independent variables:

$$Y_t = \alpha + \beta_1 X_{1t} + \beta_2 X_{1t-1} + \beta_3 X_{2t} + \beta_4 X_{2t-1} + \beta_5 Y_{t-1} + u_t \tag{6.22}$$

This ARDL structure contains contemporaneous and one lag of the explanatory variables X_1 and X_2 and one lag of the dependent variable (Y_{t-1}).

In the chapter's online resource, we provide a full account of selected distributed lag models. We discuss Koyck restrictions, the adaptive expectations model and the partial adjustment model.

• ***Online comment #6.2: Selected distributed lag models***

6.7   Attributes of a good model

In the last chapter we outlined the requirements for a good estimator. We were not in a position to consider the attributes of a good regression model since there are very few causal relationships involving a single explanatory variable. The following considerations provide a guide to obtaining a good econometric model:

(i)   Parsimony: The final regression model should only contain explanatory variables which are statistically significant. Including superfluous explanatory variables leads to imprecise estimates of the population parameters and makes it more difficult to identify the significant factors.

(ii)   Theoretical consistency: The estimates have to be plausible. This is indicated by the sign, size and significance of the estimates obtained for the population parameters. They should be consistent with the underlying theory as expressed by the PRF.

(iii)   Predictive power: The final regression model should be able to explain the variance of the dependent variable reasonably well, as indicated by the coefficient of determination (R-squared and adjusted R-squared). The model should also be able to provide reasonably accurate forecasts of future values of the dependent variable.

The attributes of a good model assume that the estimator employed in the analysis yields unbiased or consistent estimates and the correct standard errors. When applying OLS, the preceding considerations require the estimates to be BLUE, which is the same criterion as requiring the Gauss Markov assumptions of the classical linear regression model (CLRM) to be met.

6.8   Example: building a multiple regression model for Hong Kong office rents

We illustrate the fundamental concepts of multiple regression analysis with an example using data for the Hong Kong office market. We build an empirical model of office rents in Hong Kong, performing the tasks step by step. We use the same database to provide a continuation of the regression analysis in online chapter I. The steps we adopt for model building in this section include:

(i)   Existing studies on rent determination are a valuable source of ideas and information to construct an empirical model.

(ii)   Setting up (specifying) the theoretical model of rents to estimate.

(iii)   Familiarising with the data; plot the series; and discern patterns.

(iv)   Perform autocorrelation tests to assess the degree to which the values of the variable are correlated over time.

(v)   Deciding on whether the model will be estimated in levels or in growth rates/first differences. Further discussion about this can be found in Chapter 10.

(vi)   Calculate cross-correlations as an initial measure of the strength of the possible relationship between rents and the explanatory variables.

(vii)   Building the model and variable selection. Preliminary key criteria are signs of the estimated coefficients, statistical significance and evidence of model's contribution to explaining the variation of the dependent variable.

6.8.1 A general theoretical framework for office rent determination

Let us consider a theoretical framework that can be used for the study of rents.

Figure 6.2 illustrates ways in which we can set up a model of rents. We should be familiar with the data in the boxes as they were defined and discussed in Chapter 2. The solid black arrows in the figure illustrate certain paths that can be used to explain rent determination. Figure 6.2 also highlights the complex dynamics of the market. Solid grey arrows show feedback effects. There are more arrows that can be added, based on theory and our views about the workings of the market, but adding additional pathways will make this framework cumbersome. Working through the solid black arrows from left to right, a strengthening in economic activity and business conditions will impact demand, due to it being a derived demand as floor space is a factor of production, subject to floor space/employee ratio. Although employment levels increase (as business conditions improve), this ratio may fall (hence fewer squared meters per employee), reducing the effective impact of employment on the new demand for office space. The level of shadow vacancy may also affect the actual demand for space (see Chapter 2).

The demand for floor space will eventually reflect the better economic conditions with take up/absorption increasing. Demand and supply levels impact on availability and vacancy. Supply includes space from existing buildings arriving onto the market (see Chapter 2). If demand and supply conditions result in falling vacancy, the market becomes a 'landlords' market' as they will have more bargaining power in letting negotiations. Landlords will push for higher rents in new leases or possibly on rent reviews depending on lease terms. Estimated rental values are expected to rise as valuers take into account the improved market conditions into their valuations. In Chapter 2 we remarked on the notion of the natural vacancy rate and the argument suggesting that rent growth responds to the deviation of the actual from the natural vacancy rate. If the former is lower than the latter, there are upward pressures on rents. Higher rents will of course feed back into demand, as the space becomes more expensive with firms possibly economising on accommodation. Furthermore, higher

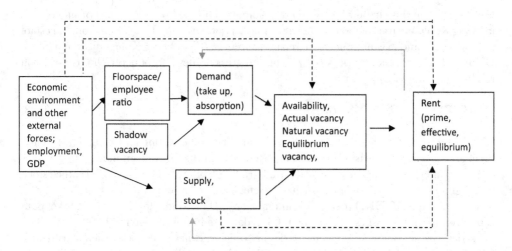

Figure 6.2 A conceptual framework for the determination of rents

Source: Brooks and Tsolacos (2010), authors.

rents are a signal to developers to consider new building starts and hence they trigger supply. This is a simple and straightforward example of how the market adjusts. Any arrow transmits influences. These influences are not instantaneous, the arrows encompass lags.

In real estate modelling the availability of data plays a significant role in the specification of the empirical model. In markets where there is an abundance of data (both in terms of existence and history of series describing the occupier and investment markets) the continuous black and grey arrows (feedback effects) show that a model of rents is part of a more general framework within which demand, vacancy, rent and supply are determined. In this system there are multiple dependent variables, some of which enter as endogenous explanatory variables in other equations. A system of equations containing endogenous explanatory variables is called a *structural* model. In markets with data limitations, it might not be possible to model all the causal relationships or include endogenous explanatory variables in particular equations. Instead *reduced form* models are estimated. The reduced form is obtained by expressing (or solving for) the endogenous variables as functions of the exogenous variables. In general, reduced models are prevalent in markets with limited data. Suppose that in a market the vacancy data is prone to measurement error (e.g. they may include space which is economically or functionally obsolete). A model using vacancy as a determinant may not be well specified and a more direct relationship between rents and economic variables may be sought. In addition, reduced models represent a good framework to examine relationships more directly and make a start with forecasting the variables of interest. The fact that they are less complex makes them a benchmark for more general models to improve upon.

Existing empirical studies offer insights for constructing the rent model. The authors explain the reasoning behind their models, the selection of variables and any data issues, all of which can provide a guide for building a model. It is important that we review past studies. For our rent model determination, relevant studies include Key *et al.* (1994), Hendershott (1996, 1995), D'Arcy *et al.* (1997), Tonelli *et al.* (2004), and Füss *et al.* (2012).

6.8.2　*Setting up the theoretical model*

Our example in this chapter focuses on the estimation of a *reduced form* model of office rents in Hong Kong. We use a subset of the office market data in Hong Kong. Based on real estate market theory and existing studies we relate real rent (*RRE*) in Hong Kong offices to (i) office vacancy (*VAC*); (ii) office stock (*STOCK*) and (iii) gross domestic product (*GDP*). Hence, real rent is a function of these drivers:

$$RRE = f(VAC, STOCK, GDP) \tag{6.23}$$

The data for rents, vacancy and stock are obtained from the Rating and Valuation Department and cover all districts of Hong Kong. Rents are in the form of an index (1999 = 100). Office vacancy is the vacancy rate (%) for all grades of office buildings. Stock is office stock for all grades and it is in millions of square meters. The economic variable GDP is a chain-linked series in 2016 HKD (hence it is in real terms). The source for the office market data is the Census and Statistics Department. GDP data are in millions of HKD. The length of the series varies though. The shortest series are office vacancy and stock. Their start date is 1985. Office rent data are available from 1981, with the GDP series even longer. Hence the shortest sample period for our analysis common is 1985 to 2017, that is 33 annual observations. The frequency of the data is annual. The data are available in the book's website as both Excel and EViews files.

The intuition behind equation (6.23) is straightforward. As we noted previously, vacancy is considered an indicator of the demand and supply balance in the real estate market. In addition, we can argue that the vacancy series on its own may not fully account for business conditions that impact on rents. For example, vacancy does not effectively allow for the impact of business profitability and affordability on rents. Measures relating to the profitability of the corporate sector can give us further insight as to the rent that firms are prepared to pay. Of course, in a market with a high vacancy rate (occupiers' market) firms will not pay the highest rent they can afford. As a measure of profitability and business conditions we use GDP. One may argue that this is not a perfect proxy and that we should use a more office-market-related economic variable such as financial and business services output. There are alternative variables to allow for business conditions and profitability. The material presented in this chapter will enable you to test the relevance of such variables on rents.

(i)   Modelling rents in levels and growth rates

A decision we need to make is whether equation (6.23) will be estimated in levels or growth rates (or alternatively in first differences). Theory might guide this decision, and we can therefore put forward a theoretical argument in real estate for modelling either in levels or in growth rates. Assuming the analyst is interested in modelling short-term cyclical movements, significant weight is given to the statistical properties of the data. Historical data which contain trends, data series that are smoothed and hence highly autocorrelated (a major area of debate about valuations data), or data which resemble random walks (see Chapter 8) may need transformation in order to perform any meaningful regression analysis. These are characteristics which are more pronounced in level series (e.g. the actual real rent series) but less so when we take their first differences or growth rates. We address this topic further in Chapter 8, where we discuss the concept of stationarity and its importance in specifying a model.

(ii)   Familiarising with the data

It is always useful to familiarise oneself with the data before proceeding with the analysis. In Figure 6.3, we plot the time-series in levels and growth rates (relative changes).

The real rent series in levels (panel (a)) exhibits long rather smooth cycles. There is a long period of declining real rents which is followed by a lengthy period of rising rents. We observe that in real terms Hong Kong office rents have not risen. When expressed as a growth rate the rents are more volatile around a mean close to zero. The vacancy rate series exhibit both short-term and long-term cycles. The volatility of this series is increased when we take the first differences as shown in panel (d). If the vacancy rate falls from 7% to 5%, panel (d) reports the absolute difference of −2%. The stock in levels is rising to denote an expansion in the office market through time. We study the topic of trended variables in more detail in Chapter 8. In terms of changes (absolute first differences) the stock series exhibits more variation. Stock changes encompass different levels of new completions each year and the possible loss of building space due to, say, demolitions. GDP displays an upward trend revealing that the Hong Kong economy became larger. The cycle pattern in GDP is captured best using its growth rate.

In Table 6.2, we present the autocorrelations of the variables in our data series. High autocorrelations are not desirable, especially for the dependent variable. If the dependent variable is highly autocorrelated, it is strongly linked to its past values and the independent

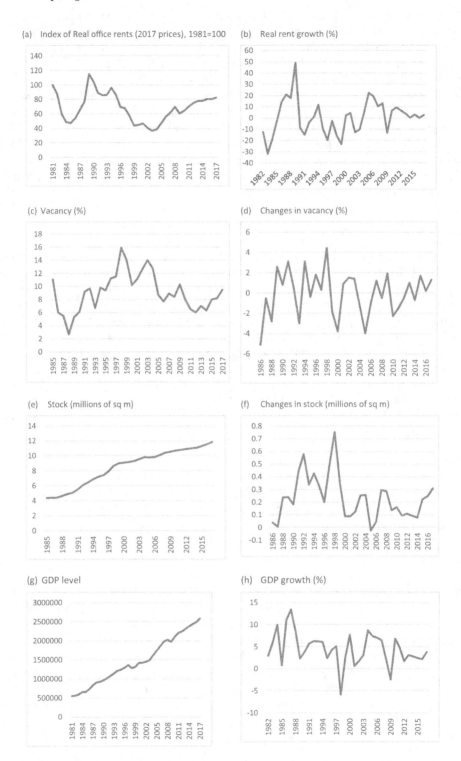

Figure 6.3 Hong Kong office market data

Table 6.2 Autocorrelation pattern of variables

Lags	RRE	RREg	VAC	ΔVAC	STOCK	ΔSTOCK	GDP	GDPg
1	0.84	0.35	0.71	−0.04	0.92	0.53	0.90	0.25
2	0.62	0.04	0.48	−0.02	0.83	0.05	0.81	−0.19
3	0.43	0.04	0.27	−0.29	0.74	0.10	0.72	0.03
Sample	1985–2017	1985–2017	1985–2017	1986–2017	1985–2017	1986–2017	1985–2017	1985–2017

Key: *RRE*: index of real office rents (level); *RREg*: real rent growth; *ΔVAC*: Vacancy; *VAC*: changes in vacancy (first differences); *STOCK*: office stock; *ΔSTOCK*: changes in office stock; *GDP*: gross domestic product; *GDPg*: GDP growth

Note: It is the practice to consider absolute changes in vacancy and stock. However, one can estimate autocorrelation patterns of these variables in growth rates. The results will be broadly similar to those reported for the changes in these variables.

Table 6.3 Cross-correlations of real rent growth with causal variables

	GDPg		VAC		ΔVAC		ΔSTOCK	
Time (t): Years	Lag	Lead	Lag	Lead	Lag	Lead	Lag	Lead
0	0.43	0.43	−0.64	−0.64	−0.27	−0.27	−0.38	−0.38
1	0.60	−0.06	−0.42	−0.55	−0.61	0.09	−0.42	−0.33
2	0.24	−0.04	0.02	−0.34	−0.19	0.27	−0.27	−0.05
3	−0.03	0.07	0.14	−0.24	−0.06	0.12	−0.17	0.21

Sample period: 1985–2017

variables may appear to be less relevant in explaining the variation in the dependent variable than in reality.

All series in levels are highly autocorrelated as denoted by the size of the first order auto-correlation coefficients. Autocorrelations remain high at the second lag and even at the third lag for *GDP* and *STOCK*. The degree of autocorrelation in the differenced series decreases across the board. The autocorrelations are confined to the first order. *ΔVAC* exhibits no autocorrelation and *Δ STOCK* only slightly higher autocorrelation, whereas *RREg* and *GDPg* show rather modest first order positive autocorrelation. *RREg*, unlike *RRE*, does not show a strong dependence to recent own past values. Its short-term variation will reflect market and economic factors.

(iii)   Cross correlations

We would like to get an initial impression of the strength of the bivariate relationships between real rental growth and the explanatory variables. We examine the cross correlations between real rent growth and the explanatory variables contemporaneously (that is at time $t = 0$) as well as using their leads and lags. The lead of an explanatory variable refers to using values of the explanatory variable one, two or more periods after the values of the dependent variable. For example, Table 6.3 reports the correlation between *RREg* at time t and *GDPg* at time $t + 1$ (correlation coefficient −0.06), $t + 2$ (−0.04) and $t + 3$ (0.07).

Table 6.3 contains the results of these cross-correlations. It includes contemporaneous correlations, that is correlations of *RREg* with the explanatory variables and correlations with their lags and leads over a 3-year horizon. An assumption that all explanatory variables affect real rents contemporaneously is restrictive even when using annual data. Recent past vacancy rates, business conditions and supply may be relevant predictors of the variation in rents. If this is the case, we should allow and test for possible influences of past values of vacancy and GDP on rents. The association of rent growth with future values of the independent variables is informative about their leading properties.

With annual data we consider a three-year window long enough to study the lag-lead relationships in this model. The strongest correlation is the contemporaneous (same year) relationship between *RREg* and *VAC* (−0.64). The correlation coefficient is negative as expected. It appears that leads of *VAC* are more strongly correlated with *RREg* than lagged values. Higher coefficients on lagged values would suggest movements of *VAC* ahead of *RREg*. This is not the case, but still the contemporaneous ($t = 0$) is the strongest. If the strongest correlation coefficient was for example at $t + 1$, then *VAC* would not be considered a strong predictive variable.

Before proceeding, it is worth referring to the possible issue of *endogeneity* between rent growth and the vacancy rate. Endogeneity would suggest that the variables determine each other simultaneously. In practice, professionals in the market use the vacancy rate to gauge the direction of rental growth. Another point to make is that the lead-lag pattern could be studied more clearly with quarterly data.

The second strongest correlation is between *RREg* and ΔVAC (−0.61). The rest of the cross-correlations are not even moderate, but this particular lag is strongly related to rent growth at time t. A similar correlation in terms of strength is observed between *RREg* at time $t = 0$ and *GDPg* at time $t − 1$. *GDPg* clearly contains leading properties. Lags of *GDPg* take the expected positive sign and show the highest coefficient values. Similarly, it is the first lag of $\Delta STOCK$ that shows the strongest association with current *RREg* and at the same time we observe a leading pattern. The negative sign on the current and lagged $\Delta STOCK$ confirms a supply side effect that has a dampening impact on rent growth. We do not show results for *GDP* and *STOCK* in levels as the results did not establish any notable association with real rent growth.

(iv)   *The theoretical model*

On the basis of the cross-correlation results in Table 6.3, we believe that real rent growth can be explained using the explanatory variables *VAC*, $\Delta STOCK$ and *GDPg*. The general form of the model we estimate at this stage is:

$$RREg_t = \beta_0 + \beta_{11}VAC_t + +\beta_{12}VAC_{t-1} + ... + \beta_{21}\Delta STOCK_t + \beta_{22}\Delta STOCK_{t-1} \qquad (6.24)$$
$$+ ... + \beta_{31}GDPg_t + \beta_{32}GDPg_{t-1} + ... + u_t$$

Equation (6.24) contains the growth rate of GDP, the change in the stock of floorspace and the vacancy rate, augmented by their lags. We do not include ΔVAC as we already have *VAC* in our model but will test the significance of ΔVAC later.

In equation (6.24), the signs on β_1s and β_2s are expected to be negative on a priori, as dictated by theory. As we expect there to be a positive relationship between $RREg_t$ and $GDPg_t$, the coefficients β_3s take a positive sign.

How many lags of each variable should be included in the model? This is mainly an empirical question. We may expect rents to reflect trends in vacancy and output within, say, 2 years. A number of metrics can be used to identify the lagged terms that are significant determinants of rent growth.

6.8.3   *Model estimation*

There are two approaches for obtaining the final specification of a model. One is the *general-to-specific* approach. This methodology involves setting up a general model and then imposing a set of testable restrictions to derive the final 'simpler' model specification. In our example in this chapter, we use a specific to general approach. An example of applying the general-to-specific approach to the same dataset in this chapter can be found in the online chapter I.

The second method is a *bottom-up approach or the specific-to-general approach,* where we start with a simple model and extend it by adding additional explanatory variables until the final and preferred model is obtained. We should note that the final model cannot solely be decided upon the material presented in this chapter. The topics in the next two chapters are important too for that purpose. In this chapter we will progress towards the final model specification, but the analysis is not complete. The model building process are guided by three important criteria: (i) the signs of the estimated values of the parameters are consistent with economic theory; (ii) statistical significance; and (iii) the explanatory power of the model. These rules apply to models derived using the general to specific approach too.

We begin by estimating a bivariate model for *RREg* and utilise the cross-correlations findings. The explanatory variable in the bivariate model is the contemporaneous vacancy rate *VAC*, which has the strongest correlation with *RREg*. The estimated bivariate model is:

$$RREg_t = 31.3 - 3.2VAC_t \qquad (6.25)$$
$$(4.8) \quad (-4.6)$$
$$[0.00] \; [0.00]$$

$\bar{R}^2 = 0.38$; *RSS* = 4014.4; S.E. of regression: 11.4; AIC = 7.76; SIC = 7.85; HQIC = 7.79
Sample: 1985–2017 (33 obs)
RSS: Residual sum of squares; S.E.: standard error;
AIC: Akaike information criterion; SIC: Schwarz information criterion; HQIC: Hannan-Quinn information criterion.
Numbers in parentheses are the *t*-ratios and in square brackets the *p*-values for the *t*-ratios.

VAC takes the expected negative sign. A rise in vacancy rate by 1 percentage point will decrease real rent growth down by 3.2 percentage points. The *t*-ratio (and the associated *p*-value) establishes the statistical significance of *VAC*. The computed *t*-value of −4.6 is in absolute terms higher than the critical *t*-value of 2.04 with 31 degrees of freedom (33 observations minus 2 parameters estimated). The null hypothesis that the coefficient on VAC_t is equal to zero is rejected. The *p*-value is even lower than 0.01, suggesting a highly significant variable.

The squared residuals (RSS) and the standard error of the regression (residuals) are also reported. We use these measures to compare model specifications and as intermediate values to calculate additional test statistics. Three new metrics accompany the results of (6.25): Akaike, Schwarz and Hanna-Quinn. These are *information criteria* metrics designed to help us select the best model by identifying variables that make contributions to the model (Akaike,

1973; Schwarz, 1978; Hannan and Quinn, 1979). As we explained earlier for the reasons for using the adjusted R square ($\bar{R}^2$) as a goodness of fit measure for multiple regression, the inclusion of an additional independent variable may lead to a small gain in the model's explanatory power that is outweighed by the loss of a degree of freedom and affect the precision of the estimates by having that extra variable, especially when the sample size is small. These criteria guide us as to whether another variable should be included in the model. The aim is to minimise the value of these tests. The online chapter presents these criteria in more detail.

- **Online comment #6.3: Information criteria**

The values reported for the information criteria on their own are uninformative. They have to be compared against a model specification that includes more or fewer explanatory variables. The aim is to minimise the reported values. If the information criteria take negative values, then a model with a more negative information criterion value is preferred.

In the next model, we include GDP growth lagged one year as per Table 6.3. Equation (6.26) is a *multivariate model* of $RREg_t$ since it contains more than one explanatory variable.

$$RREg_t = 18.3 - 2.4VAC_t + 1.4GDPg_{t-1} \tag{6.26}$$

$$(2.3) \quad (-3.5) \quad (2.6)$$
$$[0.03] \ [0.00] \quad [0.01]$$

$\bar{R}^2 = 0.48$; RSS = 3265.1; S.E. of regression: 10.4; AIC = 7.61; SIC = 7.75; HQIC = 7.66
Sample: 1985–2017 (33 obs)

The addition of $GDPg_{t-1}$ has improved the model. We make the following remarks:

- $GDPg_{t-1}$ takes the expected positive sign. Positive (negative) real rent growth is associated with positive (negative) GDP growth, ceteris paribus. This variable is statistically significant (*t*-ratio = 2.6 and *p*-value 0.01). The critical *t*-value at a 5% level of significance with 30 (33 − 3) degrees of freedom is 2.042.
- VAC_t retains its significance. We also expect GDP and VAC to have some association. If the association is strong, it means they convey similar influences into the model, and when included together one of them loses its significance. We explore this further in the diagnostics chapter. In equation (6.26) the two variables convey distinctive influences. If we run a bivariate regression with $GDPg_{t-1}$ only as an explanatory variable, we will find $GDPg_{t-1}$ statistically significant. Hence the two regressors in equation (6.26) are statistically significant on their own and when they are both included. This is a sign suggesting that these two variables convey distinct information to explain $RREg_t$.
- The slope coefficient on VAC_t is now lower, hence in the presence of $GDPg_{t-1}$ the sensitivity of real rent growth to vacancy has fallen from 3.2% to 2.4%. This is because the vacancy rate in equation (6.25) also contained the indirect influence of GDP on rent growth. The explicit inclusion of $GDPg_{t-1}$ reduces the impact VAC_t has on rent growth.
- The sum of the squared residuals is lower since equation (6.26) explains more variation of the dependent variable, indicating that the fitted values are closer to the actual or observed values of $RREg$. The standard error has fallen as well and the residuals have a smaller variance, another indicator of a better model.

- The superior explanatory power of the augmented model is reflected in the higher adjusted R-squared. The adjusted R-squared has increased to 0.48 from 0.38.
- All three information criteria reveal a better model since the values in equation (6.26) are smaller than in the bivariate equation (6.25). We have complicated the model by adding one more explanatory variable but the gain in terms of explanatory power outweighs it.
- The sample size remains unchanged when including $GDPg_{t-1}$. There is no loss of observations (and degrees of freedom) since the data for GDP are available prior to 1985.

The interpretation of the coefficients in multiple regression is subject to *ceteris paribus*. This means that an estimate reports the marginal effect of the unit change in the value of the variable holding the values of all other variables in the model constant. In the preceding equation, a 1 percentage point rise in vacancy will tend to lower rent growth by 2.4 percentage points in the presence of $GDPg_{t-1}$ and keeping its value constant.

We can also compute the regression F-statistic using formula (6.16). The null hypothesis states the proposition that the coefficients on VAC_t and $GDPg_{t-1}$ are jointly zero. In other words, they do not explain the variation around the mean of the $RREg_t$.

Null hypothesis: $\hat{\beta}_{VAC_t}(=-2.4)=0$ and $\hat{\beta}_{GDPg_{t-1}}(=1.4)=0$
Alternative: $\hat{\beta}_{VAC_t} \neq 0$ or $\hat{\beta}_{GDPg_{t-1}} \neq 0$

The unrestricted equation is equation (6.26). The RSS of unrestricted (URSS) is 3,265.1. The restricted equation is:

$$RREg_t = \beta_0 + u_t \quad \Rightarrow \quad RREg_t = 2.56 \tag{6.27}$$

RSS of restricted $(RRSS) = 6,730.4$

The sample period for the estimation of the unrestricted is 1985–2017 (33 observations), hence $T = 33$. The number of restrictions m is two ($m = 2$). In the unrestricted equation we estimate three parameters, $k = 3$. In this case, k includes the constant.

$m = 2; k = 3$ (the number of parameters in the unrestricted equation including the constant).

$$F = \frac{(6730.4 - 3265.1)/2}{3265.1/(33-3)} = 15.9$$

From Appendix A (Table A3), the critical F-value with 2 and 30 degrees of freedom at the 5% level of significance is 3.32. The computed value is larger than the critical value, hence we reject the null hypothesis in favour of the alternative. At least one of the coefficients is statistically significant. The t-tests also report that each coefficient is statistically significant. Any failure to reject this F-test implies that we have a poor model, as none of the selected independent variables explains the variation in the dependent variable.

A final observation. Equation (6.26) contains VAC which has a first order autocorrelation coefficient of 0.71. Ideally, we would have preferred a lower degree of autocorrelation. At this stage we do not envisage any problems. Tests presented in the next chapter aim to pick up misspecifications arising from the inclusion of autocorrelated variables.

• ***Online comment #6.4: Multiple regression in EViews***

We can augment the model and test the significance of a third variable. For example, suppose we include contemporaneous GDP growth ($GDPg_t$) in the equation. Although the cross correlations establish a weaker association of this variable with $RREg_t$ than between $RREg_t$ and $GDPg_{t-1}$, $GDPg_t$ may bring some additional information into the model. The results are shown. For brevity we only report the Schwarz information criterion.

$$RREg_t = 11.6 - 2.0VAC_t + 1.4GDPg_{t-1} + 0.7GDPg_t \qquad (6.28)$$
$$\quad\quad (1.2) \quad (-2.6) \quad\quad (2.6) \quad\quad\quad (1.3)$$
$$\quad\quad [0.23] \; [0.01] \quad\quad [0.01] \quad\quad\quad [0.22]$$

$\bar{R}^2 = 0.49$; RSS = 3,098.1; S.E. of regression: 10.3; SIC = 7.80
Sample: 1985–2017 (33 obs)

The contemporaneous GDP growth is not statistically significant. The RSS has decreased by adding the extra variable and although it is not statistically significant, it still explains some of the variation in $RREg_t$. The standard error is slightly reduced. This is also reflected in a marginal rise of the $\bar{R}^2$. However, the penalty that SIC imposes outweighs the marginal improvement in the explanatory power and the reduction in RSS. The increasing value of the SIC suggests GDP_t should be dropped.

We can continue the process by testing for the significance of $\Delta STOCK_t$. We find this variable to be insignificant when it is included in equation (6.26). Its influence is likely to be captured by the vacancy rate, which reflects both demand and supply conditions. The correlation of $\Delta STOCK_t$ and ΔVAC_t is 0.48, a moderate association.

From testing different specifications, we find that $\Delta STOCK_t$ is statistically significant in the absence of VAC_t, providing evidence that VAC_t embodies the impact of $\Delta STOCK_t$. But the former appears to be a better variable to use in explaining rental growth. The results are given as equation (6.29).

$$RREg_t = -0.17 + 2.3GDPg_{t-1} - 30.1\Delta STOCK_t \qquad (6.29)$$
$$\quad\quad (-0.04) \; (4.4) \quad\quad\quad (-2.7)$$
$$\quad\quad [0.97] \; [0.00] \quad\quad\quad [0.01]$$

$\bar{R}^2 = 0.45$; RSS = 3425.4; S.E. of regression: 10.9; SIC = 7.84
Sample: 1986–2017 (32 obs)

$\Delta STOCK_t$ is now statistically significant. Equation (6.29) could be seen as a competing model to equation (6.26). The latter still has the edge, though – a slightly higher explanatory power and a lower SIC value. The final choice requires diagnostic testing of each of the models for meeting the CLRM criteria, which we undertake in the next chapter. We can make a further assessment by examining how well the models forecast. Such assessments and tests are discussed in Chapter 10. In equation (6.29) $\Delta STOCK_t$ is measured in millions of square metres (e.g. 11.84 million square meters as of end of 2017). The estimate implies that if the stock rises by 1 million square metres (around 8.4% of the total stock), real rent growth falls by 30 percentage points, ceteris paribus. This is a large impact. However, the interpretation needs to be put into context. If we consider that the average completions level in the Hong Kong office market has been less than 200,000 sq m per annum in the last 15 years, 1 million sq m is more than five times that average completions level. Alternatively, if stock increases by 0.1 million (100,000) sq m, real rental growth will fall by 3 percentage points. Regression results should be interpreted and communicated in the market context, especially to a lay audience.

To complete our analysis in possible model specifications, we consider using the explanatory variable ΔVAC_{t-1} as it was found to be highly negatively correlated with real rent growth (Table 6.3). The inclusion of this variable either in equation (6.26) or in (6.29) results in an insignificant $GDPg_{t-1}$ in (6.26) and an insignificant $\Delta STOCK_t$ in (6.29). Equation (6.30) presents the results when this variable is included in the model.

$$RREg_t = 23.5 - 2.4VAC_t - 3.0\Delta VAC_{t-1} \qquad (6.30)$$
$$ (3.7) \quad (-3.6) \quad (-3.5)$$
$$ [0.00] \quad [0.00] \quad [0.00]$$

$\bar{R}^2 = 0.55$; RSS = 2786.6; S.E. of regression: 9.9; SIC = 7.67
Sample: 1987–2017 (31 obs)

Equation (6.30) includes only vacancy variables. It has the highest $\bar{R}^2$ and lowest SIC, hence it should be the preferred model among equations (6.26), (6.29) and (6.30). We remarked earlier that further diagnostic testing is pending. One issue we should highlight in equation (6.30) is that the impact of the economy (GDP) on rents is conveyed through the vacancy rate. If we are interested in the relationship between the economy and rental growth, we should also develop a model of vacancy. For example, if we wish to study the sensitivity of rent growth to GDP growth, we need at least a model of office vacancy linking vacancy to GDP growth.

To conclude this section, we can of course test for alternative specifications using more data, for example including other economic variables such as employment in finance and insurance available from Hong Kong's Census and Statistics department. Furthermore, the Hong Kong office market database contains more data such as take up.

Actual and fitted values

We can obtain the actual and fitted values for equations (6.26), (6.29) and (6.30). The actual and fitted values are generated in the same way as in the bivariate model. The fitted value for 2017 (equation (6.26)) is:

$$RREg_{2017} = 18.25 - 2.43VAC_{2017} + 1.39GDPg_{2016}$$
$$= 18.25 - 2.43 \times 9.5 + 1.39 \times 2.2 = -1.78$$

A graph of actual and fitted values along with the errors for the model represented by equation (6.26) are given in Figure 6.4.

The model appears to fit the actual data well. It does a good job replicating the volatility of the actual rent growth series. In the first few years of the sample the volatility of rent growth is excessive. Rents increased by 50% before they fell by 20% in subsequent years. The diagram reveals that since 2000 the model explains the oscillations of the actual series better than the prior periods. An explanation for this is the maturity of the market. The information used to compile the rent data in earlier periods did not have as much depth of coverage of the market as the later periods. The rent data in later years had more transactions to use

(a)   Real rent growth

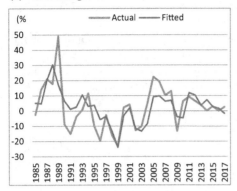

(b)   Residuals

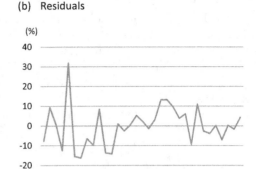

Figure 6.4 Actual and fitted values for Hong Kong office rents

in valuations, more firms compiling data and firms adopting better methods to monitor data. The Hong Kong economy was also more volatile pre-2000 (see GDP growth in Figure 6.3). Panel (b) displays the residuals, defined as actual minus fitted values (the residual in 2017 was 4.35%, the result of actual rent growth of 2.59% and a fitted value of −1.76%). In the first 10 years of the sample, there are significant fluctuations in the residuals, which dampen in later periods.

- **Online comment #6.5: Modelling the level of real rents**

6.9   Using the *F*-test to test for restrictions

Next, we provide two examples of applying the *F*-test to examine the possibility of imposing further restrictions on the model.

6.9.1   Example 1

In the context of the model (6.26) we can test whether the value of the coefficient on VAC_t is −2 and that on $GDPg_{t-1}$ is +2. This is an example of a multiple hypothesis that tests the imposition of two restrictions concurrently. The test based on the *F*-statistic is used to examine this hypothesis. Some algebra is required to obtain the restricted version of the model from equation (6.26) when we impose the restrictions:

$RRg_t = \beta_0 - 2 \times VAC_t + 2 \times GDPg_{t-1} + u_t$ which is equivalent to estimating:

$$RRg_t + 2 \times VAC_t - 2 \times GDPg_{t-1} = \beta_0 + u_t \tag{6.31}$$

Hence in this case we create a new variable $Q = RRg_t + 2 \times VAC_t - 2 \times GDPg_{t-1}$ and run the regression $Q = \beta_0 + u_t$

The estimated (restricted) equation is: $Q = -6.26$ (6.32)

RSS (restricted) = 4567.6; No of obs = 33 (1985–2017)

The estimated *F*-statistic is:

$$F = \frac{(4567.6 - 3265.1)/2}{3265.1/(33-3)} = 5.98$$

The critical *F*-value at the 5% level of significance for 2 and 30 degrees of freedom is 3.32. We reject the null hypothesis that the slopes of VAC_t and $GDPg_{t-1}$ can simultaneously be -2 and 2, respectively.

6.9.2 *Example 2*

Consider the following equation.

$$RREg_t = \alpha + \beta_1 VAC_t + \beta_2 GDPg_t + \beta_3 GDPg_{t-1} + u_t \tag{6.33}$$

We know from our earlier analysis that $GDPg_t$ is not statistically significant. However, we could examine whether $GDPg_t$ and its first lag $GDPg_{t-1}$ jointly affect rental growth and that their combined impact doubles rental growth. The argument for testing this proposition is that the economy in Hong Kong may have an impact on office rent growth over 2 years.

Our null hypothesis is $\beta_2 + \beta_3 = 2$. The unrestricted equation is equation (6.33). To obtain the restricted equation we impose 1 restriction ($m = 1$) taking the form of $\beta_2 + \beta_3 = 2$:

$$RREg_t = \alpha + \beta_1 VAC_t + 2 \times (GDPg_t + GDPg_{t-1}) + u_t \Rightarrow$$

$$Q = \alpha + \beta_1 VAC_t + u_t \tag{6.34}$$

$$\text{with } Q = RREg_t - 2 \times (GDPg_t + GDPg_{t-1})$$

The estimation of (6.34) gives:

$$Q = -6.76 - 0.92 VAC_t; \text{ RSS} = 3834.5; \text{ No of obs} = 33$$

In this example $m = 1$ and the number of parameters in the unrestricted equation is $K = 4$ (equation (6.33)). Hence,

$$F = \frac{(3834.5 - 3098.08)/1}{3098.08/(33-4)} = 6.89$$

The critical *F*-value at the 5% level of significance with 1 and 29 degrees of freedom is 4.18. We reject that $\beta_2 + \beta_3 = 2$. The joint cumulative impact of GDP growth is not 2.

6.10 Omitted variables

We can use the *F*-test to examine the impact of omitted variables. Suppose we would like to test for the joint impact of $GDPg_t$ and $GDPg_{t-2}$. If we include these variables, equation (6.26) is augmented to:

$$RREg_t = \beta_0 + \beta_1 VAC_t + \beta_2 GDPg_t + \beta_3 GDPg_{t-1} + \beta_4 GDPg_{t-2} + u_t \tag{6.35}$$

We test whether the omitted variables $GDPg_t$ and $GDPg_{t-2}$ are jointly significant with an F-test. In our preferred equation (6.26), the coefficients β_2 and β_4 are zero. In this context, equation (6.35) is the *unrestricted* equation ($K = 5$) and equation (6.26) is the *restricted equation* ($K = 3$). The number of restrictions is therefore two ($m = 2$). The *RSS* of the unrestricted equation (6.35) is 3028.35 and the number of observations is 33. The F-statistic is:

$$F = \frac{(3265.1 - 3028.4)/2}{3028.4/(33-5)} = 1.09$$

We do not reject the null hypothesis at the 0.05 significance level with 2 and 28 degrees of freedom. The coefficients on $GDPg_t$ and $GDPg_{t-2}$ are equal to zero. In a similar way we can test the impact of other variables in any equation for example in equations (6.25), (6.28) and (6.29). This test is an alternative to using the information criteria. The t-ratio is remains relevant to testing the significance of extra coefficients individually. In most cases all these alternative metrics to select variables would give us the same result.

In a similar way we can test for *redundant* variables in our model. The redundant variable test refers to testing the proposition that there are a subset of variables in your model which have no effect on determining the values of your dependent variable. For example, we can test whether VAC_t or $GDPg_{t-1}$ in (6.26) are redundant, that is they have no effect on our model. Again, the process is similar. In this case equation (6.26) is the unrestricted model.

- **Online comment #6.6: Omitted and redundant variable tests in EViews**

6.11   Standardised coefficients

Let us consider the results reported for equation (6.26). We can compare the estimated *partial regression coefficients* for VAC_t or $GDPg_{t-1}$ to determine their relative importance as they both represent the marginal impact in percentage points. VAC_t has the greater impact (2.4 compared with 1.4 of $GDPg_{t-1}$). But let us consider equation (6.29). The term $GDPg_{t-1}$ is expressed in percentages whereas the term $\Delta STOCK_t$ in thousands of square meters. The coefficients reflect the units, and their relative influence or partial effect cannot be compared unless they are *standardised*. The *standardised regression coefficients* describe what the partial regression coefficients would equal if all variables had the same standard deviation.

To standardise the coefficients in a multiple regression, we use expression (6.36).

$$\beta_n^S = \beta_n \frac{stdev(x_i)}{stdev(y_i)} \tag{6.36}$$

where β_n^S is the standardised coefficient of the *nth* coefficient (β_n), $stdev(x_i)$ is the standard deviation of the two regressors in (6.36) estimated over the shorter sample period. The sample period for $\Delta STOCK_t$ is 1986–2017 and for $GDPg_{t-1}$ it is 1985–2017. Hence the standard deviations of RRE_t, $\Delta STOCK_t$ and $GDPg_{t-1}$ will be calculated for the period 1986–2017. Table 6.4 shows the calculations.

The calculations reported in Table 6.4 indicate that $GDPg_{t-1}$ has the stronger impact. The standardised coefficients have the same sign as the original coefficients and the statistical significance of the unstandardised coefficient is not affected. The two standardised coefficients represent the effect that $GDPg_{t-1}$ and $\Delta STOCK_t$ would have if they increased by a standardised unit, their standard deviation. If $GDPg_{t-1}$ increases by 1 standard deviation (hence

Table 6.4 Estimation of standardised coefficients

Variable	Coefficient value	Standard deviation	Standardised coefficient	
$RREg_t$	–	14.706	–	
$GDPg_{t-1}$	2.29	3.756	0.58	$2.29 \times \dfrac{3.756}{14.706}$
$\Delta STOCK_t$	−30.13	0.173	−0.35	$-30.13 \times \dfrac{0.173}{14.706}$

by 3.8%), the change in the mean of $RREg_t$ will be 0.58 × 14.706 = 8.6%. The mean of $RREg_t$ will rise by 8.6% controlling for $\Delta STOCK_t$ in the model.

- ***Online comment #6.7: Time dummies in regression and EViews practice***
- ***Online comment #6.8: Background to the derivation of coefficients and standard errors in multiple regression and an example with US apartment price data***

6.12   Concluding remarks

The primary objective of this chapter is to develop the analyst's model building skills to construct, estimate and interpret a multiple regression model. It illustrates the process step by step and discusses the rationale for decisions made relating to the structure of the model and the approach to estimate it. Real estate market theory and the review of relevant empirical work are the starting point in model building. The researcher should familiarise with the dataset and data trends. Statistical tools such as autocorrelations are appropriate for this purpose. We also recommend calculating cross-correlations to gauge an initial impression of the relationships.

The chapter illustrates the model variable selection process and presents key tests and metrics such as information criteria and omitted/redundant variable tests. We note that the chapter has not presented the full process in model building. There is more material to study in subsequent chapters. Multiple regression coefficients have a unique interpretation, in that they report the marginal effect of a change in the value of an explanatory variable whilst controlling for the impact of other factors in the model. The standardised coefficient can be used to assess the magnitude of the partial effects. The rationale for time dummies in regression models is lastly discussed.

The practical example in this chapter utilised time-series data for the Hong Kong office market. The model building principles are relevant to cross-sectional studies too. We build a model with cross-section data in online chapter II. In the next chapter we study the validity of the multiple regression model with a host of tests.

Finally, an evaluation of the model's ability to forecast is not discussed in this section. We use equation (6.26) to obtain forecasts and then evaluate its forecast performance in Chapter 9.

Chapter 6 online resource

- EViews file: "ch6_eviews I" and "ch6_eviews II"
- Chapter 6 accompanying notes

Appendix 6A

F-distributions

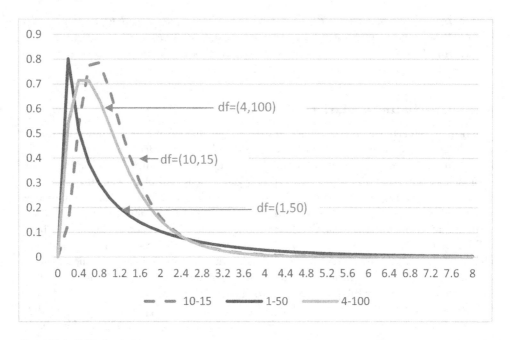

Figure 6A.1 F distributions

7   Regression diagnostics

7.1   Introduction

After building a model going through the stages followed in the previous two chapters (theory consistent signs, significance tests, minimisation of information criteria, good explanatory power), we still need to ensure we have developed a well-specified model. By a well-specified model we mean a model that satisfies the CLRM (classical linear regression model) assumptions, summarised in Table 7.1 and outlined in Chapter 5. We evaluate the quality of the model by performing *diagnostics* or *misspecification tests*. In Table 7.1 we include another misspecification test, that of parameter instability or structural breaks.

Model diagnostics are formal statistical tests that allow us to assess whether the models we have built and estimated are well-specified and robust. They are supposed to detect (diagnose) problems, which, if present, indicate that the estimates obtained from the models are biased or inefficient. If these tests raise the alarm, we may have to revisit the structure of our model and/or the variables we include. After subjecting a model to such tests, we can have confidence in the model to explain the relationships being investigated and in its use to make statistical inferences from our estimation sample for the population regression function. The model is also fit for forecasting. The diagnostics apply both to simple and multiple regression and also to other more complicated methodologies.

Table 7.1 CLRM assumptions and diagnostic tests

(i) $E(u_i) = 0$	The disturbance errors have a zero conditional mean
(ii) $Var(u_i) = \sigma^2$	Homoscedasticity: The disturbance errors have a constant variance
(iii) $E(u_i, u_j) = 0$ for $i \neq j$ or $cov(u_i, u_j) = 0$ for $i \neq j$	The disturbance errors are not autocorrelated; time independent residuals
(iv) $E(u_i, x_i) = 0$ or $cov(u_i, x_i) = 0$	The disturbance errors are not correlated with the regressors
(v) Linearity	The population regression function is linear in parameters
(vi) $u_i \sim N(0, \sigma^2)$	Normally distributed disturbance errors
(vii) No multicollinearity	Regressors are not highly correlated with one another
(viii) No structural breaks	Parameter (coefficient) stability

The CLRM assumptions are stated in terms of the theoretical properties of the disturbance error terms of the population regression function. The diagnostic tests examining whether these assumptions are met use their estimated values from the sample regression function, the residual errors. Diagnostic tests are now routinely provided in most statistical software packages. We provide a step-by-step exposition of the tests presented in this chapter as it will improve your understanding of regression analysis and give you flexibility to run these tests for other methodologies outlined in the following chapters.

The field of diagnostics testing has seen ongoing developments through devising new misspecification methods and modifications to existing tests. There are alternative versions of a given test, in general the analyst has lots of choice to test for misspecification. It is not possible to cover the whole spectrum of tests in this chapter; however, the reader will be equipped to run a battery of tests to assess models. The tests are applied to the regression models we estimated in the previous chapter (Hong Kong office rents). The calculation of the tests is illustrated with the EViews software, but the reader will also be able to carry out these tests 'manually' in Excel and make inferences.

The calculation of certain diagnostic tests involves expanded or so-called *augmented* regressions. The testing procedure requires the construction of new variables (e.g. the cross product of regressors) and more terms in expanded regressions. Furthermore, most of the tests will entail inferential analysis. We will set out a null hypothesis and seek evidence to support it or reject it. This will be the criterion to pass or fail the particular misspecification test.

7.2 $E(u_i) = 0$

The assumption of the CLRM is that the disturbance errors have a conditional mean value of zero. It implies that no relationship or correlation can exist between the disturbance error term and the explanatory variables. This assumption does not hold (i) when an incorrect functional form is used to explain the relationship between the dependent variable and the explanatory variables and (ii) a relevant explanatory variable is excluded from the model.

7.3   Homoscedastic errors

7.3.1   Definition

According to this hypothesis of the CLRM, the conditional variance of the disturbance error term in the regression is constant or homoscedastic. Mathematically it is expressed as:

$$Var\left(u_i \mid x_i\right) = \sigma^2, i = 1, 2, \ldots, n) \text{ or } E\left(X - E\left(X\right)\right) = \sigma^2 \tag{7.1}$$

The constant variance in expression (7.1) is denoted by σ^2. If the variance of the errors is not constant (unequal variance), it is referred to as *heteroscedasticity*. The errors are heteroscedastic, or they are not homoscedastic.

•   ***See online note #7.1 for an illustration of homoscedasticity and heteroscedasticity***

(a)   (b)

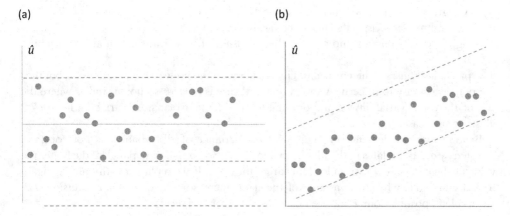

Figure 7.1 Homoscedasticity – finite error variance

Figure 7.1 presents a theoretical case of homoscedasticity. In both cases, the variance of the disturbances remains similar and within boundaries across the spectrum of disturbances.

7.3.2   Causes of heteroscedasticity

There are several reasons why the assumption of constant variance of the disturbance error is violated. Potential sources include:

(i)   Heteroscedasticity is more common in cross-sectional studies due to the nature of the data. Suppose we model commercial construction in several cities, bigger cities tend to have a greater spread of (more varied) construction levels than smaller cities. Thus, the variance of the disturbance error terms may vary with measures of city size.

(ii) A potential source for heteroscedasticity is data issues and measurement errors. The quality of data may differ across the sample (time-series or cross-sectional). For example, in time-series analysis, an index of rents may be constructed from few transactions or valuations initially making the series volatile. More recent observations of the index, though, could be based on a much larger data set of transactions/valuations. As data collection techniques improve and the index builds on more data the variance of the disturbance error of a rent model using this index should decrease.

(iii) The presence of outlier data points, which refer to a few observations that have very small or very large values in relation to the rest of the observations in the sample, can cause variability in error variance. By transforming the data, e.g. taking logs, we decrease the weight of outlier observations in our sample. The exclusion of these observations can alter the estimated output from the regression model. There is an argument suggesting that we should include outlier values in our sample and let the model learn from errors and such extreme values. We can use dummy variables (dummies) to allow for such effects. It is debatable though whether we should use dummies or remove such observations. The decision about including dummies should reflect the objectives of the study, other diagnostic tests, and whether the forecast performance of the model improves.

(iv) Skewness in the distribution of one or more regressors could lead to heteroscedasticity.
(v)   A regression model which is not correctly specified could induce heteroscedasticity. General model misspecification includes incorrect functional form applied to explaining the causal relationship and/or the exclusion of a relevant explanatory variable.

The preceding causes of heteroscedasticity could reveal themselves in different ways, but they are not always easy to discern. A straightforward case is in cross-sectional studies where the size of the error (variability) is proportional to one of the explanatory variables. Figure 7.2 illustrates this case.

In the situations depicted in Figure 7.2 the variance of disturbances is not constant, denoted by σ_i^2, as it relates to the ith observation (or its square) of x. In panel (a) the larger the value of the independent variable x, the larger the error. If we replace x with time, we again have a situation in which the variance of the errors increases with time (σ_t^2). Expression (7.2) encapsulates possible sources of the non-constant error variance:

$$\sigma_i^2 = f\left(z_1, z_2, \ldots z_K\right) \tag{7.2}$$

where $z_1 \ldots z_K$ are non-stochastic variables that could be one or all of the original regressors xs or a function of the regressors and $i = 1, 2, \ldots, K$. An example of expression (7.2) is equation (7.4) in section 7.3.4.

7.3.3   *Consequences of heteroscedasticity*

The existence of heteroscedasticity has certain consequences for the OLS estimation procedures. OLS will still be unbiased when heteroscedasticity is present. We do not need the homoscedasticity assumption to prove the unbiasedness of OLS, as the variance of the residual error does not enter the calculation of the coefficient values (see Chapter 5). Hence, as the sample gets larger, the value of the sample coefficient tends to approximate the value of the population coefficient. Heteroscedasticity does not alter the consistency property of OLS either.

The homoscedasticity assumption is needed for the efficiency of the OLS estimator. When errors are heteroscedastic, the OLS estimators are no longer the best linear unbiased estimators

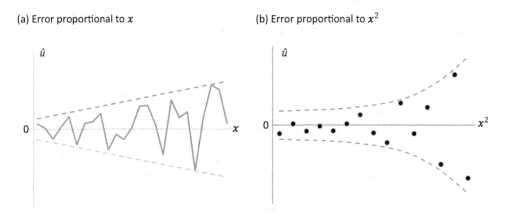

Figure 7.2 A known form of heteroscedasticity

(BLUE). The OLS is not optimal in the sense that it does not provide the smallest variance estimate among all unbiased estimators. Since heteroscedasticity renders the variance of the OLS estimators unreliable, we cannot make inferences about the standard errors, confidence intervals and t-ratios. The estimation of the variance and standard error of the coefficients is biased as the error variance that is included explicitly in the calculations is not constant. This renders the whole OLS procedure erroneous. As an example, the OLS standard error will be too low when the variance of the errors is related to the square of an independent variable. Therefore, this variable could be statistically significant whereas in reality it is not.

7.3.4   Tests to detect heteroscedasticity

The mere inspection of the residuals of estimated models may not reveal a clear pattern as in Figures 7.1 and 7.2. The variance of the errors may change through time but not in a systematic way. For this, formal tests are required. There are several tests for heteroscedasticity. One of the most commonly used tests is the White's general heteroscedasticity test, which we study in this section. All these tests come with their strengths and weaknesses. For a discussion of more tests, see Gujarati and Porter (2010).

The general test proposed by White (White, 1980) is not limited by the cause or form of heteroscedasticity which is rarely known. The White test examines the relationship of squared residuals with all independent variables x_i, their squares x_i^2 and all cross products ($x_i x_j$ for $i \neq j$). These terms are supposed to reflect the causes of heteroscedasticity. The test involves the following steps:

(i) Suppose we estimate a regression model with three explanatory variables given by equation (7.3).

$$y_i = \beta_0 + \beta_1 x_1 + \beta_2 x_2 + \beta_3 x_3 + u_i \tag{7.3}$$

We run the regression model with OLS and obtain the residuals $\hat{u}_i$.

(ii) We regress the sample squared residuals $\hat{u}_i^2$ on all explanatory variables, their squares and products. If we have many regressors and a small sample, we may exclude the cross products of variables. Following this, we run equation (7.4) which we term the *auxiliary* or *unrestricted* regression.

$$\hat{u}_i^2 = b_0 + b_1 x_1 + b_2 x_2 + b_3 x_3 + b_4 x_1^2 + b_5 x_2^2 + b_6 x_3^2 + b_7 x_1 x_2 + b_8 x_1 x_3 + b_9 x_2 x_3 + e_i \tag{7.4}$$

It is clear that with a small sample and say four explanatory variables we will run out of degrees of freedom. The terms in equation (7.4) are examples of zs in expression (7.2).

(iii) We test the overall significance of the slope coefficients $b_1, b_2, \ldots, b_9$ in the auxiliary equation (7.4). The null hypothesis is that the errors are homoscedastic or there is no heteroscedasticity. The null hypothesis requires that all terms jointly are not statistically significantly different from zero. Hence,

Null H_0: (no heteroscedasticity) $b_1 = b_2 = b_3 = b_4 = b_5 = b_6 = b_7 = b_8 = b_9 = 0$
Alternative H_1: One of the betas is not zero.

This test therefore examines whether the variance of the residuals is explained by the original regressors or a transformation of them as shown in equation (7.4).

Heteroscedasticity is seen as a function of the original regressors, their squares and their cross products (multiplicative function). The independent variables in (7.4) have a non-linear and interactive effect on the variance of the residuals. The popularity of this test partly owes to the fact that it is not constrained by whether heteroscedasticity is of linear or non-linear form as in other tests.

(iv) We run the auxiliary regression given by equation (7.5). We name this model the *restricted* model.

$$\hat{u}_i^2 = b_0 + e_i \tag{7.5}$$

This is the model when we impose the restrictions (see null above). All terms are excluded except the constant.

(v) An *F*-test or a chi-square test can be used to make inferences about the overall significance of betas.

$$\text{F-version of White's test: } F = \frac{(RRSS - URSS)/m}{URSS/(n-k)} \tag{7.6}$$

We are familiar with the elements in equation (7.6) from the discussion of the *F*-test in the previous chapter.

URSS is the sum of squared residuals of the unrestricted equation (7.4)
RRSS is the sum of the squared residuals of the restricted equation (7.5)
k is the number of parameters in the unrestricted equation (7.4) that includes the constant, hence $k = 10$ in (7.4)
m is the number of restrictions. In our example $m = 9$
n is the sample size for the unrestricted equation (7.4).

We follow the steps for inference as in previous chapters. The computed value of equation (7.6) is compared to the *F*-critical with m and $n-k$ (9 and $n-10$) degrees of freedom at the chosen level of significance (10%, 5% or 1%). If the computed *F*-statistic is greater than the critical *F*-value we reject the null hypothesis, inferring that at least one term in equation (7.4) is statistically significant and has a bearing on the variance of the errors.

Alternatively, we can run the Lagrange multiplier (LM) test, which does not require the estimation of the restricted equation and centres around the R^2 of the unrestricted equation.

$$\text{LM version of White's test } \chi^2 = n \times R^2 \text{ or } \chi^2 = nR^2 \tag{7.7}$$

R^2 is obtained from the auxiliary regression equation (7.4).

We compute the test statistic nR^2 which follows the χ^2 distribution with m degrees of freedom. m is the number of regressors in the auxiliary regression equation (7.4) excluding the constant or the number of restrictions placed under the *F*-version, in this case $m = 9$. We reject the null hypothesis of no heteroscedasticity if the computed $\chi^2(m)$ is higher than the critical $\chi^2(m)$ at the chosen level of significance. The χ^2 distribution is presented in section 4.7.6

Application

We implement the White test to examine the presence of heteroscedasticity in our Hong Kong office rent model (equation (6.26)). We go through the process step by step and we finish the application by showing how to perform the test in EViews. We will perform the test using both the F and χ^2 versions.

Equation (6.26), shown again as equation (7.8), was:

$$RREg_t = 18.3 - 2.4\,VAC_t + 1.4\,GDPg_{t-1} \tag{7.8}$$

(i) We obtain the residuals from the preceding equation and square them.

- ***The residuals and their squares are reported in online note #7.2***

(ii) We run the auxiliary or test regression equation (unrestricted equation) along with the restricted regression equations:

Unrestricted regression equation:

$$\hat{u}_i^2 = 230.45 - 29.62\,VAC_t + 1.19\,VAC_t^2 + 33.92\,GDPg_{t-1} - 1.37\,GDPg_{t-1}^2 - 2.11\left(VAC_t \times GDPg_{t-1}\right)$$
$$(0.62)\quad\;\;(0.72)\qquad\;\;(0.76)\qquad\;\;(0.52)\qquad\qquad(0.57)\qquad\qquad(0.60) \tag{7.9}$$

$R^2 = 0.11$; $URSS = 937519.8$; Number of observations $(n) = 33$; Number of regressors $(k) = 6$; numbers in parentheses are p-values.

Restricted equation:

$$\hat{u}_i^2 = 98.94 \tag{7.10}$$

$RRSS = 1,055,530$; The number of restrictions is five $(m = 5)$, that is all five slope coefficients in equation (7.9) are simultaneously zero.

(iii) *F-version:*

We test the joint significance of the five slope coefficients on the unrestricted equation (coefficients on VAC_t, VAC_t^2, $GDPg_{t-1}$, $GDPg_{t-1}^2$ and $VAC_t \times GDPg_{t-1}$) using the F approach and test. The null hypothesis is that these coefficients are jointly zero.

We substitute into the F-formula (7.6) the information from the unrestricted and restricted regressions:

$$\text{Computed } F = \frac{(1055530 - 937519.8)/5}{937519.8/(33 - 6)} = 0.68$$

At the 5% level of significance with 5 and $33 - 6 = 27$ degrees of freedom for the numerator and denominator, respectively, we get:

$$\text{Computed } F = 0.68 < \text{critical } F_{0.05}\,(5, 27) \cong 2.58$$

Since the computed F-statistic is lower than the critical value at 5%, we do not reject the assumption of constant error variance. Heteroscedasticity is not present in our model of real office rent growth in Hong Kong.

(iv) *Chi-square (χ^2) approach (LM version):*

As we noted earlier for this version, we calculate the expression nR^2 which follows a χ^2 distribution.

Computed $\chi^2 = nR^2 = 33 \times 0.11 = 0.63$. Critical χ^2 (5) at 5% = 11.07 (see Appendix A, Table A.4 for critical values or Excel '=CHISQ.INV.RT(0.05,5)' = 11.07.

We do not reject the null hypothesis that the slope coefficients are jointly equal zero in equation (7.9) or the hypothesis of no heteroscedasticity.

In the auxiliary equation (7.9) we also report the p-values for the coefficients. All p-values are higher than any of the conventionally chosen alpha levels of 0.01 (1%), 0.05 (5%) or 0.10 (10%). Hence none of these coefficients are statistically signifi-cant on the t-test. It is the same and expected result we got from the two versions of White test.

White's test can be augmented to include other variables which we may think affect the variation of the residuals. We include the candidate variables in the unre-stricted equation (7.4), estimate it and obtain the new *URSS*. The restricted equation and *RRSS* remain similar. We calculate a new F-statistic and the inference follows the same procedure.

The White test allows for a clear way to comprehend the principles of other het-eroscedasticity tests. The mechanics of the tests are similar; what differs is the auxiliary or unrestricted equation. The null hypothesis is homoscedasticity. We summarise some additional tests in the chapter's accompanying notes.

- ***Further common heteroscedasticity tests are outlined in online note #7.3***
- ***See online note #7.4 on how to run heteroscedasticity tests in EViews***

7.3.5   Actions to deal with heteroscedasticity

Knowing that the OLS standard error estimates are not reliable in the presence of heterosce-dasticity, corrective actions are sought to obtain *robust* and *heteroscedasticity consistent* standard errors. A number of options are available to deal with heteroscedasticity. The aim is to trans-form the model so that heteroscedasticity is removed or to adjust the OLS standard errors to heteroscedasticity-corrected standard errors.

(i)   Redefining the variables

Initial consideration can be given to the data. Data transformations may be enough to rem-edy the heteroscedasticity problem.

- We can moderate the impact of extreme values by taking logarithms (logs). Such values in our sample could be a source of heteroscedasticity. Logs may not be appropriate to take for all the series (e.g. when variables are expressed in percentages and have negative values).
- Choose variables that might be less likely to be a source of heteroscedasticity. This may prove a difficult task as such variables may not exist. This assumes that we know which variable causes heteroscedastic errors and that substitute variables exist.
- Create new variables in the model. Suppose we study real estate capital flows in a number of diverse countries. A regression model of foreign capital flows (*FLOW*) will include a K number of variables (z_k) such as yield differentials (between a particular

country i and other countries), expected capital value growth and the GDP growth in country i.

$$FLOW_i = \beta_0 + \beta_k z_{ki} + u_i \qquad (7.11)$$

Such a model is likely to have heteroscedastic disturbances because larger countries like US, Germany and Japan will experience greater variation of inflows into their markets than smaller countries such as the Netherlands, Austria, and Sweden. We can then express foreign capital flows in relation to the size of the market or the economy. Therefore, the equation we estimate is:

$$\frac{FLOW_i}{SIZE_i} = a_0 + a_k z_{ki} + e_i \qquad (7.12)$$

where $SIZE$ could be proxied by the level of GDP, commercial stock or an equivalent variable that represents the size of the real estate market (and could be one of the zs).

(ii) Heteroscedasticity form known: weighted least squares (WLS) and generalised least squares (GLS)

In the hypothetical case of knowing the true heteroscedasticity variance of the errors (hence σ_i^2 is observed) the remedy is easy. We can use the method known as *weighted least squares* (WLS) to estimate the model.

- **Online note #7.5: Basics of weighted least squares (WLS) estimation**

AN EXAMPLE

Assume that in our model (equation (7.8)) there is heteroscedasticity and that we know the form of heteroscedasticity, which is that the variance of the error is proportional to the square vacancy:

$$Var(u_t) = \sigma^2 VAC_t^2 \qquad (7.13)$$

Based on the preceding discussion we estimate equation (7.14). We divide the terms of equation (7.8) by $\sqrt{VAC_t^2} = VAC_t$ and apply OLS.

$$\frac{RREg_t}{VAC_t} = \frac{\beta_0}{VAC_t} - \beta_1 + \frac{GDPg_{t-1}}{VAC_t} + \frac{u_t}{VAC_t} \qquad (7.14)$$

As a further illustration, consider the case of the error variance being proportional to vacancy:

$$Var(u_t) = \sigma^2 VAC_t \qquad (7.15)$$

VAC_t is positive, therefore, the variance will be positive as expected. Now the equation we estimate is:

$$\frac{RREg_t}{\sqrt{VAC_t}} = \frac{\beta_0}{\sqrt{VAC_t}} - \beta_1 + \frac{GDPg_{t-1}}{\sqrt{VAC_t}} + \frac{u_t}{\sqrt{VAC_t}} \tag{7.16}$$

In practice, it is difficult to know the exact source of heteroscedasticity. We may have some insight as to which term or influence is responsible. This insight is primarily the result of our understanding of the market. We can investigate by weighting the terms of the original model with the variables we speculate to be a source of heteroscedastic errors. Even if we are successful, we encounter another issue, that of coefficient interpretation. The meaning of the coefficients will not be as straightforward as for the original regression.

(iii)   Unknown forms of heteroscedasticity

If the type of heteroscedasticity is not known, we can use the feasible GLS (FGLS) estimator. An in-depth theoretical exposition is not attempted in this text (see online note #7.6).

•   **Online note #7.6: Key features of FGLS estimator**

A comprehensive discussion is presented in Wooldridge (2012), chapter 8. Fortunately, econometric software has built-in heteroscedasticity methods, such as White's, to calculate robust standard errors for the regression coefficients.

•   **See online note #7.7 on heteroscedasticity consistent estimation in EViews**

Table 7.2 shows two sets of results for the Hong Kong office rent growth model (equation (7.8)): the OLS estimates and the results when the model is estimated with the Huber-White robust technique (performed in EViews).

Table 7.2 Comparison of OLS and White estimation for heteroscedasticity

Dependent variable: RREgt

Independent variable	Ordinary OLS				White			
	Coeff	*St error*	*t-stat*	*Prob*	*Coeff*	*St error (robust)*	*t-stat*	*Prob*
C	18.25	7.82	2.34	0.03	18.25	7.34	2.45	0.02
VAC_t	−2.43	0.70	−3.45	0.00	−2.43	0.65	−3.71	0.00
$GDPg_{t-1}$	1.39	0.53	2.62	0.01	1.39	0.47	2.98	0.01
Adj. R^2	0.48				0.48			
SSR	3265.10				3265.10			
F-statistic	15.92				15.92			
p-value (*F*-statistic)	0.00				0.00			
Wald *F*-statistic					13.16			
p-value Wald *F*-statistic					0.00			

Earlier we found that the White test showed no evidence of heteroscedasticity in the model. Hence, we would not expect that heteroscedasticity robust estimation (applying the Huber-White or similar techniques) would yield different regression results. The value of the coefficients remains the same (expected), but what has changed marginally is the magnitude of the standard errors. The coefficients in the model remain statistically significant and this is important. Coefficients can become insignificant when there is heteroscedasticity and we estimate the model with a heteroscedasticity-robust technique. The Wald *F*-value is the appropriate regression *F*-statistic to consider when we estimate a GLS or feasible GLS model.

A final note: we can directly report heteroscedasticity-robust standard errors using one of the methods in-built in statistical software. However, by testing for the presence of heteroscedasticity we are studying the dynamics of the relationship more closely, and we understand data issues and other important features of the relationship we examine.

7.4   Uncorrelated error terms: $E(u_i u_j) = 0$

7.4.1   Nature of the problem and notation

The third assumption of CLRM states that the disturbances are independent from one another, that is, they are not correlated. We say that the disturbance errors are *orthogonal*.

More formally this condition is expressed as:

$$E\left(u_i u_j\right) = 0 \text{ or } cov\left(u_{ij}\right) = 0 \text{ or } correl\left(u_i u_j\right) = 0 \text{ for all } i \neq j \tag{7.17}$$

The assumption of independent disturbances applies both to cross-section and time-series models. We are more concerned with this assumption in the context of time-series data. More specifically for time series errors we can write expression (7.17) as:

$$E\left(u_i u_{t-i}\right) = 0 \text{ or } cov\left(u_i u_{t-i}\right) = 0 \text{ or } correl\left(u_i u_{t-i}\right) = 0 \text{ for all } i = 1, 2, \dots \tag{7.18}$$

Expression (7.18) states that the covariance or correlation between successive errors (one-, two-, . . . periods apart) is zero, meaning that the errors are not autocorrelated at these lag lengths. The magnitude and sign of the error at some observation *i* does not bear any relationship to any other errors. The violation of this assumption in a time-series application of OLS results in the misspecification problem of *residual autocorrelation* or *serial correlation*. The problem of serial correlation is only encountered in models with time-series data and the analysis in this section is presented with such models in mind.

Table 7.3 outlines forms of serial correlation in time series analysis and defines the order of residual autocorrelation. In first order serial correlation, the error is generated by the following autoregressive AR(1) equation:

$$u_t = \rho u_{t-1} + e_t \tag{7.19}$$

where $-1 \leq \rho \leq 1$ and e_t is an *iid* disturbance term.

In equation (7.19), the error at any time is a proportion of the error in the previous period plus another random disturbance e_t. Serial correlation can be positive or negative, that is the residuals are correlated positively or negatively with their own past values. This will

Table 7.3 Forms of serial correlation in time series

Classical assumption: No autocorrelation	Violation: Evidence of ith order serial correlation	
$correl\,(u_t u_{t-1}) = 0$	$correl\,(u_t u_{t-1}) \neq 0$: 1st order serial correlation	AR(1) errors
$correl\,(u_t u_{t-2}) = 0$	$correl\,(u_t u_{t-2}) \neq 0$: 2nd order serial correlation	AR(2) errors
$correl\,(u_t u_{t-3}) = 0$	$correl\,(u_t u_{t-3}) \neq 0$: 3rd order serial correlation	AR(3) errors
. . .	. . .	
$correl\,(u_t u_{t-r}) = 0$	$correl\,(u_t u_{t-r}) \neq 0$: rth order serial correlation	AR(r) errors

determine the sign ρ takes. The coefficient ρ measures the strength of the correlation among residuals. If $\rho = 0$ there is no serial correlation; the term u_{t-1} has no effect on u_t. The closer the value of ρ to 1 or -1, the stronger the correlation.

When $0 < \rho \leq 1$, we have positive serial correlation. Successive errors are positively correlated that is positive errors are followed by positive errors and negative errors are followed by negative errors on average. If $-1 \leq \rho < 0$, then we have negative autocorrelation in the errors. Mostly successive errors are followed by opposite sign errors. The error term switches signs from observation to observation. Equation (7.20) represents the case of first order negative serial correlation.

$$u_t = -\rho u_{t-1} + e_t \tag{7.20}$$

In the rth order serial correlation, the residuals bear a relationship with residuals r periods ago.

$$u_t = \rho_1 u_{t-1} + \rho_2 u_{t-2} + \ldots + \rho_r u_{t-r} + e_t \tag{7.21}$$

We cannot determine a priori the order of residual autocorrelation. Therefore, the value of r will depend on the frequency of data. It is likely to be up to two periods for annual data, four periods for quarterly data, and 12 periods for monthly data. For quarterly data, we will usually test for up to fourth order serial correlation. It is an empirical matter, though, whether an empirical model with data of quarterly frequency will exhibit first, second, third, or fourth order autocorrelation in the residuals. The properties of the dependent variable (e.g. how autocorrelated the dependent variable is) and other factors will depend on the model itself. In real estate analysis with time-series data, first order serial correlation is the prevalent form. The preceding discussion might give the impression of always experiencing the incident of residual autocorrelation in models using time-series data. However, this is not the case. Many models we estimate do not exhibit any serial correlation problems.

In Figure 7.3, panel (a) gives a graphical representation of first order positive serial correlation. We can observe two clusters of positive errors and one cluster of negative errors. Errors remain positive or negative for some time before they change sign. If we calculate the first order autocorrelation coefficient of the errors shown in Figure 7.3a we get +0.68, confirming positive first order autocorrelation. This means that successive errors have a fairly strong association. The errors are described by equation (7.19). Panel (b) illustrates

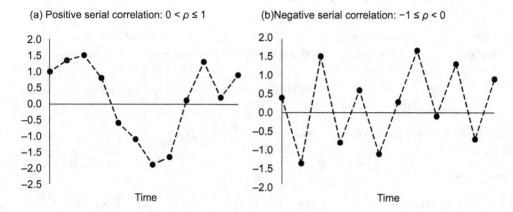

Figure 7.3 Graphical illustration of serial correlation (first order)

negative serial correlation. The first order autocorrelation coefficient in this set of errors is -0.60. The errors in this case follow the pattern that equation (7.20) defines. The graphical representation gets more complicated when higher-order residual autocorrelation is plotted. It is difficult to discern patterns, and therefore have to rely on formal tests to detect such problems.

Positive first order autocorrelation in the residuals is the most common situation when we run regressions with time-series data in real estate. We should expect some residual autocorrelation in our models, although for particular classes of models, such as VARs studied in Chapter 11, serial correlation is not expected. We should be concerned when there are strong autocorrelation patterns in the residuals (as they are detected by appropriate tests) due to the consequences and implications for our models.

7.4.2 Consequences

(i) The presence of autocorrelation in the residuals of a regression model results in OLS estimates that are not BLUE (see Chapter 5). The OLS estimator is still linear and the parameter estimates remain unbiased provided there are no violations of other classical assumptions. As in the case of heteroscedasticity, the variances of the parameter estimates are affected. Standard errors, t-ratios, F-tests and in general statistical inference are not reliable. Although OLS estimators will be unbiased and consistent they will not be efficient, meaning they will not have the lowest standard errors in this class of estimators. This is important in small samples where inefficient estimators do not perform well. For example, positive serial correlation (a common case in real estate analysis) underestimates the size of the standard errors resulting in higher t-ratios. We may accept a coefficient (variable) as being statistically significant (type one error) whereas in reality it is not.

(ii) The R^2 could be inflated.

(iii) Forecasts from models with serial correlation will tend to have inflated variances.

(iv) We over-reject the true null hypothesis H_0: $\beta = 0$.

7.4.3   Causes of serial correlation

Of the several causes of serial correlation discussed in econometric textbooks, omitted variables and data smoothness are referred to as the most common causes of the problem within real estate. However, we should not disregard other important sources of serial correlation, such as the impact of shocks.

(i)   Omitted variables

Suppose the true specification we should estimate is given by:

$$y_t = \beta_0 + \beta_1 x_{1t} + \beta_2 x_{2t} + \beta_3 x_{3t} + u_t \tag{7.22}$$

Instead, we estimate the following model that excludes variable x_3:

$$y_t = \beta_0 + \beta_1 x_{1t} + \beta_2 x_{2t} + e_t \tag{7.23}$$

The omission of x_3 may be due to inadequate theory, lack of knowledge, and unawareness of available data. Suppose we model take up and do not include a variable that allows for the influence of the floor space/employment ratio due to the fact that we may not have thought of this variable or may not have been aware that this data exists. In theory, take-up requirements from additional employment will depend on how much floor space each extra employee will use. We may knowingly omit a variable due to lack of data and attempt estimation. If x_3 has a significant impact on y, then the errors e_t will embody the influence of x_3. The errors will be determined by x_3 (see equation (7.24)), which is not desirable.

$$e_t = \gamma_0 + \gamma_1 x_{3t} + \varepsilon_t \tag{7.24}$$

It is easy to understand why the errors are autocorrelated. In this instance, in every period our model makes a persistent error in the same direction since the error reflects the same omitted influence on the dependent variable. If in a rent model we only include demand-side variables, the omission of a supply-side variable and its expected inverse impact on rent growth will be reflected in the errors for each time period. The result is correlation among successive errors. Hence, a missing variable can create an upward or downward bias in the error term for several periods.

The discussion extends to situations of not including the dependent variable as a lagged regressor when needed. For example, we estimate equation (7.25) without the term y_{t-1}. The error series e_t will reflect a systematic pattern due to the omitted influence of y_{t-1} on y_t.

$$y_t = \beta_0 + \beta_1 x_{1t} + \beta_2 x_{2t} + \beta_3 y_{t-1} + e_t \tag{7.25}$$

The lagged dependent variable may reflect persistence and momentum in the dependent series or indeed other influences as we discuss next.

(ii)   Smoothed dependent variable

When the dependent variable is highly smoothed, serial correlation will arise. This is common with particular real estate series (mainly private market data) and when levels of the series are used. We may have to estimate equation (7.25), as the term y_{t-1} will convey the influence of past values on the current value. The lag can be extended to longer than one period. When we model in growth rates or in first differences, serial correlation from this source of misspecification bias is reduced.

Smoothness in the data could be the result of the way the data is generated (see Chapter 2) or manipulated. For example, various forms of interpolation such as creating quarterly data from annual observations or taking averages of neighbouring observations to fill omitted data or other manipulation.

Further, there can be inertia in the real estate market. Prices may not change or firms may not adjust space they demand even if economic conditions change for a variety of reasons, such as testing the resilience of the market. Even in growth rates, real estate data can be pretty smooth and autocorrelated.

(iii)   Incorrect functional form

The true relationship may not be linear in the variables as we assume. We estimate equation (7.26) whereas the true model is equation (7.27), which can also be estimated with OLS.

$$y_t = \beta_0 + \beta_1 x_{1t} + \beta_2 x_{2t} + u_t \tag{7.26}$$

$$y_t = \beta_0 + \beta_1 x_{1t} + \beta_2 x_{2t}^2 + v_t \tag{7.27}$$

In the next section, we study the functional form of a model and the problem of linearity in parameters.

(iv)   Shocks

Any shocks occurring in the market will have an impact on the variable we study. A shock may not be absorbed quickly and its impact may take several periods to work through the market. This means that in any one period, the error term will reflect the current period's shock and will partially be carried over from the previous shock. If the shock dissipates at a slow speed, successive errors will be correlated reflecting the remainder of the (declining) impact from the shock.

(v)   Simultaneity

This is a situation in which the dependent variable is determined along with the independent variable(s) from a common third influence. Further, the independent variables are influenced by the dependent variable (a kind of feedback or bilateral effect). The point is that the dependent and independent variables are determined concurrently (simultaneously). In such a case, the independent variables will be correlated with the error term. Hence, if we have a model that relates rents to vacancy, both variables may be determined simultaneously by an economic variable. If not, there may be contemporaneous effects from rents to vacancy and from vacancy to rents, a point we raised and discussed in the previous chapter. This is a more complex situation in modelling and requires specific treatment since OLS is not the appropriate estimator.

(vi)   Seasonality

A model estimated with quarterly or monthly data may be subject to seasonal influences in a particular quarter or month. For instance, seasonality might arise in the last quarter/

month of the year when the quarterly/monthly valuations will reflect not just the quarterly/monthly sample of properties but the larger annual sample. If seasonality is ignored, the seasonal factors will cause correlated errors of order 4 or 12 for quarterly and monthly data respectively. Therefore, the pattern of the error will respectively be described by:

$$u_t = \rho u_{t-4} + e_t \tag{7.28}$$

$$u_t = \rho u_{t-12} + e_t \tag{7.29}$$

The errors could still be subject to first or other order autocorrelation patterns. A good example of seasonality affecting the value of dependent variable can be found in construction, especially in colder climates when activity slows down in the winter months and picks up again in the spring and summer months.

7.4.4   Tests to detect serial correlation

In practice, a plot of the residuals and a visual inspection is unlikely to give a clear picture of serially correlated errors (positive or negative) as in Figure 7.3. But, it is always a good idea to plot the residuals to observe the general patterns. We can also do some preliminary analysis and run simple correlations between residuals at various periods apart for example, t and $t-1$, t and $t-2$ and so forth. High-correlation coefficients are indicative of significant residual autocorrelation.

The literature proposes several tests for residual autocorrelation. We review the classic Durbin Watson (DW) d-test (Durbin and Watson, 1951), reported almost universally in diagnostics outputs, and one of the most commonly used general tests, the Breusch-Godfrey test.

(i)   The DW test

The DW d-test (Durbin and Watson, 1951) is a test for just first order autocorrelation and despite its several limitations the majority of software packages report it routinely. This is a test of whether $\rho = 0$ in equation (7.30).

$$u_t = \rho u_{t-1} + e_t \tag{7.30}$$

The d-value is computed from the following statistics:

$$d = \frac{\sum_{t=2}^{n} \left(\hat{u}_t - \hat{u}_{t-1} \right)^2}{\sum_{t=1}^{n} \hat{u}_t^2} \tag{7.31}$$

It can be shown that:

$$d \approx 2\left(1 - \hat{\rho}\right) \tag{7.32}$$

where $\hat{\rho}$ is the estimated ρ in equation (7.33). Table 7.4 gives the range of values the DW d-statistic takes and the corresponding form of autocorrelation.

Table 7.4 Form of autocorrelation and DW *d*-statistic values

No first order autocorrelation	$\hat{\rho} = 0$	$d = 2$
Positive autocorrelation	$0 < \hat{\rho} < 1$	$0 < d < 2$
Perfect positive autocorrelation	$\hat{\rho} = 1$	$d = 0$
Negative autocorrelation	$-1 < \hat{\rho} < 0$	$2 < d < 4$
Perfect negative autocorrelation	$\hat{\rho} = -1$	$d = 4$

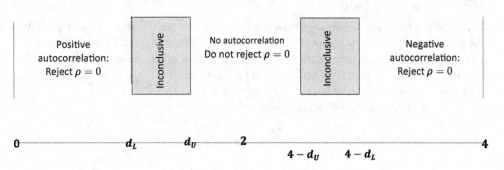

Figure 7.4 Decision rules for the DW statistic

Table 7.5 Durbin Watson critical values (5% level of significance)

T	k = 1		k = 2		k = 3		k = 4		k = 5	
	dL	dU	dL	dU	dL	dU	dL	dU	dL	dU
...	...	...	...	...	...	...	...	...	...	...
15	1.08	1.36	0.95	1.54	0.81	1.75	0.69	1.98	0.56	2.22
20	1.20	1.41	1.10	1.54	1.00	1.68	0.90	1.83	0.79	1.99
25	1.29	1.45	1.21	1.55	1.12	1.65	1.04	1.78	0.95	1.89
30	1.35	1.49	1.28	1.57	1.21	1.65	1.14	1.74	1.07	1.83
32	1.37	1.50	1.31	1.57	1.22	1.65	1.17	1.73	1.11	1.82
40	1.44	1.54	1.39	1.60	1.34	1.66	1.39	1.72	1.23	1.79

The DW statistic takes values between 0 and 4. There is no exact distribution for the DW statistic, and upper and lower limits are computed.

Table 7.5 contains selected critical values for the Durbin Watson d_L and d_U statistics (from which we can compute $4 - d_L$ and $4 - d_U$). The critical values assume that the residuals are homoscedastic and normally distributed, there is no autocorrelation and that all regressors are exogenous (no simultaneity). There are different tables for different levels of significance. k refers to the number of parameters in the model. In some tables, such as Table 7.5, k excludes the constant.

Suppose that we estimate a model with two explanatory variables with 30 observations. A DW statistic of 1.45 falls into the inconclusive area ($d_L = 1.28$ and $d_U = 1.57$). We cannot tell whether there is first order positive serial correlation. The application of more general tests will rectify this shortcoming. The DW test is not appropriate for models that have the lagged dependent variable as one of the regressors. Another version of the DW test, the

DW h-statistic is a test for residual autocorrelation in models containing a lagged dependent variable as a regressor. This test is not discussed in this section as the more general test presented next is appropriate for any model specification. The DW d-test can be used to study other forms of pairwise correlations, for example u_t and u_{t-2}. In practice, though, it is not used for this purpose. In statistical software, the DW test is only reported for first order serial correlation.

(ii)   The Breusch-Godfrey LM test

A general test for higher-order autocorrelation that overcomes the limitations of the DW's test is the Breusch-Godfrey LM test (Godfrey, 1988, 1978b). This test allows us to perform a joint test for residual autocorrelation of different order, that is first, second and higher order. This test encompasses the DW d and h tests. This test requires the estimation of an auxiliary regression that involves the residuals of the original model.

Suppose we estimate the following model with n observations.

$$y_t = \beta_0 + \beta_1 x_{1t} + \beta_2 x_{2t} + u_t$$
(7.33)

The model for the errors or the auxiliary regression in this test is:

$$\hat{u}_t = b_0 + b_1 x_{1t} + b_2 x_{2t} + \rho_1 \hat{u}_{t-1} + \rho_2 \hat{u}_{t-2} + \ldots + \rho_r \hat{u}_{t-r} + e_t \text{ with } e_t \sim N\left(0, \sigma^2\right)$$
(7.34)

We regress the estimated residuals on all explanatory variables in the model and successive values of lagged residuals up to lag r. The auxiliary regression is estimated with $n - r$ degrees of freedom.

We test in the auxiliary regression whether the coefficients on the lagged residual terms are jointly zero. Therefore, the hypotheses are:

Null hypothesis (H_0): $\rho_1 = \rho_2 = \ldots = \rho_r = 0$.
Alternative H_1: $\rho_1 \neq 0$ or $\rho_2 \neq 0$ or ... or $\rho_r \neq 0$

We reject the null if one of the autocorrelation coefficients $\rho_1, \rho_2, \ldots, \rho_r$ is not zero. We can compute two forms of this test, an F-test and a χ^2-test, as we did for heteroscedasticity.

F-VERSION

(i)   We estimate equation (7.33) and obtain the residuals $(\hat{u}_t)$.
(ii)   We run equation (7.34) and obtain the RSS (*URSS*). We cannot predetermine the order of r theoretically. We noted that first order residual autocorrelation is most common in real estate analysis, but it is typical to test for higher order depending on the frequency of the data. The auxiliary equation (7.34) is the unrestricted equation in this test.
(iii) We run the restricted equation (7.35) and obtain the RSS (*RRSS*):

$$\hat{u}_t = c_0 + c_1 x_{1t} + c_2 x_{2t} + v_t$$
(7.35)

(iv)   We calculate the F-test (expression 7.6). k is number of regressors in the unrestricted equation (7.34) including the constant, m is the order of autocorrelation (hence $m = r$).

(v)   We do not reject the null hypothesis if the calculated F-test is smaller than the critical F-value at the preferred level of significance α with m and $n - k$ degrees of freedom or equivalently, when the probability of the estimated F-value is lower than α.

χ^2-VERSION

The test statistic for this test is $(n - r)R^2$, where R^2 is obtained from the auxiliary regression. We reject the null hypothesis (as above) if the calculated χ^2 value from $(n - r)R^2$ is higher than $\chi^2_{critical}$ at the selected significance level a. Alternatively, the null is rejected if the probability associated with the computed χ^2 test is lower than the preset significance level (usually 5%).

Therefore, the Breusch-Godfrey test can be used for higher-order serial correlation and applies to models which contain lags both of the dependent and independent variables. The two versions will generally produce the same results, but in some cases, they may produce conflicting evidence. This might be the case in small samples in which case the F-version is considered more reliable. Should the test results differ, we can state that there is evidence of serial correlation, but the tests collectively do not indicate significant serial correlation in the residuals.

APPLICATION

We apply the preceding tests to our Hong Kong office price model. For the Durbin Watson statistic, we estimate the first order autocorrelation coefficient:

$\hat{u}_t = -0.109\hat{u}_{t-1}$ (number of observations $(n) = 32$).
$DW - d = 2 \times (1 - 0.109) = 2.22$

The first order autocorrelation coefficient is negative, which is not common in real estate analysis using private market data. The absolute value of the correlation coefficient 0.11 (0.109) is small perhaps indicating that it might be insignificant. This is confirmed by the $DW - d$ statistic as the value of 2.22 is falling into the non-rejection region. With $k = 2$ (number of regressors in equation (7.8)) and $T = 32$ from Table 7.5 we get $d_L = 1.31$ and $d_U = 1.57$. Therefore, $4 - d_L = 2.69$ and $4 - d_U = 2.43$. On these results, we do not reject the null that $\check{p} = 0$. There is no evidence of first order serial correlation. The value of the $DW - d$ test in the EViews output (online note #6.4) is slightly different ($d = 2.19$), as its calculation is based on formula (7.31) and not formula (7.32).

We apply the Breusch-Godfrey test to examine both first order and higher order autocorrelation in the residuals of (7.8). We illustrate both versions of the test.

F-VERSION

We apply the LM test to detect first order serial correlation in our model:
    Auxiliary regression:

$$\hat{u}_t = -1.90 + 0.13VAC_t + 0.21GDPg_{t-1} - 0.12\hat{u}_{t-1}$$
$$\phantom{\hat{u}_t = }(0.82)\ (0.86)\ (0.71)\ \phantom{+0.21GDPg_{t-1}}(0.53) \tag{7.36}$$

$R^2 = 0.017;\ URSS = 3150.3;\ n = 32\ (1986\text{--}2017)$

The restricted equation is:

$$\hat{u}_t = 0.00 - 0.00VAC_t + 0.00GDPg_{t-1} \tag{7.37}$$

$$RRSS = 3224.7$$

We calculate the *F*-statistic with $k = 4$ as per equation (7.36) and $m = 1$ since we test whether the coefficient on $\hat{u}_{t-1}$ is not statistically different from zero.

$$\text{Computed } F = \frac{(RRSS - USRR)/m}{USRR/(n-k)} = \frac{(3224.7 - 3150.3)/1}{3150.3/(32-4)} = 0.66$$

The critical *F*-value with 1 and 28 degrees of freedom at 5% $\approx$ 4.19. We do not reject the null of no first order serial correlation.

$\chi 2$ VERSION

The estimated χ^2 is $(n - r)R^2 = (32 - 1) \times 0.017 = 0.53$; critical $\chi^2(1)$ at 5% $= 3.84$.

This version of the test also does not reject the null hypothesis. There is strong evidence that equation (7.8) does not exhibit symptoms of serial correlation.

The process is similar when we test for second-degree serial correlation or higher order. If we test for second-degree residual autocorrelation, the unrestricted equation is:

$$\hat{u}_t = -2.33 + 0.18VAC_t + 0.14GDPg_{t-1} - 0.08\hat{u}_{t-1} + 0.02\hat{u}_{t-2} \tag{7.38}$$

$$R^2 = 0.012$$

The restricted equation remains equation (7.37).

The decision process is similar in applying either of the tests, $k = 5$ and $m = 2$ (two lagged residual values). The number of observations is now 31 (1987–2017). We should note that EViews maintains the sample period 1985–2017 by setting the values for $\hat{u}_t$ in 1984 and 1983 to zero (see online note #7.8) so that the sample begins in 1985.

- **Online note #7.8: Breusch-Godfrey test in EViews**

7.4.5 Remedies

Since one of the causes of serial correlation is omitted variables, we should think of additional variables that might explain the dependent variable. We may need to add more lags in the model as effects may not be contemporaneous (see discussion on lagged effects in real estate in Chapter 6).

We noted that the features of real estate data, in particular data from the private or direct market, can be a source of serial correlation. Autocorrelation tests will be informative of the smoothness of our variables, in particular the dependent variable. If serial correlation is present, we will have to transform highly autocorrelated variables by taking first differences or growth rates. In Chapter 8 we will also suggest such transformation for other reasons (e.g. non-stationary variables).

Another transformation that resembles the one for first differences is to use the estimated value of ρ from equation (7.30). Assume that our Hong Kong model exhibits serial correlation with $\rho = 0.46$. Then, we would transform the model in the following way:

$$RREg_t - 0.46RREg_{t-1} = \beta_0(1-0.46) + \beta_1(VAC_t - 0.46VAC_{t-1})$$
$$+ \beta_2(GDPg_{t-1} - 0.46GDPg_{t-2}) + (u_t - 0.46u_{t-1}) \tag{7.39}$$

The model is estimated with OLS and the new error term $(u_t - 0.46u_{t-1})$ is now free from serial correlation.

The use of the lagged dependent variable could account for omitted variables or other systematic influences on the dependent variable. In the previous chapter, we noted that a significant lagged dependent variable can again be the result of a strongly autocorrelated dependent variable. The use of lags (dependent or independent variable) may not remove serial correlation from our model.

The Newey-West method

The Newey and West (1987) estimator produces consistent standard errors for OLS regression coefficient estimates in the presence of serial correlation and heteroscedasticity. As with White's test, an adjustment is made to the variance covariance matrix. We should note that this test applies to large samples. The sample size in our Hong Kong office rent example is just over 30 observations, which is considered small. For this method a sample of over 50 observations would be preferable.

- **See online note #7.9 to run the test in EViews**

We discussed autocorrelation of the disturbance error terms in the context of time-series data and models where such model misspecification is more likely. Autocorrelation could also be an issue in cross-section studies, especially those involving location. The dependent variable may be correlated among cross-sections. Spatial autocorrelation is a measure of the relationship between the value of the dependent variable at a location and the same dependent variable but at another location separated by some specified distance. Positive spatial autocorrelation describes the tendency for locations separated by a specified distance (or 'lag') to have similar values of the dependent variable, while negative spatial autocorrelation is the tendency for the values to be dissimilar. The presence of spatial autocorrelation in the data is often indicative that there is something of interest that calls for further investigation in order to understand the reasons behind it. For example, house prices within a city often exhibit positive spatial autocorrelation. The topic of spatial autocorrelation is a much more complex one, though, and beyond the scope of this book.

7.5 Regressors not correlated with disturbances: $E(u, x_i) = 0$

This assumption states that none of the explanatory variables is contemporaneously related to the error term. Statistical inference becomes difficult if one or more of the regressors is correlated with the error term. Consider our Hong Kong office rent model.

Table 7.6 Correlation between regressors and residuals

	Residuals	VAC_t	$GDPg_{t-1}$
Residuals	1.00		
VAC_t	0.00	1.00	
$GDPg_{t-1}$	0.00	−0.41	1.00

If in a particular period rent growth accelerates (due to a large number of rent reviews in under-rented leases), the influence will be conveyed to rent growth through the residual error (the error takes a high value). Our model does not contain variables that capture this particular 'shock' to rents and rent growth. If vacancy is negatively correlated with the error, the higher error will be associated with lower vacancy. The OLS estimator will erroneously attribute the stronger rent growth to the fall in vacancy when in reality rent growth is stronger due to the shock from rents adjusting to their ERVs (market values). This is also termed an endogeneity problem (which we have referred to in earlier discussions).

There are some unwelcome repercussions. The estimated coefficients are biased and inconsistent. The fitted values will fit the actual data much better than they do in reality. In cross sectional studies, *unobserved heterogeneity* may be encountered. It means we have no obvious reasons to explain the correlation of a regressor with the disturbances. In this case and in practice, we had better be aware of the potential problem even if we are unable to fully explain why this happens.

In addition, if we measure one of the independent variables with error, this error will become part of the overall model error. Consider availability for office space: its measurement may exclude availability in small buildings or have limited coverage of grade B buildings. But in our investigation we require an overall market vacancy measure. The measurement error in the actual regressor will also be reflected in the regression model error as the relationship between the regressor and the dependent variable is not fully accounted for, resulting in correlation between the model error and the regressor. The larger the measurement error, the stronger the association.

It is therefore useful to know whether the independent variables have an association with the estimated model residuals. A simple test will be to run correlations between the error term and the regressors. Table 7.6 presents the cross-correlation results for the Hong Kong office rent growth model. More formal tests can be found in Wu (1974), and for a detailed treatment of this topic the reader is referred to Hill *et al.* (2001).

None of the two regressors of equation (7.7) shows any correlation with residuals. The regressors are moderately correlated (−0.41). The associations of regressors is a topic we discuss in the section on multicollinearity later.

7.6   Inappropriate functional form (non-linearities)

7.6.1   Defining the problem

A functional form misspecification is caused by the presence of non-linear relationships between the dependent and independent variables. An assumption of the CLRM is that the functional form of the model is linear in parameters. Thus, in the bivariate case, the relationship between y and x is best represented by a straight line and a linear equation. Suppose we estimate the following model:

$$y_i = \beta_0 + \beta_1 x_{1i} + \beta_2 x_{2i} + u_i \tag{7.40}$$

However, the true relationship is a complex one and described by a polynomial:

$$y = \beta_0 + \beta_1 x_{1i} + \beta_2^2 x_{2i} + u_i \tag{7.41}$$

The CLRM assumption of linearity in OLS refers to linearity in parameters and requires that we test for non-linearities in our model, and if a problem is detected, remedial action should be undertaken through variable transformation and model modifications. There are some forms of non-linear relationships such as equation (7.41) which cannot be addressed using OLS and we have to deploy estimators that are designed for the estimation of non-linear models, for example non-linear squares.

7.6.2 Causes and consequences of ignoring non-linearities

Non-linearities arise because real-world relationships are complex, and real estate is not an exception. This holds true even in very large samples. Ignoring or not treating non-linearities results in poor fits and if non-linearity in parameters is present, the ordinary least squares estimation will have major defects. In real estate, we will not encounter non-linear functions that are too complex and as a result some non-linearities can be treated and OLS apply.

In Figure 7.5, panel (a) illustrates the poor fit of the linear regression line when the true relationship is a quadratic one. Panel (b) presents another source of functional form misspecification. It shows the impact of outlier values to the regression line. More correctly, the two values which are out of proportion with the rest of the data points are said to exert

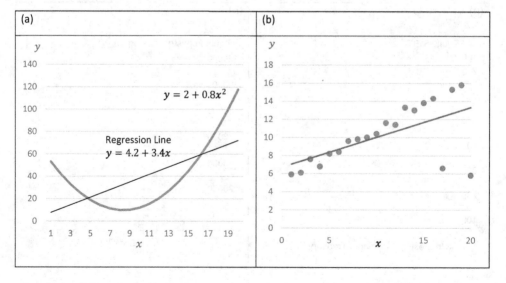

Figure 7.5 Non-linearity and outliers

high *leverage*. The regression line is pulled towards these observations. As can be seen, the regression line does not fit well with the rest of the observations as the slope is distorted by the high leverage of the two observations. At the same time, large errors are made for these observations making them an *outlier* in our sample. Outlying values can cause functional form problems. In addition, in small samples their presence can result in heteroscedasticity and non-normally distributed residuals.

Figure 7.6 illustrates two cases of non-linearities in the real estate market. Figure 7.6(a) plots the all equity REIT price index against the short-term interest rate in the US using annual data. By observing the data points (the dots in the graph) it is difficult to observe a relationship. Fitting a straight line is not that informative. It results in an *R*-squared of 2% (not shown in the graph) and it suggests a really low sensitivity of the price index to interest rates. Fitting a non-linear line reveals a distinctive pattern. At very low interest rates (less than 1%), the value of the index fluctuates in a wide spectrum. As interest rates rise, there is a cluster of index values, suggesting a falling tendency for the price index. For interest rates over 2%, a positive relationship is emerging. To generalise, many relationships in practice could be difficult to characterise on theoretical grounds. It is advisable to run a scatter plot and check for the best fit too (which is easily done in Excel). Sometimes fitting a polynomial (for the possibility of a non-linear relationship) will not differ much from a straight line.

Panel (b), in Figure 7.6, relates construction activity to GDP growth. At low rates of GDP growth, designating a weak economy, construction is not profitable or it does not respond as the economy is weak. As GDP growth gets stronger, construction activity picks up. At the other end, strong GDP growth would be expected to trigger high levels of construction. However, there are constraints such as labour, equipment and raw material shortages. Hence, the construction industry cannot be as responsive to stronger economic growth as before. This is the *Weibull* growth pattern, and a linear approximation does not reflect the true relationship between commercial construction and the economy.

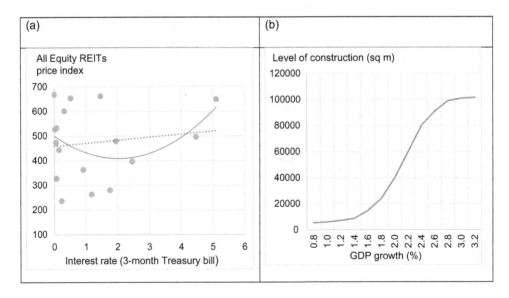

Figure 7.6 Illustration of non-linearities in real estate

It is evident from these examples that adopting the wrong functional form and not accounting for non-linearities will result in a sub-optimal model that will not capture the true relationship (which has implications for parameter estimates and inferences). The problem might not correct even if the sample gets larger.

7.6.3 Detecting functional form problems – the RESET

In practice, it is difficult to know a priori whether our model is subject to non-linearities and of what particular form. It is customary, however, to test for non-linearities since such mis-specification could be treated. A common test for linearity is the RESET (regression speci-fication error test) proposed by Ramsey (1969). In this test we run the model again and we add powers of the fitted values, usually the quadratic and cubic powers. If the powers of the fitted values are not statistically significant, the model is correctly specified.

Suppose the regression model is:

$$y = \beta_0 + \beta_1 x_1 + \ldots + \beta_k x_k + u_t \tag{7.42}$$

To apply the RESET we estimate:

$$y = b_0 + b_1 x_1 + \ldots + b_k x_k + \gamma_1 \hat{y}^2 + \ldots + \gamma_p \hat{y}^p + e_t \tag{7.43}$$

It is common to test for $\hat{y}^2$ and $\hat{y}^3$.

By introducing $\hat{y}^2$ or $\hat{y}^3$ we attempt to account for unknown non-linear relationships. For instance, when we test for the statistical significance of $\hat{y}^2$, we test for some complex dynam-ics as these expressions illustrate:

$$\hat{y}^2 = \left(\hat{\beta}_0 + \hat{\beta}_1 x_1 + \ldots + \hat{\beta}_k x_k\right)^2 \tag{7.44}$$

$$\hat{y}^3 = \left(\hat{\beta}_0 + \hat{\beta}_1 x_1 + \ldots + \hat{\beta}_k x_k\right)^3 \tag{7.45}$$

The expansion of equations (7.44) and (7.45) results in powers and cross terms that are sup-posed to capture non-linearities. Even if the form of non-linearities is unknown, the expansion of equations (7.44) and (7.45) will reflect the existence of non-linear elements in our equation.

The RESET is run along the same lines as the tests earlier.

(i) Estimate equation (7.42) and obtain the fitted values $\hat{y}$. This is the restricted equation in the RESET.
(ii) Re-estimate the model including powers of the fitted values $\hat{y}$ as per equation (7.43). The quadratic power is usually sufficient in less complex models. This is the unrestricted equation.
(iii) Null hypothesis: $\gamma_1 \ldots = \gamma_2 = 0$.

The alternative hypothesis is that one of the coefficients on the powers of the fitted values is not zero. The original model is not correctly specified due to non-linearities.

(iv) The hypothesis test can run with an F-test and a t-test. For the F-test, the number of observations and number of parameters is that of the unrestricted (7.43). m in the

F-formula is the number of powers of the fitted values that are included in the unrestricted equation.
(v)   Compare the estimated *F*-test statistic with the critical value at the chosen level of
significance or observe the associated *p*-value to reject or not reject the null hypothesis
and make inferences about possible functional misspecification problems.

Application to the Hong Kong office rent growth model

The *RRSS* of the restricted equation in the RESET (equation (7.8)) is 3,265.1 ($n = 33$,
1985–2017).

The unrestricted equation (with $\hat{y}^2$):

$$RRg_t = 18.48 - 2.44VAC_t + 1.40GDPg_{t-1} - 0.001\widehat{RRg}_t^2$$
$$(0.03)\quad(0.00)\qquad(0.02)\qquad\quad(0.89)$$

p-values are shown in parentheses: *URSS*: 3262.9; $n = 33$; $k = 4$ ($n - k = 29$); $m = 1$;
$R^2 = 0.515$

$$\text{Computed } F = \frac{(3265.1 - 3262.9)/1}{3262.9/(33-4)} = 0.02$$

Critical *F* at the 5% level of significance: $F_{0.05}(1,29) = 4.18$
Since computed $F <$ critical F, we do not reject the null hypothesis. There is no evidence
of misspecification owing to possible non-linearities. The *t*-test can alternatively be used. The
p-value of 0.89 (way above 0.05, the chosen level of significance) tells us that the square of
fitted values is not statistically significant.
The process is similar when we add $\widehat{RRg}_t^3$ to the unrestricted equation. In that case, *k* will
be 5 and *p* (or *m* the number of restrictions in the *F*-test) is 2.
If we fail the RESET test, we know that the assumption of linearity is violated, but the
RESET gives us no guide as to what a better specification may be. We will need to look at
data issues and try transformations that are presented next.
In the same way, we can apply the RESET to examine the presence of linearities embodied
in $\hat{y}^2$ or $\hat{y}^3$ in models with cross-section data.

•   ***Online note #7.10: Conducting the RESET in EViews***

7.6.4   Remedies

When the RESET test picks up the presence of non-linearities, we can estimate the model
in different forms. We transform the data by taking logs and investigate relationships that
can be log-log linear (logs taken for both the dependent and independent variables) or
log-linear (dependent variable is logged only). We give indicative examples of transformations in Table 7.7. We estimate models involving different functional forms in online
chapter I. When we model in percentage growth rates, the logarithmic transformation is
not appropriate.

Table 7.7 Possible transformations to induce linearity

Original model	Transformation
(i) $y = \beta_0 + \beta_1 x_1 + \beta_2 x_2^2 + u$	$y = \beta_0 + \beta_1 x_1 + \beta_2 z + u$ where $z = x_2^2$; model still linear in the parameters
(ii) $y = \beta_0 + \beta_1 x_1 + \beta_1^2 x_2 + u$	Cannot be linearised. It is not linear in the parameters. Non-linear estimator required.
(iii) $y = \beta_0 x \beta_1 u$	OLS can apply to: $\ln(y) = \ln(\beta_0) + \beta_1 \ln(x) + \ln(u)$

Outliers could be responsible for the presence of non-linearities especially in small samples (say, up to 50 observations). The outliers will be absorbed as the sample size lengthens. Taking the logarithm of the original series before calculating growth rates may resolve the problem. If an outlier is responsible for failing the RESET, we can use a dummy variable for that observation. A significant dummy variable could restore the appropriate functional form. Another suggested method is to create and include in our model an *interactive term*. Functional problems may also reflect the omission of needed interactive terms (such as in the White test). If we have prior knowledge that the impact of variable x_2 on y depends on the association of x_2 with x_1, we can include the interactive term $x_2 \times x_1$ in the equation, check its statistical significance and assess whether it rectifies the misspecification.

But when the source of the problem concerns non-linear parameters, OLS cannot be used. A more appropriate estimator has to be used instead.

7.7 Residuals are normally distributed: $u_i \sim N(0, \sigma^2)$

7.7.1 The problem, causes and testing

The assumption of normality in the disturbance error distribution is necessary for OLS to estimate the standard errors accurately and obtain confidence intervals for the coefficients, allowing us to make statistical inferences about their significance. Disturbances may not be normally distributed due to an inadequate model or outlier values particularly in small samples. We noted that if all other assumptions of the CLRM are satisfied, the property of normal distribution of the residuals may not be necessary. A violation of this assumption may be an indication of problems with the model, and it is worthwhile to undertake further investigation.

The dependent and independent variables in a regression model do not have to be normally distributed themselves. It is the disturbance errors of the regression model we focus on. However, we cannot disregard significant deviations from normality exhibited by the variables in the model. It could be a cause for the violation of the normality assumption in the disturbances. Non-normality of the dependent variable can cause non-normality in the errors even if the independent variables are normally distributed. Lack of normality in the errors can still occur even though the dependent variable is normally distributed. A wrongly specified model can lead to non-normal disturbances.

We assess the assumption of normally distributed errors using the Jarque-Bera (JB) test (Jarque and Bera, 1980), which is one of several tests for normality. Another popular test is the Anderson-Darling test. The Jarque-Bera test statistic is given by (see Chapter 4):

$$JB = n\left[\frac{S^2}{6} + \frac{(K-3)^2}{24}\right] \tag{7.46}$$

S is skewness, K is kurtosis ($K - 3$ is excess kurtosis) and n is the number of observations. The JB test statistic follows the χ^2 distribution with 2 degrees of freedom. The null hypothesis is that the disturbance errors are normally distributed.

If the null is rejected, a first response is to look at observations with high leverage and outliers. In a small sample, the problem may be caused by just one or two outlying values. If the model satisfies all other properties, we can overlook the non-normality of the residuals. As the sample becomes larger the problem will be rectified. However, if we observe concurrent outlier values as the sample increases, then we need to rethink the specification of our model. A transformation of the dependent or independent variables may be necessary such as taking logs. If we fail normality due to heavily skewed residuals, it is a sign of model misspecification.

Note that the normality assumption may not be required if our aim is to estimate the best linear unbiased coefficients. Under the assumption of uncorrelated errors and homoscedasticity, OLS will produce the best linear estimates for the population coefficients.

7.7.2 *Example*

We calculate the JB test statistic value for our Hong Kong model:

$S = 0.733$ (sample skewness in Excel); excess kurtosis $K - 3 = 1.689$ (kurtosis calculation in Excel); $n = 33$.

$$JB = 33\left[\frac{0.699^2}{6} + \frac{(1.689)^2}{24}\right] = 6.61$$

The $\chi^2(2)$ at the 5% and 1% level of significance are 5.99 and 9.12, respectively. The computed value is higher than the critical value at 5%. Hence, we reject the null hypothesis of normality in our regression at the 5% level. At the 1% level of significance, we do not reject normality. What disturbs normality is kurtosis, skewness or both, as their values differ from those of the normal distribution. This discrepancy reflects model issues or outlier values. We should also note that the JB test is an asymptotic test for normality and 33 observations may be too small.

- **Online note #7.11: Testing for normality in EViews**

In the plot of the actual and fitted values we observe an outlier in 1989. Let us address the problem with a dummy variable. The dummy we create takes the value zero for every year except year 1989 when it takes the value 1. An event occurred in 1989 that pushed prices higher than implied by the model. It is our job to have an explanation about what happened to justify the inclusion of the dummy variable. Good market knowledge comes in handy here. The results of the model estimation including the dummy variable ($D1989$) is:

$$RRg_t = 14.45 - 2.04VAC_t + 1.21GDPg_{t-1} + 35.07D1989$$
$$(0.03)(0.00)(0.01)(0.00) \tag{7.47}$$

Adj. $R^2 = 0.65$ *(0.48)*; DW = 1.45 *(2.19)*; p-value White F-test (cross terms) = 0.85 *(0.64)* p-value Breusch-Godfrey F-test for first order serial correlation (F-version) = 0.13 *(0.55)*; RESET p-value = 0.57 *(0.89)*; p-value for JB normality test = 0.67 *(0.09)*

In brackets we show the same statistics for the original model (7.8) for direct comparisons. As expected, the dummy variable is highly significant and takes a positive sign. This suggests that the event in 1989 pushed rents higher over and above the value predicted by vacancy and GDP growth. Both VAC and $GDPg_{t-1}$ retain their significance, a welcome result. In small samples, the addition of a dummy or dummies could impact on the significance of the original regressors. The impact on the model is also seen in the changing values of the coefficients. The slope coefficients have fallen in value but again not considerably. As the dummy takes away explanatory power from the coefficients, the sensitivity of RRg to VAC and $GDPg$ is now lower. For VAC it is -2.04 compared with -2.43 in the original regression. In the case for $GDPg_{t-1}$, it is 1.21 down from 1.39. The explanatory power has risen as expected to 0.65. When we add a dummy, we should re-run the diagnostics. The model with the dummy passes all tests and the JB statistic has improved. If we look at the actual and fitted values (not shown here) there is no error in 1989; the actual value is fully replicated by the inclusion of the dummy by definition.

7.8 Multicollinearity

7.8.1 Nature of the problem

A basic assumption in the successful multiple linear regression model is that the regressors are independent, that is there is no linear relationship among the explanatory variables. In our example, it is desired that VAC_t is independent or orthogonal to $GDPg_{t-1}$, meaning there is no linear association among them or the two regressors do not correlate. Multicollinearity materialises when two or more explanatory variables are highly correlated. There are consequences for the regression model from a strong dependence among regressors.

In practice, it is difficult to have orthogonality among explanatory variables in multiple regression. The world is complex and economic; finance and real estate series or data could correlate. Consider our example, without even conducting any tests should we expect some association between VAC_t and $GDPg_{t-1}$? GDP growth at time t is expected to affect vacancy at time t but it could affect it at period $t + 1$ too, as some firms may have reacted with a delay to GDP growth in taking on more people and reducing vacant space. In general, different degrees of association among regressors is likely when working with real estate data from the private market. True independence among regressors is difficult to observe. When pairs of variables in our model are highly correlated, it could lead to possible model misspecification and flawed inferences.

7.8.2   *Consequences*

In general, when explanatory variables exhibit a high degree of dependence the following problems might occur:

(i)   In the extreme case when the variables are perfectly correlated (correlation coefficient ± 1) we obtain an ill-conditioned $X'X = 1$ (see online note #6.8). Estimation is not possible.
(ii)   The variance of the estimators becomes large, which results in unreliable estimates. With large standard errors, the t-ratios become small indicating non-significant variables, which can be a false inference. Thus, important variables may be excluded in the final model specification.
(iii)   It follows that the confidence intervals for the coefficients will be too wide and this will transpire into forecasts obtained from the model.
(iv)   The problem of multicollinearity is so complex that, in practice, two collinear variables could appear statistically significant. If we run univariate regressions, we may observe that one of the variables is not statistically significant. It is significant only when it is combined with another variable or other variables in the model.
(v)   The standard errors are sensitive to even small changes in the values of the explanatory variables. When we add or exclude variables the results change significantly, that is the existing variables in model may change sign or have their significance affected.

7.8.3   *Causes*

(i)   A model contains explanatory variables that are highly correlated. For example, in a model of demand for commercial space model (e.g. a model take up) we include both GDP and employment. Economic output and employment are bound to correlate; they broadly capture the same influence. A model with both may render one or both of the variables insignificant. Furthermore, the signs may not conform to our prior expectations. The problem also arises when, say, two of the regressors are jointly determined by a third variable.
(ii)   An overdetermined model (overfitting the model). We may have too many variables (e.g. many lags of quarterly or monthly data) in the model, increasing the possibility of associations among the independent variables.
(iii)   Lack of variability in the explanatory variables. Since the regressors do not show distinct variability, they appear as correlated.
(iv)   Generating or using an independent variable that is based on one or more of the existing regressors.

7.8.4   *Detection of multicollinearity – multicollinearity diagnostics*

(i)   The model has a high R^2 with insignificant coefficients. This is evidence of multicollinearity. The variance of the coefficients becomes large, resulting in small t-ratios and insignificant coefficients.
(ii)   Inspection of cross correlation among the independent variables. If pairwise correlation tests reveal strong correlation, multicollinearity is present in the model. There is no value for the correlation coefficient that is indicative of harmful multicollinearity. We can test the significance of the correlation coefficient. However, values less than

0.30 could be considered acceptable with values of over 0.60, indicating the presence of strong multicollinearity.

(iii) Running regressions among the independent variables. In a model with two regressors, x_1 and x_2, we run the following model:

$$x_2 = \alpha_0 + \alpha_1 x_1 + e_t \tag{7.48}$$

There is evidence of multicollinearity if α_1 is statistically significant and when the adjusted R^2 is high. This shows a strong and significant linear dependency among the two variables.

(iv) Estimating the variance inflation factor (VIF). The VIF is an extension of the previous approach. VIF is a metric calculated for every regressor in the model. A high VIF for an explanatory variable means that it is explained by other variables in the model. The metric is calculated as:

$$VIF = 1/\left(1 - R^2\right) \tag{7.49}$$

where R^2 is obtained by running an auxiliary regression of an explanatory variable against the other explanatory variables in the model. Hence in a model which has three regressors x_1, x_2 and x_3, the R^2 that enters the VIF formula for x_2 (7.49) is the R^2 of equation (7.50):

$$x_2 = \alpha_0 + \alpha_1 x_1 + \alpha_2 x_3 + e_t \tag{7.50}$$

Consider the values of *VIF* when is R^2 is high, say 0.8, and when it is low, say 0.25:

$$R^2 = 0.8 \rightarrow VIF = 1/\left(1 - 0.8\right) = 1/0.2 = 5$$
$$R^2 = 0.25 \rightarrow VIF = 1/\left(1 - 0.25\right) = 1/0.75 = 1.33$$

There is no distribution or critical values for this metric. It is the convention to consider acceptable values less than four ($VIF < 4$). If we are strict about multicollinearity, in order to dismiss harmful multicollinearity $VIF < 2$ is expected.

VIF is therefore an alternative to cross correlations as a test for multicollinearity.

7.8.5 *Application*

We apply the multicollinearity diagnostics to our Hong Kong office rent model. The correlation between VAC_t and $GDPg_{t-1}$ is −0.41. It is not strong but it is certainly not negligible. We also run the regression as per (7.48) and calculate the VIF:

$$VAC_t = 10.4 - 0.31 GDPg_{t-1} \tag{7.51}$$
$$(1.0) \quad (0.02)$$

p-values in parentheses; $R^2 = 0.17$; $VIF = 1/(1 - 0.17) = 1.20$

Equation (7.51) shows a significant $GDPg_{t-1}$ in explaining VAC_t, a sign of multicollinearity. The explanatory power of 0.17 is, however, on the low side. The VIF takes a value suggesting that multicollinearity is not an issue despite a cross correlation of moderate strength (-0.41). We further run a univariate regression between $RREg_t$ and each of the two regressors. The latter appears significant when they are included on their own in the respective univariate regressions. This is a desired outcome implying that these two regressors convey different influences to $RREg_t$ in (7.8). In general, if the significance of two independent variables x_1 and x_2 in a model is due to these variables being collinear, they may appear not significant in univariate regressions.

7.8.6   Remedies

(i)   Increase the sample size. Additional data may break up multicollinearity, hence the variance and standard errors of the parameter estimates might decrease. There is no assurance, though, that larger samples will remove multicollinearity from the model.

(ii)   Drop collinear variables. In a model with collinear variables, x_1 and x_2, drop the one which has the smaller t-ratio. In a model with more than two variables, identify which pair of variables is most collinear and drop the one which has the smaller t-ratio.

(iii)   Use real estate theory to determine which variables should be included in the model. For example, in a retail rent model you contemplate the inclusion of consumer spending and retail sales. You may believe that retail sales are a better indicator since general consumer spending may be too broad as a measure (e.g. includes spending on services). You can include both in the regression model and assess their estimates. If the variables are collinear, then drop the one with the lower t-ratio.

(iv)   Transformations. Suppose you estimate a model of retail rents $(RETR)$ that includes consumer spending (EXP) and online sales $(ONLINE)$:

$$RETR_t = \alpha_0 + \alpha_1 EXP_t + \alpha_2 ONLINE_t + e_t \qquad (7.52)$$

Further, assume that EXP_t and $ONLINE_t$ are strongly collinear and the two variables are not statistically significant. $ONLINE_t$ has the smaller t-ratio but we still wish to keep this variable in the model, as intuitively it has become more relevant in recent years. We have a couple of options: first to get the ratio of the two variables and estimate the following model:

$$RETR_t = Y_0 + Y_1 \frac{EXP_t}{ONLINE_t} + e_t \qquad (7.53)$$

This model addresses multicollinearity. The loss is in the interpretation of Y_1, which is not as straightforward any longer.

Another option is to extract from consumer spending the part that is spent online since we take that explicitly on board via online sales $(ONLINE)$. We run the following regression:

$$EXP_t = b_0 + b_1 ONLINE_t + e_t \qquad (7.54)$$

The residuals or unexplained part of this equation will represent consumer spending after online sales are taken into account, hence consumer spending in actual shops in shopping centres, the high street or elsewhere.

Now we run this model:

$$RETR_t = \alpha_0 + \alpha_1 EXPRES_t + \alpha_2 ONLINE_t + e_t \tag{7.55}$$

where *EXPRES* is expenditure net of online sales.

(v) Finally, scholars also suggest the use of panel analysis (online chapter IV).

7.9   Structural breaks and parameter stability

7.9.1   *Structural change and model implications*

A linear relationship and parameter (coefficient) stability in the classical regression analysis are assumed throughout the sample period. There may be disruption, though, from events that can have a permanent impact on the relationship between the dependent and independent variables. If such events affect the model, the slope coefficients can change significantly, or even one or more of the explanatory variables may lose their significance after an event.

Structural change is primarily a source of misspecification in time series models. The top panel of Figure 7.7 shows that the fitted values after a certain point in time fail to track the actual values as closely as before. There must have been a change in the original relationship and most likely in the parameters (coefficients). Panel (b) of Figure 7.7, gives a graphical illustration of a change due to an event that results in a distinctive effect on the relationship between *y* and *x*. This situation is also relevant to cross-section models, in that the impact of a change of the explanatory variable on the dependent variable may differ between cross-section units. For example, household incomes may differ in magnitude in different regions affecting house prices. Overall, the dotted line depicts a positive relationship between *y* and *x*. A close observation reveals that at smaller values, the regression line is steeper. Therefore, *y* is more sensitive to lower than higher values of *x*, where the relationship gets flatter. Irrespective of the cause, this is the result of a structural change and parameter value shift. Structural change refers to permanent effects in the relationship or effects that last for a significant length of time.

Events that cause structural breaks include a major policy change (e.g. new regulation, tax policy, planning policy), a significant economic shock, a stock market crash, a banking crisis and natural disasters. The major stock market crash in October 1987, the financial crisis in East Asia in 1997, the global financial crisis in 2008 and the eurozone crisis in 2010 are events that can cause permanent changes to relationships. The UK's departure from the EU could be seen at the time and subsequently as a structural event slowly manifesting itself and leading to a structural change in the relationship between the economy and the real estate market in the UK. Another example is online shopping and the impact on retail models that link the retail property sector (demand, rent, prices, other) to the consumer economy.

Suppose the general model describing the relationships in Figure 7.7 is:

$$y_t = \alpha_0 + \alpha_1 x_t + e_t \tag{7.56}$$

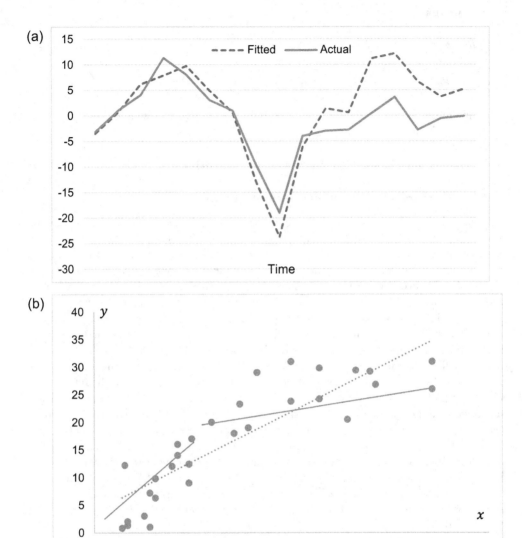

Figure 7.7 Graphical illustrations of a structural break

We observe that the relationships do not remain constant across the sample. This is more apparent in panel (b), whereas in panel (a) the changing relationship is reflected in the fitted values of the model. In both cases, equation (7.56) does not seem to hold constant throughout the sample – in particular, before and after structural events. We can write (7.56) for the two sub-samples as follows:

Before the event: $y_t = \beta_0 + \beta_1 x_t + e_{1,t}$ (7.57)

After the event: $y_t = \gamma_0 + \gamma_1 x_t + e_{2,t}$ (7.58)

What we would like to investigate is whether the coefficients are different in the three models (7.56 through 7.58). Minor differences are to be expected. But if they are too large, it is a sign of a structural change in the relationships being modelled. We will perform basic tests to examine the constancy of coefficients or parameter stability throughout the sample.

The pervasiveness of structural changes and breaks in the relationships has long been a concern in econometric analysis. In the extensive research in economics and finance, uncertainty and parameter instability are found in testable relationships. For example, Rapach and Wohar (2006) and Paye and Timmermann (2006) provide empirical evidence of structural breaks in equity premium models. These findings are of interest and of relevance in modelling property risk premia.

The key implications from the presence structural breaks in a model are:

- Instability of parameters leads to inaccurate sensitivities (betas)
- Large forecasting errors due to changing parameter estimates (see Goyal and Welch (2008), Rapach *et al.* (2010)
- In the presence of a structural break (or breaks), the wrong inferences can be made. The results are not representative for the whole sample and the population.

7.9.2 *Chow tests and applications*

A range of tests for structural breaks is available in the econometrics of structural change. Part of the literature is concerned with the timing of the structural break and whether the breakpoint date is known a priori. When there is a structural change and the breakpoint date is known with certainty, there are two classical Chow tests available for the analysis (Chow, 1960). The two versions of this test are the breakpoint test and the predictive or forecast failure test. We discuss these tests as they are customary in econometrics software. Chow tests are popular in empirical real estate modelling.

(i) *Chow's breakpoint test*

We are interested in testing the presence of a structural break in the following linear regression model estimated with time-series data:

$$y_t = \beta_0 + \beta_1 x_{1t} + \beta_2 x_{2t} + u_t \tag{7.59}$$

Suppose we have in mind an obvious breakpoint date or a date when a major event occurred which we believe might have caused a structural break in the model. Chow's breakpoint test involves the estimation of (7.59) over the full sample period, the pre-break or pre-event period, and the post-event period. We set the breakpoint date around the date we believe a break might have occurred.

We estimate three equations and calculate the sum of the squared residuals (*RSS*) for each of them.

Pre-breakpoint date: $y_t = a_0 + a_1 x_{1t} + a_2 x_{2t} + u_{1t}$ (7.60)

Post-breakpoint date: $y_t = \gamma_0 + \gamma_1 x_{1t} + \gamma_2 x_{2t} + u_{2t}$ (7.61)

The assumptions are that the error series in the sub-sample periods are homoscedastic and not serially correlated. The variance of the errors in the two periods is equal:

$$u_{1t} \sim N\left(0, \sigma^2\right), \; u_{2t} \sim N\left(0, \sigma^2\right)$$

Further, u_{1t} and u_{2t} are independent.

The null hypothesis is of no structural break or parameter stability:

$$H_0: a_i = \gamma_i \text{ for all } i$$
$$H_1: a_i \neq \gamma_i \text{ for all } i$$

We compute the following F-test statistic:

$$F = \frac{\left(RSS - \left(RSS_1 + RSS_2\right)\right)/k}{\left(RSS_1 + RSS_2\right)/\left(n - 2k\right)} \sim F\left(k, n - 2k\right) \tag{7.62}$$

where RSS, RSS_1 and RSS_2 are the residual sum of squares of the model estimated over the whole sample period, the pre-event period and the post-event period, respectively. n is the number of observations in the full sample and k is the number of parameters including the constant. A decision can be made by comparing the computed F from equation (7.62) to the critical value of F with k and $n - 2k$ degrees of freedom at the chosen level of significance a.

At its simplest, when testing for a structural break and possible parameter instability, a common breakpoint date researchers adopt is the sample mid-point. The full sample is split into two equal parts; if a break has occurred in the first or second half it would be picked up by this straightforward test. This test is conducted when we have a breakpoint date in mind and we undertake a basic structural change test.

(ii)   *Chow's forecast or predictive failure test*

The breakpoint test requires a large sample. In small samples, such as in our example (33 observations), splitting the full sample results in small sub-samples. This is a shortcoming of this test. Hence, if we wish to test, say, the impact of the global financial crisis on Hong Kong's office market, it is not possible since in the post-event sample there will be only 8–9 annual observations (depending whether we set the breakpoint date as 2008 or 2009). An alternative formulation of the Chow test is the *forecast* or *predictive failure test* (see also Pesaran *et al.*, 1985). It is designed to address situations in which we do not have a large sample after the break. The nature of this test differs from the breakpoint test. It examines whether the parameters estimated over the first sub-sample continue to predict over the second and usually smaller sub-period. If there is a structural break, the errors and the RSS will be inflated showing a higher outcome of what would be expected under no structural break in the second sub-period.

The formula for the test statistic that follows the F distribution is:

$$F = \frac{\left(RSS - RSS_1\right)/n_2}{RSS_1/\left(n_1 - k\right)} \sim F\left(n_2, n_1 - k\right) \tag{7.63}$$

where RSS_1 is the sum of squared residuals of the large sub-period, n_1 is the number of observations in the large sub-sample and n_2 the number of observations after the break (the observations that are predicted in this test).

This statistic is useful when breaks occur towards the end of the sample, that leaves us with just a few observations to apply Chow's breakpoint test. It is also useful to start monitoring the effect of a possible break. Consider the EU membership referendum of June 2016 (2016Q2) in the UK. The referendum result could be a possible source of a structural change that would impact the relationship between the economy and the real estate market in the UK. Chow's forecast test could be applied to real estate models. For quarterly data, we set the breakpoint date in 2016Q2 with the sub-sample period starting in 2016Q3. The computation of the test is repeated as subsequent data become available. If we start with post-event sample period 2016Q3 to 2017Q3, as data become available the post-event sample becomes 2016Q3–2017Q4, 2016Q3–2018Q1 and so forth.

An equivalent test is based on examining the joint significance of the dummies in the model but it is not often used in real estate. The model specification is:

$$y_t = \beta_0 + \beta_1 x_{1t} + \beta_2 x_{2t} + c_1 D_1 + \ldots + c_{n2} D_{n2} + u_t \tag{7.64}$$

The full sample is n with the break occurring at n_1, $(n_1 < n)$. The dummy variable D_i takes the value 1 in period $n_1 + i$ $(i = 1 \ldots n_2)$ and zero in all other periods. $n = n_1 + n_2$ and the number of dummies we include is n_2. We test the joint significance of the dummy variables with an F-test. The preceding model is the unrestricted model. The null hypothesis is: $c_1 = \ldots c_{n2} = 0$. The alternative is that one of the dummies is statistically significant.

EMPIRICAL EXAMPLE

We perform these tests in our Hong Kong office price model, starting with Chow's breakpoint test for parameter stability. Initially, we split the sample roughly into half. Our full sample period is 1985–2017 and roughly, 2001 is taken as the midpoint. We estimate two models for the sub-periods 1985–2001 (17 observations) and 2002–2017 (16 observations).

1st sub-period: 1985–2001 (17 observations):

$$RRg_t = 19.58 - 2.77VAC_t + 1.29GDPg_{t-1} \tag{7.65}$$
$$(0.16) \quad (0.02) \qquad (0.15)$$

Adj. $R^2 = 0.47$; $RSS = 2{,}352.7$; S.E. of regression: 13.0; standard deviation of the dependent variable 18.6; p-values in brackets

2nd sub-period: 2002–2017 (16 observations):

$$RRg_t = 9.94 - 1.49VAC_t + 2.04GDPg_{t-1} \tag{7.66}$$
$$(0.25) \quad (0.08) \qquad (0.01)$$

Adj. $R^2 = 0.54$; $RRS = 618.6$; S.E. of regression: 6.9; standard deviation of the dependent variable: 10.2; p-values in brackets

Our sample is small and estimating individual regressions with 16 to 17 observations may not lead to reliable results. We should keep this in mind in the discussions that follow – the results may reflect a small sample.

We make the following observations:

(i) The results for the two models do not look similar. The estimate of the constant is different and the estimate of VAC_t is less negative. In contrast, the estimate of $GDPg_{t-1}$ has risen implying that it has a bigger impact on RRg_t in the second period. Moreover, $GDPg_{t-1}$ is not statistically significant in the first period. Hence, the relationship between the dependent variable and explanatory variables seems to be different between the two periods. Do these differences suggest a structural break and parameter instability? Is the model different between the first and second half of the sample period? We formally test this with Chow's breakpoint test.

(ii) The standard error in the first half of the sample is larger than the second half. This partly reflects the greater volatility of RRg_t in the first half of the sample (see the differences in standard deviations in the two sub-samples). The model is unable to explain this volatility resulting in a larger standard error (volatility of residuals).

It is not uncommon in real estate models with time-series data to observe weaker explanatory ability at the beginning of the sample. Data such as rents and prices can be more volatile in the early part of the sample largely reflecting reduced quality and limited depth of databases.

We calculate the breakpoint test statistic and compare it to the critical value at say $\alpha = 5\%$.

$$\text{Computed } F = \frac{\left[3265.1 - (2352.7 + 618.6)\right]/3}{(2352.7 + 618.6)/(33 - 2 \times 3)} = 0.89$$

The critical $F(3,27)$ at 5% level is approximately 2.97 (Appendix A, Table A3).

We find that the calculated F-statistic is lower than the critical value, hence the null hypothesis of structural change cannot be rejected. The estimated coefficients are considered stable, despite the differences we observe in the first and second half of the sample. Still, these differences are within the acceptable boundaries.

Instead of splitting the sample into two halves, let us look at the impact of the global financial crisis in the Hong Kong office market. More specifically, what happened after the Lehman's collapse in September 2008. Using annual data, we attempt to trace the impact from 2009. If we had quarterly data, we would have studied the implications from the fourth quarter of 2008. The application of Chow's breakpoint test is not appropriate in this instance since the post-break period is even shorter now, 9 years (9 observations). Due to this fact, the predictive failure test is more appropriate. We populate (7.63) by estimating the model over the first sample period:

1st sub-period: 1985–2008 (24 observations):

$$RRg_t = 17.67 - 2.41VAC_t + 1.54GDPg_{t-1}$$
$$\quad\;\; (0.15)\;\;\; (0.02)\qquad (0.05)$$

$$(7.67)$$

Adj. $R^2 = 0.49$; $RRS = 2951.9$; the RRS for full sample period is 3,265.1 (see equation (6.26)).

$$\text{Computed } F = \frac{(3265.1 - 2951.9)/9}{(2951.9)/(24 - 3)} = 0.25$$

Critical $F(9,21)$ at 5% level is approximately 2.38.

The estimated *F*-test does not exceed the critical value. Hence, we do not detect a break in the Hong Kong office market owing to Lehman's collapse and subsequent global financial crisis. To get further assurance that this is a correct statement, we can run the test and set the breakpoint date as 2010 to allow for slow responses in the market. We re-run the test with the results revealing once again no structural break.

- ***Online note #7.12 illustrates the detection of a structural change with the use of dummy variables***
- ***Online note #7.13: Chow's breakpoint test in EViews***
- ***Online note #7.14: Chow's forecast (predictive failure) test in EViews***

In the preceding tests the breakpoint dates are known. This is not unrealistic in real estate. Real estate professionals have good market knowledge and can be insightful in identifying major events that have occurred in the market and provide information of the potential sources of a structural break. Chow tests present the first port of call. But the dates for structural breaks may not always be known. There is a large and technical literature concerned with the detection of unknown breakpoints and the ongoing development of methodologies to identify breakpoint dates. Several of these tests are now customary in econometrics software. However, the first stage is always to test the break at dates using market knowledge from real estate professionals.

7.9.3 Tests based on recursive estimates

(i) Recursive estimates of coefficients

This approach to parameter instability tracks the evolution of coefficient values as we add more data. We use the first few observations to run the OLS. It is expected that initially the coefficients will be unstable as the sample is small. Suppose we start the recursive estimates from the beginning of the sample (which is the approach in EViews for these tests). If the number of regressors is k, the first three observations are reserved and estimation begins at $t = k + 1$. In our example $k = 3$, hence, the first estimate of coefficient values is obtained from the sample 1985 to 1988. The second estimate of values from 1985 to 1989 and so forth. The final sample will be 1985 to 2017 and the coefficient values are those reported in equation (7.7).

Initially the coefficient is expected to be unstable and volatile as it is estimated from small samples. As the sample increases though, the coefficient value should stabilise. This is an indication of a stable coefficient and a good model. In the case of a structural break, the coefficient values will shift after the breakpoint date. This test captures parameter instability. If a coefficient is not stable, we have less confidence in its ability to capture the influence of an explanatory variable as it might exhibit further instability in the future. This test is also useful to trace the impact of structural change that gradually and steadily affects the market. A coefficient value will progressively increase or decrease as a result of new market regime. In the same way, we estimate the model recursively and track the residuals. These recursive residuals provide the basis for two further tests of structural stability described next.

- ***Online note #7.15: Recursive coefficients in EViews***

(ii)   Cumulative sum of recursive residuals (CUSUM) test

This test for parameter change or instability proposed by Brown *et al.* (1975) is based on the cumulative sum of the recursive residuals. The cumulative sum of the recursive residuals is expected to vary randomly around a mean of zero. This is because random positive and negative errors will offset each other, resulting in the cumulative sum to swing around zero. A misspecification of the model will quickly result in an upward or downward trend in the cumulative sum. CUSUM charts include critical boundaries. The test detects parameter instability if the cumulative sum goes outside the 5% critical boundaries.

(iii)   Cumulative sum of squares (CUSUMSQ) test

The CUSUMSQ test (also proposed by Brown et al., 1975) is a similar procedure to CUSUM but is grounded on the cumulative sum of the squared recursive residuals. The CUSUMSQ graph in EViews also contains 5% critical bounds. The sum of the squared recursive residuals from a model with stable parameters will stay within the bounds throughout the sample.

- **Online note #7.16: CUSUM and CUSUMSQ tests in EViews**

Turner (2010) finds that the power of the CUSUM test is high (that is the probability of concluding there is no structural break when there is one is high) but falls if the break is in the intercept of the regression model. In general, since the intercept controls for the mean of the dependent variable, a structural break owing to the changing intercept reflects changes in the mean of the dependent variable after the event. The CUSUMSQ has higher power when the structural break is caused by a slope coefficient.

The literature on structural breaks extends to study the possibility of a number of unknown points of structural change through so-called *sequential* testing procedures. Bai and Perron (1998, 2003) propose tests for multiple unknown structural breaks and a further discussion, and application of multiple breakpoint tests can be found in Jouini and Boutahar (2005), who outline the workings of these tests. The maths behind these tests is beyond the objectives of this book. In principle, breaks are added and the constancy of the parameters is tested in each sub-sample. Bai and Perron (2004) discuss the application of the sequential process in some detail.

- **Online note #7.17: Sequential structural break tests in EViews**

7.9.4   Remedies

In the simplest case of a structural break, which is not transitory and has permanent effects on the relationship, we cannot estimate the model over the full sample period as this ignores the break. The explanatory or predictive ability of regressors will vary markedly. If we have enough observations in the post-break period, we could estimate the model over the second sub-sample. If the model has changed much (e.g. notably lower explanatory power), we may have to reformulate it. This may not always be possible, though. Structural breaks can be treated in an ad hoc fashion. We can include an intervention dummy taking the value 1 after the breakpoint date. This may result in too many dummies that might pick up other effects as well. This is not an ideal treatment of a structural break.

If testing has found one or more breaks, we should use estimators that allow estimation and forecasting in the presence of structural breaks. The discussion of these estimators is technical, and the reader is referred to Clements and Hendry (2011) and Perron (2006) for a comprehensive analysis of the subject.

7.10   Concluding remarks

We present in this chapter key tests to detect a range of possible misspecifications in a regression model. These tests are necessary to assess the reliability of regression models and are referred to as *robustness tests*. Given the current computing capabilities these tests are performed speedily. Hence it is important that we have a good understanding of the nature of these tests and we are able to make appropriate interpretations of the test outputs obtained. The chapter illustrates the computation of diagnostics in EViews, but the exposition of most of these tests allows you to perform these tests in Excel.

We study the diagnostics with reference to OLS regression models. Some of the tests are only applicable for OLS estimates (e.g. recursive tests in section 7.9.3). We noted that these tests can be applied both to models with time-series data (as in our examples in this chapter) and cross-section data. They are also relevant to more complex methodologies, although there are additional tests specific for other methodologies.

Running diagnostics serves the purpose of assessing the model but also finding out more about the relationship, the data and the impact of shocks. Further, they can be used for the selection of the preferred model(s) among competing models. Small samples can be an issue for these tests. A single outlier observation can lead to the detection of misspecification. We will link the material in this chapter to Chapter 8 on stationarity. Problems may be due to non-stationary data. By using stationary data, we could potentially eliminate some of the diagnostics problems.

Chapter 7 online resource

- Chapter 7 accompanying notes
- EViews file: "ch6_eviews": Work from EViews files in Chapter 6.

8 Stationarity

8.1 Introduction

It is now generally recognised that a major requirement in regression analysis with time series data is that the series are stationary. In this section we study additional features of time-series data and address the topics of *unit roots* and *stationarity* in time series. These are important topics in regression analysis involving time-series data and highly relevant to the validity of empirical models we estimate. As such, an extended literature has focused on the theory of unit roots, stationarity and associated testing procedures that have been evolving through time, resulting in more efficient tests. Major problems in empirical work arise from trended variables and persistent innovations (shocks) to the variables we include in our models. Trends and shock persistence are characteristics of the so-called *non-stationary* variables. Modelling with non-stationary variables leads to model misspecification, invalid inferences and *spurious regressions*, which we discuss later. The aim of this chapter is to explain the notions of stationarity and unit roots, present and run commonly used tests for the detection of unit roots, outline data transformations to remove unit roots and obtain stationary data series, and discuss the key problems when we include non-stationary variables in our models.

Stationarity relates to the dynamic properties of variables that we need to understand before estimating models and forecast. Do our data follow a random walk process or do our time-series data contain trends? If so, we need to transform the data so that our regression analysis and results are reliable.

Furthermore, stationarity in time series is associated with mean reversion. This property in time-series data has useful implications for investment and portfolio construction. When a series exhibits mean reversion, the series will tend to revert to its long-term average when it moves away from it. This provides the basis for the construction of trading rules. Further mean reversion is part of assessing market risk. A class of investment funds constructs portfolios on the basis of mean reversion. It is apparent that the topic of stationarity has wide-ranging applications and uses.

8.2 Stationarity

Stationary time series are characterised by time invariant moments of their distribution. We distinguish between two types of stationary processes.

8.2.1 Covariance stationary or weakly stationary processes

A random variable is covariance or weakly stationary if its mean, variance and autocovariance are constant through time. In this weak form of stationarity, the first and second moments of the series are time invariant, that is the series has a constant mean and variance throughout the time. More formally a weakly stationary series satisfies the following:

$$E(y_t) = E(y_{t-s}) = \mu \tag{8.1}$$

$$Var(y_t) = E(y_t - \mu)^2 = (y_{t-s} - \mu)^2 = \sigma^2 \tag{8.2}$$

$$Cov(y_t, y_{t-k}) = Cov(y_{t-s}, y_{t-s-k}) = \gamma_k \text{ or}$$

$$E(y - \mu)(y - \mu) = E(y - \mu)(y_{t-s-k} - \mu) = \gamma_s \tag{8.3}$$

The first two conditions suggest that the first two moments, mean and variance, do not depend on the actual location in time or the time origin of the data. y_{t-s} in the first two conditions represents a sub-sample of the full sample y_t. Through y_{t-s} we shift/lag the sequence of values within the same time series.

Suppose we have a total return series for the period 1990Q1 to 2018Q4. According to the first two conditions the mean and variance of the series is constant for sub-periods such as 1990Q1–2002Q4, 1990Q1–2010Q4 or similar. The sub-samples are denoted by y_{t-s}. The mean and variance of the original series (over the full sample) is the same as lagged time series by any number of lags s are considered. Further, this condition should hold for any other sub-samples, e.g. 1998Q1–2012Q4.

The third condition states the autocovariances (autocorrelations) only depend on time decay or on the time interval between the terms. The autocovariance coefficients depend on whether we examine first, second or higher order autocovariances but not in the time (sample) itself when autocovariances are estimated. Therefore, autocovariances are a function of the lag k and do not depend on the time at which the series is observed. That is, if k is 1, then the covariance between y_t and y_{t-1} $\left(cov(y_t, y_{t-1})\right)$ should be constant irrespective of the time period we examine this order of covariance. $cov(y_t, y_{t-1})$ refers to whole sample period. Another time period can be defined if, say, s equals 10. We still consider first order autocovariance but over a different time period, that is $cov(y_{t-10}, y_{t-10-1})$. The first order autocovariance can be examined over the full sample period 1991Q4 to 2018Q4 or 1990Q4 and 2015Q2 (to denote y_{t-10}). The autocovariance coefficient should be similar in the two sample periods. The interpretation of the third condition remains similar if instead of first order we examine the second or higher order autocovariances.

8.2.2   Strong stationarity

In weak stationarity we implicitly assume a normal distribution, which is defined by the mean and standard deviation. In the case of strong stationarity, the distribution of the series is unspecified but it remains the same in any sequence of values within the series. The distribution of the original time series is exactly the same as lagged time series (by any number of lags) or even sub-segments of the time series. Hence, all moments of the probability distribution are time invariant irrespective of sub-sample. Our focus in this section is weakly stationary processes.

8.2.3   Properties of stationary series

A stationary series has a number of appealing properties.

- It has a constant long-term mean and the series fluctuates around this mean. When it drifts away from the mean, the series reverts to it. This is an appealing property in empirical modelling and also in forecasting. Further, it has implications for investment portfolios. A number of funds follow mean-reverting strategies. They consider assets for inclusion in the portfolio that exhibit mean reversion in performance, hence returns from these assets are stationary. For a discussion on the mean reversion of stock returns and portfolio implications see Akarim and Sevim (2013), Pastor and Stambaugh (2012), Spierdijk *et al.* (2012), and Kim *et al.* (1991).

- Shocks dissipate quickly. The impact of shocks on a stationary time series dies away as the time goes on. For example, the global financial crisis (GFC) had a major impact on the real estate market in 2009. The impact of GFC on the market would fade away in 2010, 2011 and eventually vanish.
- There are no trends or seasonality in the data. These are desirable features in regression analysis. Trends and seasonality will affect the value of the time series at different times.
- In general, stationary time series do not have predictable patterns. One may argue that this contradicts the property of mean reversion. This is not so, as cycles may be aperiodic. The series reverts to the mean but the timing of the cycle phases is not predictable. It is difficult to time the cycle.

Figure 8.1 gives a graphical representation of two stationary series drawn from real estate market data. In case (i) the series has a mean of 0.9. The series wanders around its mean, however, it tends to move back to it. The volatility of the series remains largely constant throughout the sample period. This series exhibits positive first order autocovariance or autocorrelation. The first order autocorrelation coefficient is 0.48. Case (ii) presents a more volatile series with a mean of 4.3. This series too clearly fluctuates around its mean. In both cases the time period (in the preceding examples the number of years) that the series remains above or below their mean varies, there is no pattern. In a stationary process, the series cuts through its mean several times. The variance appears to be constant throughout the sample. In reality, the pattern of the series (the original series or after transformation) may not be as clear-cut as in Figure 8.1, particularly when we work with small samples. Formal tests will apply to decide whether a series has no unit roots and is stationary. The series in case (ii) has a negative first order autocorrelation of −0.15 and it is less autocorrelated than the series in case (i). Two series can be stationary but their strength of autocorrelation (autocovariance) could vary in degree.

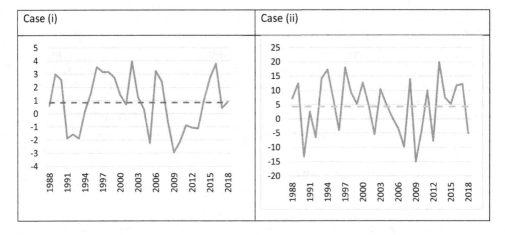

Figure 8.1 Stationary series

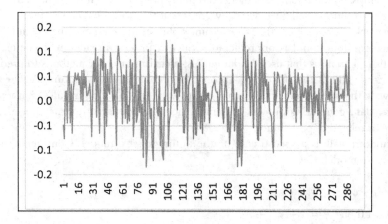

Figure 8.2 A white noise series

8.2.4   *White noise process*

This is a particular case of a stationary process. A series is white noise if it has a zero mean and constant variance irrespective of sequence and elements which are independent, hence the covariance is zero.

$$E(y_t) = E(y_{t-s}) = 0 \tag{8.4}$$

$$Var(y_t) = E(y_t - \mu)^2 = (y_{t-s} - \mu)^2 = \sigma^2 \tag{8.5}$$

$$Cov(y_t, y_{t-k}) = Cov(y_{t-s}, y_{t-s-k}) = \gamma_k \text{ or}$$

$$E(y_t - \mu)(y_{t-k} - \mu) = E(y_{t-s} - \mu)(y_{t-s-k} - \mu) = 0 \tag{8.6}$$

Figure 8.2 illustrates a white noise sequence, which is highly volatile and erratic. White noise processes are encountered in many contexts. In regression analysis the residuals of a model are often described as white noise. White noise residuals are independent and identically distributed (*iid*) usually normally distributed with zero mean. White noise residuals are mostly seen as a representation of shocks or unpredictable events in the model allowing for stochastic fluctuations. The predictability of most series is aided by the presence of autocorrelation. Autocorrelation is non-existent in white noise processes. A white noise process can still be predicted through non-linear techniques and as a part of a general mathematical model that describes broader phenomena. A white noise process is part of this bigger system. White noise has a place in simulation work whereby we can create many realisations and generate useful statistics.

8.3   Random walks

8.3.1   *Definitions*

A *random walk* or a *stochastic process* is a sequence in which the value of the series at time $t + 1$ is the value at t plus a random shock. The series takes a random and unpredictable path. The series jumps in a random direction at each subsequent time period and this jump is independent of earlier jumps. The random walk model has many financial and non-financial applications. In finance the random walk theory simply states that the stock market is unpredictable. The price of a stock at time $t + 1$ is the price at time t plus a random move, thereby it is therefore determined randomly and is unpredictable. Such a process in stock price determination defies statistical analysis that uses fundamentals, chartism or other methods to predict stock market prices. This is the subject of an ongoing debate in the literature. Apparently, the random walk theory is relevant to the real estate market. Do random walks characterise the underlying data generating process of rents, prices, returns or other meaning that future values are unrelated to past values?

The random walk is a special case of the autoregressive model. Consider a first order autoregressive model:

$$y_t = \alpha + \beta y_{t-1} + u_t \tag{8.7}$$
$$\text{If } \beta = 1 \text{ then } y_t = \alpha + y_{t-1} + u_t \tag{8.8}$$

The implication of this is that the series equals the previous value and an error term which is random. Hence the move from y_t to y_{t+1} is due to the random error (u_{t+1}):

$$y_{t+1} = \alpha + y_t + u_{t+1} \tag{8.9}$$

The value of y_{t+2} is y_{t+1} plus another random shock u_{t+2}. The series moves due to random shocks (random error). This is the simple random walk also termed a *stochastic trend* series. There is a number of forms of random walks summarised in Table 8.1.

In all forms of the random walk model the error term u_t is independent and identically distributed (*iid*) with zero mean and constant variance. In Figures 8.3 and 8.4 we illustrate various forms of random walks.

Figure 8.3 contains the three types of random walks with stochastic trend we presented in Table 8.1. We observe that the stochastic trend in all cases follows different patterns, it rises or falls at varying intensity. These stochastic trends (random walks) represent unpredictable series. In Figure 8.4 the series wanders around but it reverts to a deterministic trend (an upward-sloping straight line that cuts through the original data). All these cases signify non-stationary processes.

Table 8.1 Random walk models (forms of non-stationarity)

Stochastic trend	
$y_t = y_{t-1} + u_t$	The current value is equal to the previous value (since the coefficient on y_{t-1} is one) with a random impact up or down from u_t. This model has no constant or intercept.
$y_t = \alpha + y_{t-1} + u_t$	Random walk model with a drift (or a constant) α. It adds a permanent drift to the data generating process.
$y_t = \alpha + y_{t-1} + \gamma t + u_t$	Random walk model with a drift and a trend. In addition to the above, a deterministic trend component is included in the random walk.
Deterministic trend	
$y_t = \alpha + \gamma t + u_t$	Deterministic trend process. Series fluctuates around a deterministic trend.

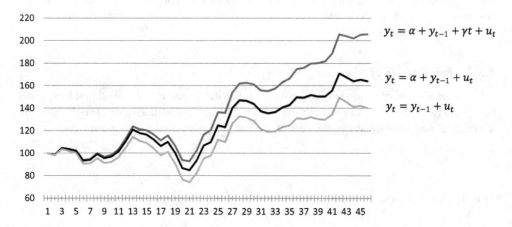

Figure 8.3 Random walks – stochastic trends

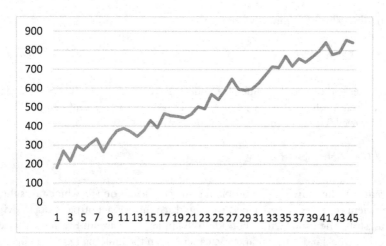

Figure 8.4 Random walks – deterministic trend

Of these cases the random walk with a drift is the most common form in real estate data. Let us consider a rent series in levels as a random walk. Each time the rent equals the previous rent level and a shock in the current period. Unpredictable random events can push the rent higher or lower. This is plausible. The drift in the random walk could reflect inflation if the rents are linked to inflation or to a constant uplift of say 2% or other to reflect long-term inflation. Rent at time $t + 1$ will be rent at time t, plus 2% and a random shock. The relevance of this discussion in modelling rents (level) is that if the rent level path is unpredictable, the results of the rent regression model would be unreliable. We therefore need to test whether the rent level behaves as a random walk with a drift.

A deterministic trend process is not too common in real estate data. It is hard to hypothesise why a deterministic trend is relevant to the rent level series on the top of a drift. We can think of such a trend for the building stock. The office stock increases to reflect higher levels of employment in financial and business services (the result of a larger local economy and real estate market). But again there is no strong theory to rationalise why the office stock increase along a deterministic trend.

8.3.2 Persistence of shocks

A key characteristic of a non-stationary series is that shocks are not absorbed and have non-decaying effects on the series. We illustrate how shocks are not dissolved through time but affect the series permanently in the case of the random walk model with a drift.

$$y_t = \alpha + y_{t-1} + u_t; \ t = 1,2,\ldots,T \tag{8.10}$$

From the general form we write the model for two particular periods 1 and 2.

$$y_1 = \alpha + y_0 + u_1 \tag{8.11}$$

$$y_2 = \alpha + y_1 + u_2 \tag{8.12}$$

Substituting the value of y_1 into y_2 we get:

$$y_2 = \alpha + \alpha + y_0 + u_1 + u_2 = 2\alpha + y_0 + u_1 + u_2 \tag{8.13}$$

Hence in the second period the value of y_2 reflects fully the shock in the first period (u_1). This influence will remain for $y_3, y_4, \ldots y_T$. The shock has a permanent influence. We say that y_t has a *stochastic trend*. We can generalise with successive substitutions to get:

$$y_t = T\alpha + y_0 + u_1 + u_2 + \ldots + u_t \tag{8.14}$$

It is apparent that the source of the problem is the coefficient on y_{t-1}, which takes the value 1. If the coefficient on y_{t-1} was less than 1, we would see the impact of past shocks dying away.

   Just to illustrate the issue further, consider the impact of Lehman Brothers' shock in 2008 on the London Docklands office market (vacancy) when the building Lehman's occupied had all of a sudden became vacant. If vacancy in this London sub-market is a random walk (not stationary), this shock will perpetuate. As an illustration, this is what would happen in the 3 years following the shock.

$$VAC_{2008} = \alpha + VAC_{2007} + u_{2008} \ \text{(vacancy in 2008 reflects the shock through } u_{2008})$$

$$VAC_{2009} = \alpha + VAC_{2008} + u_{2009} = \alpha + \alpha + VAC_{2007} + u_{2008} + u_{2009}$$

$$VAC_{2010} = \alpha + VAC_{2009} + u_{2010} = \alpha + \alpha + \alpha + VAC_{2007} + u_{2008} + u_{2009} + u_{2010}$$

The Lehman's shock is still affecting the level of vacancy in 2010. The impact of u_{2008} (representing the shock) remains unchanged. Consider a model of London Docklands office vacancy on, say, employment (EMP) and supply of space ($SUPPLY$) (equation (8.15)). The implication of a non-stationary vacancy in equation (8.15) can lead into erroneous regression results.

$$VAC_t = \alpha + \beta_1 EMP_t + \beta_2 SUPPLY_t + u_t \tag{8.15}$$

Assume that in our example of vacancy in the London Docklands market the vacancy series is described by expression (8.16), which is the underlying data generating process.

$$VAC_t = \alpha + 0.5VAC_{t-1} + u_t \qquad\qquad (8.16)$$

In this case we have:

$$VAC_{2008} = \alpha + 0.5VAC_{2007} + u_{2008}$$

$$VAC_{2009} = \alpha + 0.5VAC_{2008} + u_{2009} = \alpha + 0.5\left(\alpha + VAC_{2007} + u_{2008}\right) + u_{2009}$$

$$= \alpha + 0.5\alpha + 0.5VAC_{2007} + 0.5u_{2008} + u_{2009}$$

$$VAC_{2010} = \alpha + 0.5VAC_{2009} + u_{2010}$$

$$= \alpha + 0.5\left(\alpha + 0.5\alpha + 0.5VAC_{2007} + 0.5u_{2008} + u_{2009}\right) + u_{2010}$$

$$= \alpha + 0.5\alpha + 0.25\alpha + 0.25VAC_{2007} + 0.25u_{2008} + 0.5u_{2009} + u_{2010}$$

We observe that the impact of the 2008 shock gradually fades away. The coefficient on u_{2008} is declining $(0.5, 0.25, \ldots)$ in each round, signifying a decaying effect on vacancy. VAC_{2009} partially reflects the Lehman's shock in 2008 $(0.5u_{2008})$ and the impact is dissipating. In 2010 the Lehman's shock has even a lower weight $(0.25u_{2008})$. It is apparent that the impact of the shock decays faster the smaller the value of β is (that is the β in equation (8.7)).

8.4 Implications of non-stationarity

Models estimated with non-stationary variables run into a number of problems. There are major implications from non-stationary variables, which econometrics textbooks highlight. Most variables in real estate, economics and finance are non-stationary in levels. Hence, a rent series or a total return index are unlikely to be stationary. Mostly real estate series will contain a stochastic trend. Particular series such as take up or net absorption may not be random walks in levels, however.

8.4.1 Spurious regressions

The problem of *spurious* or *meaningless regression* arises when a non-stationary dependent variable is regressed on one or more non-stationary independent variables. If two variables are trending over time, a regression of one on the other could have a high R^2 and a significant regression F-statistic even if the two are totally unrelated. Since the persistence of shocks is infinite, the variable we explain will not be driven by current developments in the explanatory variables. The latter too will reflect past shocks. It is a cumbersome and certainly unwelcome situation in regression analysis.

8.4.2 Unreliable inferences

If the dependent variable is not stationary, then it can be proved that the standard assumptions for asymptotic analysis will not be valid and we cannot apply standard inference procedures. The distribution of the t-statistic is affected, and it may not follow the t distribution. The usual inferences are invalidated and we cannot validly carry out hypothesis tests about the regression parameters. The least squares estimator will not have its usual properties (BLUE estimator).

8.4.3　*Limited value in forecasting*

Gujarati (2014) states that non-stationary time series are of little practical value for forecasting purposes. It is difficult to generalise a random walk beyond the period with the realised data. We can study the behaviour of the random walk only for the period with available data, but it is not possible to predict it.

8.5　Inducing stationarity

8.5.1　*Data transformations*

In Table 8.1 we presented three versions of the random walk (stochastic non-stationarity) and the trend-stationary process. These processes require transformation to obtain a stationary process. The treatment is different in the two cases. Detrending the random walk with or without a drift and a trend is achieved by taking the first differences of the series. Therefore, for the simple case of a random walk with no drift and no trend the treatment is as follows:

$$y_t = y_{t-1} + u_t \Rightarrow y_t - y_{t-1} = y_{t-1} - y_{t-1} + u_t \Rightarrow \Delta y_t = u_t \tag{8.17}$$

If Δy_t is a stationary series, we have induced stationarity by differencing once. If Δy_t is not stationary, we have to difference one more time, hence difference twice (Δy_t or $\Delta^2 y_t$).

If y_t is the log of a series, by taking the first difference of logs we obtain the growth rates. Another way to induce stationarity is to take the growth rates of a series. It is the convention, though, to difference the data. Inducing stationarity by taking growth rates is appealing in a number of situations. The interpretation of coefficients in models involving growth rates of series is more straightforward than using first differences.

Similarly, if the random walk incorporates a drift or both a drift and a trend we get:

$$y_t = \alpha + y_{t-1} + u_t \Rightarrow \Delta y_t = \alpha + u_t \tag{8.18}$$
$$y_t = \alpha + y_{t-1} + \gamma t + u_t \Rightarrow \Delta y_t = \alpha + \gamma t + u_t \tag{8.19}$$

We illustrate the transformation of an original series to a stationary sequence with two real estate data series: (i) prime retail rents in Central London (annual data 1990–2018) compiled from several sources and (ii) all property total returns in the US (quarterly data 1977Q4–2017Q4). The latter series is NCREIF's all property index (NPI).

Figure 8.5 plots the level of retail rents. The visual inspection of the plot suggests a process that resembles the theoretical case of a random walk. The stochastic trend process is apparent with time varying trends. In Figure 8.6 we show the first differences and growth rates of the rent series. Both transformations result in series that appear to revert to a long-term mean. However, there are spikes in the transformed data which might disturb stationarity. In first differences or growth rates, there is mean reversion but the volatility tends to differ in sub-periods. In any case, decisions cannot be made through observing the plots of the data, rather we would have to run appropriate tests which we discuss later.

Figure 8.7 plots the log level and growth rates (changes in logs) of the NCREIF all total return index at national level in the US. Again the log-level series appears to be stochastic process. We can discern the impact of the global financial crisis in 2008. The series fell and subsequently grew along a new trend. It seems that the global financial crisis had a

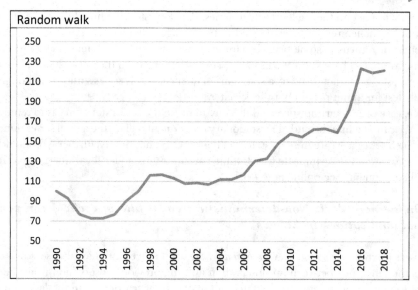

Figure 8.5 Central London headline retail rents (index, nominal)

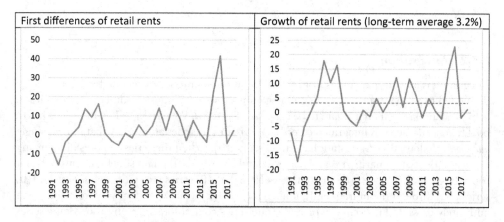

Figure 8.6 Making the retail rent series stationary

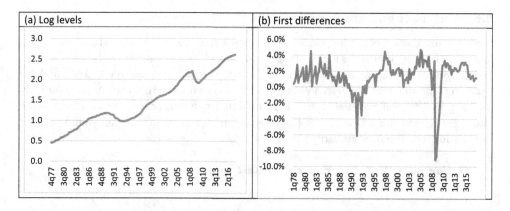

Figure 8.7 All property total returns (real) in the US

permanent impact on the log level return series. Panel (b) plots the first differences of the log return index, hence returns in expressed in percentage. We also observe declining and rising trends for returns in some periods. However the series exhibits a tendency to revert to a long-run average. The series (returns in % terms) recovered from the fall in 2008. The impact was transitory, a characteristic of a stationary sequence. Is this series stationary, though? We formally test for a unit root in both series later.

In the case of a deterministic trend the way to make the series stationary is to remove the trend. We can fit a linear trend. The stationary component will be the actual series minus the trend. Such series are rare in real estate and hence we do not discuss further. We should note that we can fit non-deterministic trends (time variant trends) to level variables to extract the cyclical component. See online note #8.1.

- **Online note #8.1: Non-deterministic trends, filters, cyclical components and applications in EViews**

Some series exhibit seasonality. A *seasonal difference* will induce stationarity. A seasonal difference is the difference between an observation and the previous observation from the same season. Quarterly or monthly real estate data may exhibit seasonality in the last quarter of the year. For quarterly data the seasonal difference is:

$$\Delta^4 P_t = P_t - P_{t-\Delta^4} \tag{8.20}$$

8.5.2 Integration

Integration is a term that comes from the fact that some time series need to be differenced or detrended to become stationary. If a non-stationary series, y_t, must be differenced d times to achieve stationarity, then it is said to be *integrated* of order d. We write $y_t \sim I(d)$. If a series has to be differenced once it is integrated of order 1. It is an $I(1)$ series and we say that it has a single *unit root*. It follows that Δy_t is stationary, that is $\Delta y_t I(0)$. When the series requires differencing twice to induce stationarity it is an $I(2)$ series and $\Delta^2 y_t \sim I(0)$. This series contains two unit roots. In general, if $y_t \sim I(d)$ then $\Delta^d y_t \sim I(0)$. An $I(0)$ series is a stationary series. An $I(0)$ series goes through mean frequently, unlike $I(1)$ or $I(2)$ which wander away from the mean and rarely cross it. The majority of real estate series contains a single unit root. Some series however are $I(0)$.

8.6   Unit root and stationarity tests

Since the recognition of the importance of stationarity in regression analysis, a voluminous literature on testing time series for unit roots has developed. This literature has become increasingly technical aiming to construct more efficient tests to address the so-called *size* and *low power* problems that unit root tests are subject to. In this section we review the most commonly used tests in empirical real estate research. More technical testing procedures are based on the same general principles. Unit root tests will determine the order of integration and whether differencing is required. Stationarity tests, another family of tests, complement unit root tests.

- **Online note #8.2: The concept of unit roots**

8.6.1   Background to unit root tests

Consider the general form of a random walk model with no drift. φ is common notation for the coefficient on y_{t-1} in the unit root literature.

$$y_t = \varphi y_{t-1} + u_t \tag{8.21}$$

We can write:

$$y_{t-1} = \varphi y_{t-2} + u_{t-1} \tag{8.22}$$
$$y_{t-2} = \varphi y_{t-3} + u_{t-2} \tag{8.23}$$

We substitute y_{t-1} from (8.22) into (8.21):

$$y_t = \varphi\left(\varphi y_{t-2} + u_{t-1}\right) + u_t = \varphi^2 y_{t-2} + \varphi u_{t-1} + u_t \tag{8.24}$$

A further substitution for y_{t-2} from equation (8.23) into (8.24) yields:

$$y_t = \varphi^2\left(\varphi y_{t-3} + u_{t-2}\right) + \varphi u_{t-1} + u_t = \varphi^3 y_{t-3} + \varphi^2 u_{t-2} + \varphi u_{t-1} + u_t \tag{8.25}$$

With successive substitutions we get:

$$y_t = \varphi^T y_0 + \varphi^T u_0 + \ldots + \varphi^2 u_{t-2} + \varphi u_{t-1} + u_t \tag{8.26}$$

By rearranging the order of past shocks we obtain equation (8.27):

$$y_t = \varphi^T y_0 + \varphi u_{t-1} + \varphi^2 u_{t-2} + \varphi^T u_0 + u_t \tag{8.27}$$

We distinguish three cases with regard to equation (8.27):

(i) $\varphi < 1$. In this case $\varphi^T \to 0$ as $T \to \infty$

   If $\varphi = 0.9 \to \varphi^2 = 0.81$, $\varphi^3 = 0.73$; If $\varphi = 0.4 \to \varphi^2 = 0.16$, $\varphi^3 = 0.06$

(ii) $\varphi = 1$. In this case $\varphi^T = 1$ for any value of T

(iii) $\varphi > 1$. In this case $\varphi < \varphi^2 < \varphi^3 \ldots$ as $T \to \infty$

In case (iii), the magnitude of the shock rises and becomes more influential as time goes on. Hence the impact of the initial shock u_0 becomes stronger with time. This is a condition that does not describe data series in the real estate market and indeed in economics and finance. Hence we ignore this possibility and focus on the other two cases, whether $\varphi = 1$ or $\varphi < 1$. As noted earlier, $\varphi = 1$ characterises a non-stationary process.

8.6.2 The Dickey-Fuller (DF) and augmented DF test for unit roots

Dickey-Fuller tests

In the founding work on testing for a unit root in time series by Dickey and Fuller (Dickey and Fuller, 1979; Fuller, 1976) the basic objective is to examine the null hypothesis that $\varphi = 1$ against the one-sided alternative $\varphi < 1$ in equation (8.28). The case $\varphi > 1$, as we mentioned earlier, is not realistic, hence the one-sided alternative.

$$y_t = \varphi y_{t-1} + u_t \tag{8.28}$$

The null and alternative hypotheses are:

$H_0: \varphi = 1$ (series contains a unit root)
$H_1: \varphi < 1$ (series is stationary)

The DF test does not apply directly to equation (8.28). The regression equation for the DF test is obtained following a simple transformation of equation (8.28). Subtracting y_{t-1} from both sides of equation (8.28), we get:

$$y_t - y_{t-1} = \varphi y_{t-1} - y_{t-1} + u_t \rightarrow \Delta y_t = (\varphi - 1) y_{t-1} + u_t \rightarrow$$
$$\Delta y_t = \psi y_{t-1} + u_t \text{ where } \psi = \varphi - 1 \tag{8.29}$$

Testing for $\varphi = 1$ is equivalent to a test of $\psi = 0$ and $\varphi < 1$ is equivalent to $\psi < 0$.

We summarise the null and alternative hypotheses for the three forms of the random walk in Table 8.2.

In all three cases and DF regressions the null and alternative hypotheses are similar. If $\psi = 0$ is not rejected, we conclude that y_t contains a unit root (it is not stationary). This is the

Table 8.2 The Dickey-Fuller test

		Hypotheses	
Case (i): No intercept (drift), no time trend ($\alpha = 0, \gamma = 0$)			
Random walk	$y_t = \varphi y_{t-1} + u_t$	$H_0 : \varphi = 1$	$H_1 : \varphi < 1$
DF regression	$\Delta y_t = \psi y_{t-1} + u_t$	$H_0 : \psi = 0$	$H_1 : \psi < 0$
Case (ii): Intercept, no time trend ($\alpha \neq 0, \gamma = 0$)			
Random walk	$y_t = \alpha + \varphi y_{t-1} + u_t$	$H_0 : \varphi = 1$	$H_1 : \varphi < 1$
DF regression	$\Delta y_t = \alpha + \psi y_{t-1} + u_t$	$H_0 : \psi = 0$	$H_1 : \psi < 0$
Case (iii): Intercept, time trend ($\alpha \neq 0, \gamma \neq 0$)			
Random walk	$y_t = \alpha + \varphi y_{t-1} + \gamma t + u_t$	$H_0 : \varphi = 1$	$H_1 : \varphi < 1$
DF regression	$\Delta y_t = \alpha + \psi y_{t-1} + \gamma t + u_t$	$H_0 : \psi = 0$	$H_1 : \psi < 0$

familiar *t*-test for $\beta = 0$ in regression analysis. We test for $\psi = 0$ in a similar way. The test statistic therefore is:

$$DF\ test\ statistic = \frac{\hat{\psi} - 0}{SE(\hat{\psi})} = \frac{\hat{\psi}}{SE(\hat{\psi})} \tag{8.30}$$

Following the convention in hypothesis testing, the calculated DF statistic is compared with a tabulated critical value. Dickey and Fuller (1979) shows that this test statistic does not follow a conventional *t* distribution under the null hypothesis of a unit root. They derive asymptotic results from a non-standard distribution. Critical values for the various versions of the test (inclusion of a constant, trend or both) and different sample sizes are derived from Monte Carlo simulations. Each version of the test has its own critical value that depends on whether there is a constant with or without a trend in the test equation. It also depends on the size of the sample. Since the DF-test statistic does not follow a *t* distribution, it is called the DF *tau* (τ) statistic. This is a one-sided test and the negative sign indicates that the rejection region is on the left. Examples of critical values for the tau statistic are given in Table 8.3.

We compare the calculated tau statistic with the critical values in Table 8.3. The null hypothesis of a unit root is rejected in favour of the stationary alternative in each case if the computed tau statistic is more negative than the table value at the chosen level of significance. The more negative the tau statistic, the stronger the sample evidence for rejecting the null hypothesis of a unit root in the time series. If the computed DF test statistic in a test regression with a constant but no trend estimated with 50 observations is −3.16 we reject the null hypothesis of unit root at the 5% level of significance (computed DF-statistic is more negative than the critical value −2.93) but not at the 1% level (computed DF tau statistic is less negative than −3.58). The estimated coefficient ($\hat{\psi}$) in equation (8.29) is statistically different from zero (negative) and therefore ($\hat{\varphi}$) in equation (8.28) is different from 1 (less than 1).

Econometric software are most likely to report critical values from subsequent studies. It is common to use McKinnon critical values (MacKinnon, 1996) as for example in EViews. The inference process remains similar. We reject the null of a unit root if the DF test values are more negative than McKinnon's critical values (or higher in absolute terms). Econometric packages will provide the corresponding *p*-values. That makes the decision process much easier as we do not have to find critical values. A *p*-value that lies below the chosen significance level of 10%, 5% or 1% suggests rejection of the null hypothesis.

Table 8.3 Critical values for the Dickey-Fuller test

Sample size	Constant but no trend		Constant and trend	
	5%	1%	5%	1%
T = 25	−3.00	−3.75	−3.60	−4.38
T = 50	−2.93	−3.58	−3.50	−4.15
T = 100	−2.89	−3.51	−3.45	−4.04
T = ∞	−2.86	−3.43	−3.41	−3.96

Source: Fuller (1976).

The augmented Dickey-Fuller test

The Dickey-Fuller test is valid if u_t in equation (8.29) is white noise. In particular, u_t will be autocorrelated if there was autocorrelation in the dependent variable of the regression (Δy_t). The solution is to *augment* the test using p lags of the dependent variable in order to make the residuals (u_t) independent (hence the augmented Dickey-Fuller test). The lag length should be chosen so that the residuals aren't serially correlated. The alternative augmented model in case (i) with no constant or trend is now written:

$$\Delta y_t = \psi y_{t-1} + \sum_{i=1}^{p} a_i \Delta y_{t-i} + u_t \qquad (8.31)$$

The same critical values from the DF tables or MacKinnon (1996) are used. The null and alternative hypotheses and the conditions for rejecting the null hypothesis are similar to the DF test. The optimal number of lags of the dependent variable in equation (8.30) is determined by one of the information criteria.

8.6.3   The Phillips-Perron (PP) test

Phillips and Perron (1988) have developed a more comprehensive set of tests for unit roots. The tests are similar to ADF tests, but they incorporate an automatic correction to the DF procedure to allow for autocorrelated and heteroscedastic residuals. The Phillips-Perron (PP) test regression with no constant or a trend is similar to the DF test regression:

$$\Delta y_t = \pi y_{t-1} + u_t \qquad (8.32)$$

u_t in (8.32) is serially autocorrelated. The hypothesis $\pi = 0$ is tested with the computation of two statistics. The calculation of the modified test statistics is complex and not presented here. We use the critical values suggested by Phillips-Perron (routinely reported by statistical software) to make inferences. An advantage of the PP approach is that we do not need to include lagged terms of Δy_t in the test regression. If the residuals of the ADF regression fail normality, autocorrelation or heteroscedasticity the PP approach should be adopted.

Under the null hypothesis that $\psi = 0$ (in equation (8.31)) or $\pi = 0$ (in equation (8.32)) the ADF tau and PP test statistics have the same asymptotic distribution and the tests should give similar results. The two tests are asymptotically equivalent. In finite (limited) samples though the results from these tests may differ owing to the different treatment of serial correlation in the test regressions. The PP test is not recommended when the autocorrelation in the residuals is strongly negative in small samples. On this issue, Schwert (1989) finds that if Δy_t is an ARMA process (ARMA models are studied in Chapter 10) with a large and negative moving average (MA) component, then the ADF and PP tests are severely distorted in finite samples. They reject the null of a unit root (reject an I(1) process) much too often when it is true in favour of an I(0) process. Schwert (1989) states that the PP tests are more size-distorted than the ADF tests. Perron and Ng (1996) modify the PP tests to mitigate this size related bias.

Both the ADF and PP tests have low power if φ is close to 1. The *power* of any test of *statistical* significance is *defined* as the probability that it will reject a false null hypothesis. That is they fail to distinguish between if $\varphi = 1$ or $\varphi = 0.95$, especially in small samples. Hence if the true generating process is:

$$y_t = 0.95 y_{t-1} + u_t \qquad (8.33)$$

ADF and PP tests may not reject the null hypothesis of a unit root. Both tests have low power against $I(0)$ as y_t is close to being an $I(1)$ process. The tests struggle to distinguish highly persistent stationary series (such as equation (8.33)) from non-stationary sequences. This is the so-called *near observation equivalence* (see Campbell and Perron, 1991).

8.6.4 ERS tests

A wrong inference can therefore be made if the root is close to unity. Tests proposed by Elliott *et al.* (1996) – often labelled as ERS – and Ng and Perron (2001) increase the power of the test against highly persistent alternatives (e.g. $\varphi = 0.95$ or $\varphi = 0.98$). ERS tests enhance the power of a unit root test by estimating the test regression using so-called fractional differencing. This is something like first differencing: $\Delta y_t = y_t - \alpha y_{t-1}$ with $0 < \alpha < 1$. We are familiar with first differencing when the value of $\alpha = 1$. Hence we get $\Delta y_t = y_t - y_{t-1}$. The *point optimal test* developed by ERS is superior to conventional unit root tests, when a time series has an unknown mean or a linear trend. In making inferences, the test statistic calculated is compared with the simulation based critical values estimated by Elliot *et al.* (1996). Both the Elliot *et al.* (1996) and Ng and Perron (2001) tests are customary tests in most econometric software or can be downloaded from the software's library.

8.6.5 Stationarity tests

One way to get around the problem of low statistical power – in that the unit root tests often cannot distinguish between true unit process and near unit root process – is to use a stationarity test to complement unit root tests. One such stationarity test is the KPSS test (Kwiatkowski *et al.*, 1992). The null hypothesis is that the series is stationarity. The alternative hypothesis is of non-stationarity. Again we skip the exposition of the KPSS specification and test statistic and focus on the interpretation of the test. The hypotheses of the KPSS test are stated in Table 8.4 along with the testable hypotheses of ADF and PP tests.

We can compare the results of the three tests to infer whether a series has a unit root. There are four possible outcomes, summarised in Table 8.5.

Table 8.4 Hypotheses for ADF, PP and KPSS tests

ADF/PP	*KPSS*
$H_0 : y_t \sim I(1)$ – y_t has a unit root	$H_0 : y_t \sim I(0)$ – y_t is stationary
$H_1 : y_t \sim I(0)$ – y_t is stationary	$H_1 : y_t \sim I(1)$ – y_t has a unit root

Table 8.5 Outcomes of ADF, PP and KPSS tests

ADF/PP H_0	*KPSS* H_0	*Comment*
Reject	Do not reject	Both tests point to stationarity
Reject	Reject	Inconclusive evidence from the two tests
Do not reject	Do not reject	Inconclusive evidence from the two tests
Do not reject	Reject	Both tests point to non-stationarity

The KPSS test also suffers from problems identified for unit root tests. Caner and Kilian (2001) have found small size distortions similar to those in ADF and PP tests.

The power of traditional tests diminishes as deterministic terms are added to the test regression. That is, tests that include a constant and a trend in the test regression have less power to differentiate between $I(1)$ and $I(0)$ series than tests that only include a constant in the test regression. This is another weakness that advancements in the unit root literature attempt to address. Ng and Perron (2001) among many authors offer modified tests to overcome the size and power shortcomings of traditional tests.

An analyst has the choice of a range of formal statistical tests to detect non-stationarity in time series. The array of ADF, PP and KPSS tests, despite their shortcomings as we described, have found wide applicability in empirical real estate research. With econometric software offering more advanced and efficient tests as a standard, their use will become greater.

- **Online note #8.3: Unit root tests and higher order integrated series**

In online note #8.4 we run the tests discussed in section 8.6 in EViews and we elaborate further on the outputs of these tests.

- **Online note #8.4: Unit root and stationarity tests practice, running the tests in EViews**

Online note #8.5 extends the discussion of unit root tests in the presence of structural breaks.

- **Online note #8.5: Structural breaks and unit roots**

8.7   Practical considerations in real estate analysis

In empirical work in real estate, the ADF and PP unit root tests remain the most commonly used and to a degree, the KPSS test. With software packages now offering more options more efficient tests are likely to be used. We often work with small samples in real estate and we should remember that these tests have low power and can be problematic when the number of datapoints is small. What we mean by small is that we do not have enough observations and cycles in the series. For example, 40 observations could be a satisfactory sample for annual data but a small sample of quarterly frequency. Sometimes we need to make a decision about the model specification based on theory. Suppose we model vacancy and we use employment growth (*empl*) as the explanatory variable. More specifically we estimate the following model with 35 annual observations:

$$vacancy_t = \alpha + \beta empl_t + u_t a$$

$$(8.34)$$

Further, assume we run ADF tests and get the following probabilities for the ADF test statistics:

vacancy: p-value $= 0.15$; *dvacancy*: p-value $= 0.02$ *empl*: p-value $= 0.08$

Since vacancy has a unit root we cannot run (8.34), we will opt out to run *dvacancy* on *empl*. The following situation might arise. In the *dvacancy* model *empl* may not be significant. Perhaps *dvacancy* has become too volatile (over differenced) now, or indeed employment growth determines the level of vacancy and not changes in vacancy. On the other hand, *empl*, is significant in the vacancy model when *vacancy* is the dependent variable. In such a case we can proceed with the estimation of equation (8.34). We have strong theoretical backing for this relationship. It cannot just be spurious. We appreciate and acknowledge of course that the final results may be biased and hence we need to take a close look at all diagnostics. After all, *vacancy* does not fail the stationarity test too badly and we have a workable model for vacancy. The non-rejection of the unit root null for *vacancy* may be the result of the small sample or indeed a structural break. In such situations consider how badly the series fails the unit root test. Strong stochastic processes will badly fail the test. Theoretical insight as to how the market works is always helpful. A good understanding of the market workings could explain for example persistence in shocks (see Voith and Crone, 1988).

In our example earlier, using retail rents in the UK, is an extension of the previous line of thinking. A single test (ADF) suggested the use of second differences of retail rents. Two other tests suggested the use of first differences. It is therefore useful to consult an array of tests given the low-size power of these tests (our sample is viewed small). Using the 2nd differences of real retail rents in our model may have resulted in difficulties to construct a rent model. A reason is the over-differencing of the rent series.

Further consider our results for retail capital values. The series can be modelled in levels, but we need to allow for the structural break in 2009Q4. We can include a dummy that takes the value 1 in 2009Q4 and zero otherwise for one-time break; or create a dummy that takes the value zero before the break and 1 from 2009Q4 to denote an intercept break. If we do not include the dummy, the effect of the break will be passed on to the residuals, which could be I(1). A unit root test on the residuals will require a dummy to allow for the known structural break.

8.8   Concluding remarks

The subject of stationarity and unit roots is of high relevance to empirical work in real estate involving time-series data. We highlighted the fact that non-stationary data in our model can be a source of misspecification and the so-called spurious regression. Having a good grasp of the concept of stationarity is useful to understand the dynamics of time-series data and interpret regression outputs. A high explanatory power of a model may reflect non-stationary variables.

The chapter gave an overview of testing procedures for unit roots and stationarity, which continually evolve. Of the many tests available it highlights the most prevalent tests in real estate research. Our examples are illustrated in EViews. However, the specification of the tests and interpretations of the outputs are relevant to carrying out the tests in other statistical packages.

There are instances in real estate when we will have to work with small samples of fewer than 40 observations. In small samples these tests may struggle to give reliable evidence on whether a series is stationary. We discuss some practical matters when the tests suggest to over-difference the series. The use of non-stationary series in a model may be dictated by the need to construct and estimate a model. In that case we have to make sure that the models are strongly supported by theory and do well in diagnostics tests.

Chapter 8 online resource

- EViews file: "ch8_eviews"
- Excel file: "ch8_excel"
- Chapter 8 accompanying notes

9 Forecast evaluation

9.1 Introduction

Of the uses of applied quantitative analysis and econometric modelling work, a main one is forecasting. There is a plethora of forecasting methodologies ranging from simple naïve methods and judgemental forecasts to general and complex models. It follows that forecast evaluation is necessary to assess how accurate forecasts produced by these methodologies are and to establish the features of the errors they make in order to provide guidance into the characteristics of likely errors. The literature on the subject of forecast evaluation is vast, and a profusion of tests exists to assess the forecast performance of statistical and other methodologies. Models and econometric methodologies are also many and diverse. Forecast evaluation tests and metrics provide a common denominator to assess the forecast accuracy of any of these models.

With forecasting being an integral part in real estate research, in this book we use the statistical models we study for the purpose of forecasting and evaluate their forecast accuracy. Forecasting models should be robust and reliable. Tests of forecast accuracy complement the diagnostics tests we performed in Chapter 7 and can further inform the selection of competing regression models. We may have situations in which a few contesting theory-consistent models for a real estate series have all passed a set of diagnostics tests. The relative forecasting ability of these models is an additional criterion we can use to make the final selection, especially if the objective of the empirical analysis is forecasting. We should also note that forecast evaluation extends beyond quantitative forecasting to cover judgmental forecasting.

The literature on forecast accuracy is large and expanding. It can get pretty technical too. In this chapter we present conventional forecast accuracy tests. These tests are commonly used in real estate to assess the forecasting ability of models and should be seen as the starting but essential step in forecast evaluation. The understanding of basic forecast testing procedures is important for the implementation of more complex tests of forecast adequacy.

Prior to presenting and applying the forecast evaluation tests, we outline the objectives of forecasting and forecasting approaches and discuss the sources of the forecast error. Forecast evaluation is conducted in different ways and through a variety of tests. The forecast objectives will determine to a degree the format and particular tests in the evaluation of forecasts. We illustrate the key aspects of the forecast assessment exercise with detailed examples drawn on the office rent models we estimated with Hong Kong data and a model of all property prices (capital values) for the US.

The examples illustrate different ways to construct the sample of the forecasts and forecast errors. We explain the concepts of *ex-post* and *ex-ante* forecasting and carry out perfect foresight and dynamic forecasts. Emphasis is placed on the forecast performance of a model in relation to a benchmark or naïve model and we highlight situations where a regression model is fitter to forecast the market.

This chapter is an adaptation of the online version. The latter extends the analysis to cover qualitative or judgmental forecasting, which is a key component in real estate forecasting. The need for judgmental forecasting, common errors and forecasting in practice are additional topics in the online chapter.

9.2   Objectives in real estate forecasting

Empirical models can be used to achieve a number of forecasting objectives. These objectives are not completely distinct; rather they are supplementary. The particular forecast goal can determine the empirical methodology to adopt. Objectives in real estate forecasting, especially for time-series data, include the following:

(i)   *Point forecasts*: The purpose of this objective is to forecast the exact value of a real estate market variable over a specific horizon. For example the volume of take up or capital value growth in the next four quarters or 3 years. Hence, the forecast output is an exact figure, say 180,000 sq m of take up activity or capital growth of 6.5% is predicted for next year.

(ii)   *Directional forecasts*: The goal of the analyst is to predict the direction of the movement of the target series. Would rent growth be more or less positive in the next period than the current period? Would net absorption be positive or negative in the coming quarters?

(iii)   *Turning point forecasting*: The aim in this case is to identify forthcoming turning points or make predictions about the probability of a turning point occurring. This goal is related to directional forecasting. For an individual variable, say prices, turning points may simply be thought of prices moving from phases of positive to negative growth and vice versa. Identifying turning points is a much more complex process and involves both simple rules and sophisticated methodologies (for a review of methods, see Chauvet and Piger, 2008; Krystalogianni *et al.*, 2004).

(iv)   *Forecast horizon*: The forecast horizon is another goal in forecasting. Is the interest in short- or long-term forecasts? In the private (direct) real estate market, short-term forecasting is mostly a year but could extend to over a year (four to eight quarters), with long-term forecasting referring to a 5-year period and occasionally longer. In the securitised market the short-run is of course much shorter, up to 6 months.

There are further goals an analyst can pursue in forecasting work, including the estimation of confidence intervals for the forecast and conducting scenario analysis.

(i)   *Confidence boundaries or intervals.* Based on the forecasts and analysis of forecast errors, a base case or central forecast can be accompanied by confidence boundaries. These boundaries aim to capture uncertainty surrounding the forecast. A forecast is an expected value and most likely it will over- or under-estimate future realised values. We can therefore state how much the forecast can vary and at what probability.

(ii)   *Scenario analysis.* This is the 'what-if' analysis in forecasting. The analyst studies the sensitivity of the forecast to the changing values of the drivers of the model. Sensitivity analysis is important for stress testing and estimating downside risks in the case of future major adverse market developments.

• *See online note #9.1 for additional discussion on forecast objectives*

9.3   Forecast approaches

Forecasting methods used in real estate do not differ from those in economics or other assets classes. They fall into two broad categories: *quantitative* and *qualitative* forecasting methods. In turn, these categories comprise many alternative methodologies and approaches. Figure 9.1 provides a high-level summary of forecast approaches that are used in real estate.

The left side of Figure 9.1 (quantitative methods) summarises econometric or statistical approaches that have been used to a different extent in the real estate field. It by no means exhausts the full set of approaches in empirical forecasting work in real estate. In this book we cover econometric and time-series techniques commonly applied in forecasting real estate markets, but there are many more.

The right side of Figure 9.1 brings a different dimension to forecasting. It draws attention to the qualitative approach to forecasting. These approaches may offer alternative options for the forecast goals we have set. There may be good reasons to consider judgmental forecasts on their own or alongside quantitative forecasts. The complementarity of these approaches should always be considered. However, the bottom line for all these approaches is that they should be subjected to forecast accuracy evaluation with regard to the forecast goals we pursue.

9.4   Sources of error in real estate forecasting

Forecasts represent an expected value for the series we forecast. Rarely will they be totally accurate, hence small or large errors are expected. Basic forecast assessment is based on the size and in general the behaviour of forecast errors. In this section we review the sources of the forecast error defined in this book as realization (or actual value) minus the forecast. The discussion has similarities with the analysis of the regression error we studied in Chapter 5.

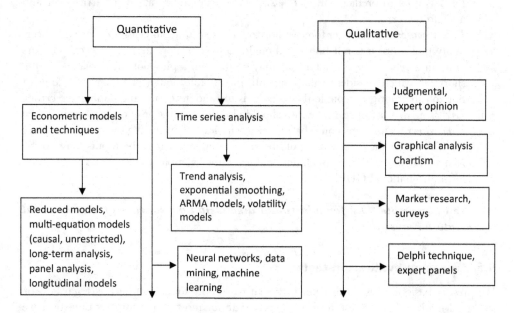

Figure 9.1 Summary of forecasting approaches

We noted in Chapter 5 that the error of a model reflects the goodness of the model, issues with data and small samples and random unpredictable events. A detailed account of the classification of model-based forecast errors can be found in Clements and Hendry (1998) and Diebold (1998). Based on the forecasting literature and taking into account the particular features of the real estate market and data, we categorise the sources responsible for the error in real estate forecasts. The discussion focuses on errors from quantitative analysis and later in the chapter we refer to the sources of error for judgmental (qualitative) forecasts.

(i)   *Poor specification of the model*: This situation could be the result of inadequate theory, lack of real estate market knowledge or adopting a model which is not robust.

(ii)   *Structural events*: These are major events that change the structure of a good model permanently, as we discussed in the previous chapter.

(iii) *Random events*: Unpredictable market circumstances in the forecast period will impact on the accuracy of the forecast even if they do not cause a structural break.

(iv)   *Data problems*: The data series we forecast may not be of good quality. Data issues, in terms of measurement and quality, are more likely to arise in the direct real estate market. Such data issues can affect the predictive ability of the model.

(v)   *Inaccurate input predictions*: When third-party forecasts are used for the future path of inputs in the model, our forecasts can be as good as these third-party forecasts are. This is a particular issue when we use inputs which have been obtained/forecast from auxiliary models. For example a model which has GDP growth as one of the independent variables will make a large error if the forecasts for GDP growth prove to be highly inaccurate.

(vi)   *Cumulative errors*: In some models the forecast process is such that the error is accumulating over time. This is particularly the case when we use forecasts as an input into the next stage in order to extend the forecast one step further. For example we use rents at time t in the prediction of rents at $t + 1$. The predicted value for rent at $t + 1$ is used to predict rent at $t + 2$. A large error made at $t + 1$ carries over and adds to the error at $t + 2$.

(vii) *Forecast horizon*: For a longer forecast horizon a less accurate forecast is expected. Uncertainty builds up. It is not unlikely, though, to have a more accurate forecast 5 years on than 3 years. The series forecast may go through a phase of high volatility in the short run which subsides in the long run. In general, though, the longer the forecast horizon the more susceptible the forecast is to random events and cumulative errors.

(viii) *Level of aggregation*: In real estate we should also think of the level of aggregation. The forecast errors for rents could be higher at city level than the regional or the country level. Series at the local level could be more volatile whereas the inputs may not be as accurate as for the regional/national level, for example accuracy and volatility of local and national GDP.

•   ***See online note #9.2 for additional discussion on sources of error in real estate forecasting***

9.5   Forecast evaluation tests

We present basic forecast evaluation tests that can apply to assess forecasts regardless of statistical methodology. These tests are appropriate for most of the forecast objectives we discussed earlier. An object of crucial importance in measuring forecast accuracy is the so-called *loss function* $L(A_{t+n}, F_{t+n,t})$ or $L(\hat{e}_{t+n,t})$, where A are the realizations or actual values of a series, F are the forecasts for the series, n is the forecast horizon and $\hat{e}$ is the forecast error defined as $A - F$. Hence $A_{t+n,t}$ is the realization of the series at time $t + n$ and $F_{t+n,t}$ is the forecast for time $t + n$ made at time t (n periods before) (see Diebold and Lopez, 1996; Diebold, 1993). In the forecasting literature there have been several measures which describe the loss function. These measures of forecast quality are grouped into a number of categories. In our treatment of forecast evaluation these are forecast bias, sign predictability, forecast accuracy with emphasis on large errors, forecast efficiency and encompassing and directional forecasting. The evaluation of the forecast performance on these measures takes place through the computation of appropriate statistics. These metrics represent different ways to quantify forecast errors and obtain useful information about their behaviour.

9.5.1 Mean error

Frequently the question arises as to whether there is systematic bias in a forecast, that is whether the forecast *consistently* over- or under-predicts the actual or realised values. A desirable property is that the forecast is not biased. That is, the forecasts on average are not consistently too high or too low compared to actual values. A measure of forecast bias is the mean error.

The mean error (ME) or mean forecast error (MFE) is defined as:

$$ME = \frac{1}{n}\sum_{i=1}^{n}\hat{e}_i \tag{9.1}$$

where n is the number of periods in the forecast (forecast horizon) and $\hat{e}_i$ is the estimated forecast error in the ith period.

We can run a significance test on the bias of a forecast. The null hypothesis is that the model produces forecasts and forecast errors ($\hat{e}_i$) with a zero mean: $E(\hat{e}_i) = 0$. The alternative is that the mean error can be negative or positive $E(\hat{e}_i) < 0$ or $E(\hat{e}_i) > 0$. A t-test can be calculated to test whether there is statistically significant negative or positive bias in the forecasts (see Davidson and MacKinnon (1993). This will require a large number of observations (errors) that should be independent and approximate the normal distribution.

9.5.2 Mean absolute error

Another important metric in forecasting is the mean absolute error (*MAE*), which is the average of the differences between the actual and forecast values in absolute terms. An error of -2% or $+2\%$ has a similar absolute value of 2%. The calculation of *MAE*, given in formula (9.2) is based on the absolute values of the errors $\left(|\hat{e}_i|\right)$.

$$MAE = \frac{1}{n}\sum_{i=1}^{n}|\hat{e}_i| \tag{9.2}$$

This is a true measure of the error of the forecast as positive and negative errors do not cancel each other out. Offsetting positive and negative errors would result in underestimating the size of the error and a far better forecast performance than actually exists. The smaller the size of *MAE*, the better the forecast.

Since both *ME* and *MAE* are scale-dependent measures, a variant test often reported is the mean absolute percent error (*MAPE*):

$$MAPE = \frac{1}{n}\sum_{i=1}^{n}|p\hat{e}_i| \quad \text{with} \quad p\hat{e}_i = \frac{A_i - F_i}{A_i}$$

(9.3)

The *MAPE* measure is more appropriate when the series under investigation is not expressed in percentage terms. For example the forecast error of a take-up model is market-size specific. In the Central London office market the error would be higher in actual terms than, say, the error in a smaller office market, for example Geneva. In Central London the take-up forecast error is likely to be in the hundreds of thousands of sq m per annum whereas in Geneva it is likely to be in the tens of thousands. By calculating the error in percentage terms (*MAPE*) we get a better idea of the relative error in the two markets. The take-up model makes an average error of 110,000s sq m or 6% in London, whereas the Geneva model an average error of 32,000 sq m or 8%. The smaller the *MAPE* value, the more accurate the forecast. If the series we forecast is price growth or expressed in percentage (e.g. vacancy), the *MAE* criterion is sufficient.

The preceding measures are used to assess how close individual variables track their corresponding real data and in particular the mean of the actual values. Another set of statistics is based on the variance of the forecast errors and is concerned with the ability of the model to forecast the variability of the actual data. We discuss the key statistics in this category next.

9.5.3   *Mean squared error (MSE) and root mean squared error (RMSE)*

The mean squared error, also termed the mean squared forecast error (*MSFE*), is a measure that sums up the square deviations of the forecast from their actual values. By squaring the deviations the *MSE* gives a higher weighting to larger errors than smaller ones. The objective of the forecaster is to assess the performance of the model or models with respect to large forecast errors made. Using the terminology earlier, the loss is expressed in a quadratic form. The *MSE* measures accuracy by the quadratic loss function.

$$MSE = \frac{1}{n}\sum_{i=1}^{n}\hat{e}_i^2$$

(9.4)

The *MSE* will have the units of the square of the data. In order to produce a statistic that is measured on the same units as the data series we forecast, the root mean squared error (RMSE) is proposed:

$$RMSE = \sqrt{MSE}$$

(9.5)

The *MSE* and *RMSE* measures have been popular methods to compute the volatility of the forecast around the actual values trajectory. The smaller the values of the *MSE* and *RMSE*, the more accurate the forecasts. Due to its similar scale with the dependent variable, the

RMSE of a forecast can be compared to the standard error of the model. An *RMSE* higher than say twice the standard error of the model is not indicative of a good forecast. The *RMSE* and *MSE* are useful when comparing different methods to the same set of data, but should not be used when comparing forecasts across data series that have different scales (see Chatfield, 1988; Armstrong and Collopy, 1992). The MSE can be decomposed into three parts – see online version.

We noted earlier that the *MSE* (and as a consequence, the *RMSE*) impose a penalty for large errors. The *RMSE* is a better performance criterion than measures such as *MAE* and *MAPE* when the studied variable undergoes fluctuations and turning points. If the forecast misses these fluctuations and turning points, the *RMSE* will heavily penalise the larger errors. If the variable follows a steadier trend, the evaluation could rely on the mean absolute error. It follows that the *RMSE* penalises disproportionately forecasts with a few large errors, relative to forecasts with a large number of small errors. This is an important matter in smaller samples we work with in real estate. A few large errors will produce higher *RMSE* and *MSE* statistics and may lead us to conclude that the model is less fit for forecasting. Since these measures are sensitive to outliers, some authors have recommended caution in their use in forecast accuracy evaluation (Armstrong, 2001).

Given that the *RMSE* is scale-dependent, the root mean squared percentage error can also be used (*RMSPE*):

$$RMSPE = \sqrt{\frac{1}{n}\sum_{i=1}^{n} p\hat{e}_i^2} \tag{9.6}$$

As per *MAE* versus *MAPE*, if the series we forecast is expressed in percentage terms, the *RMSE* is sufficient to compare forecasts.

9.5.4   Theil's U1 and U2 statistics

Theil's U1 statistic

Theil (1966, 1971) utilises the *RMSE* metric to propose an inequality coefficient that measures the difference between the predicted and actual values in terms of change. An appropriate scalar in the denominator restricts the value of the coefficient between zero and 1:

$$U1 = \frac{RMSE}{\sqrt{\frac{1}{n}\sum_{i=1}^{n} A_i^2} + \sqrt{\frac{1}{n}\sum_{i=1}^{n} F_i^2}} \tag{9.7}$$

Theil's $U1$ coefficient value ranges between zero and 1. The closer the computed $U1$ value is to zero, the better the prediction. Pindyck and Rubinfeld (1998) note that a value of less than 0.20 indicates a robust forecast. A variant of this test is given by Winker (2002). It is defined as the *RMSE* divided by the standard deviation of the variable we forecast. If it is less than 1, it means that the forecast error is smaller than the actual variation of the actual values – a desirable feature.

- ***Online note #9.3: Forecast bias decomposition of U1***

Theil's U2 statistic

The second coefficient by Theil, the $U2$ coefficient, assesses the contribution of the forecast against a naïve rule (such as 'no change', that its future values remain unchanged from the last available observed value) or an alternative model (naïve or otherwise). The $U2$ metric is defined as:

$$U2 = \left(\frac{MSE}{MSE^{ALT}}\right)^{1/2} \tag{9.8}$$

Another common formula for $U2$ is:

$$U2 = \sqrt{\frac{\sum_{t=1}^{n-1}\left(\dfrac{F_{t+1} - A_{t+1}}{A_t}\right)^2}{\sum_{t=1}^{n-1}\left(\dfrac{A_{t+1} - A_t}{A_t}\right)^2}} \tag{9.9}$$

where MSE^{ALT} is the MSE of the alternative model. Based on MSE, Theil's $U2$ statistic also measures accuracy by the quadratic loss criterion, hence the focus is on the variability of the forecast from the actual values. Its value will be lower than one if the model outperforms the naïve model (since the MSE of the naïve will be higher than the MSE of the model). If the naïve model produces a more accurate forecast the value of the $U2$ metric will be higher than one. Of course the naïve approach does not need to be the 'no change' extrapolation or a random walk (which we studied in Chapter 8) but other naïve methods such as an exponential smoothing and an autoregressive model. This criterion can be generalised in order to assess the contributions of an alternative model to a base or an existing model. Again, a $U2$ value of less than 1 indicates that the base model is doing better than the alternative model. Variants of Theil's $U2$ statistic can be found in Collopy and Armstrong (1992) and Thompson (1990).

- ***Online note #9.4: C-statistic***
- ***Online note #9.5: Optimal forecasting – tests for unbiased forecasts, forecast efficiency and forecast encompassing***

9.5.5 Directional forecast tests

Directional forecasts refer to the ability of the model or forecast methodology to predict the direction of future change. Many scholars have argued that solely studying the exactness of the error magnitude is not sufficient to assess forecast capacity (Blaskowitz and Herwartz, 2011). The loss function in the realised signs supplements the error magnitude (e.g. *MAE*) and quadratic loss functions (*MSE*, *RMSE*). In finance, directional forecasts are popular as the accuracy of forecasts using traditional criteria and the emphasis on the magnitude of the errors may give little guide to the profitability of trading strategies. The ability of the model to predict direction changes is important (Gerlow *et al.*, 1993; Leitch and Tanner, 1991). The assessment of the success of directional forecasts can simply be the percentage of correct predictions. More formally, the methodology used for evaluating the directional accuracy of the forecast that is whether the change in direction predicted by the model is the same as the actual one, is to estimate the probability of correct sign prediction (see Pons, 2000; Brooks and Tsolacos, 2000, 2001) using expression (9.10).

$$prob\left(\tilde{z}_i = 1\right) = \frac{\sum_{i=1}^{n} \theta\left[A_i * F_i > 0\right]}{n} \tag{9.10}$$

In (9.10), if $\tilde{z}_i = 1$, it indicates the success of correctly predicted signs (forecast and actual movement in the same direction); A_i is the actual change, and F_i is the forecast change; $\theta(\cdot)$ is a function where $\theta(\cdot) = 1$ if $A_i * F_i > 0$ (referring to the same sign, hence the same direction) and $\theta(\cdot) = 0$ if $A_i * F_i < 0$; and n is the forecast evaluation sample. Refenes (1995) and Pesaran and Timmerman (1992) have proposed other but related indicators to measure the percentage of correct sign predictions and the percentage of correct direction change predictions. In practice the estimation of (9.10) and other formulae is not as complicated as it appears. We illustrate the assessment of directional forecasts applying (9.10) with a detailed example in section 9.8.

9.6 Application of forecast evaluation tests – ex-post forecasts

The following three sections offer examples for the calculation of the forecast evaluation tests. We show different ways to create the sample of forecast errors. We calculate the tests and discuss practical aspects in forecast evaluation. We begin with the evaluation of the forecast performance of the Hong Kong office rent model (estimated in Chapter 6) with perfect foresight for inputs. In the next section we use US data to build a model of real estate prices and practice more tests.

9.6.1 Forecast evaluation of Hong Kong office rent growth model

(i) Forecast period and sample

A comprehensive forecast evaluation requires a large dataset of forecast errors for in-depth testing. This may not be feasible in several markets where we have limited data. In our Hong Kong rent model in Chapter 6, we have 33 annual observations. Even with such small samples, we can still perform a forecast evaluation although we are constrained to using basic tests. Understanding the application of basic tests with a simple example is essential, since the same steps are followed for larger sets of forecast errors.

A typical forecast period in the real estate (private) market is 5 years. We will use equation (6.26), now shown as (9.11) in this section to forecast real rents over this common horizon. With small samples, as in our case, it is common in empirical work and in practice to evaluate the forecast at the end of the sample. This allows sufficient observations for the estimation of the model and coefficients to settle. In order to have more forecast errors for the forecast evaluation, in our example we add another year and we do the evaluation over a 6-year period. Equation (9.11) was estimated over the sample period 1985–2017 (33 observations). We reserve the last six observations to perform the forecast evaluation. Hence we estimate (9.11) for the period 1985–2011 (27 observations) – that is equation (9.12) and forecast for the 6-year horizon 2012–2017, obtaining six forecast data points and errors. The section of the sample reserved to forecast and calculate forecast errors 2012–2017 is termed the *hold-out* period. Let us first look at the model estimations over the two samples as the structure of the model may have changed considerably.

Estimation in full sample period 1985–2017:

$$RREg_t = 18.3 - 2.4VAC_t + 1.4GDPg_{t-1} \qquad (9.11)$$
$$\quad\ \ [0.03]\ [0.00] \qquad [0.01]$$

$\bar{R}^2 = 0.48$; No of observations: 33

Estimation period 1985–2011:

$$RREg_t = 20.8 - 2.6VAC_t + 1.3GDPg_{t-1} \qquad (9.12)$$
$$\quad\ \ [0.05]\ [0.00] \qquad [0.05]$$

$\bar{R}^2 = 0.49$; No. of observations: 27

Estimating the model over the shorter period does not result in any major changes in the structure of the model, which is a good sign indicating model stability. The slopes in the two models are pretty similar – the sensitivity of *RREg* to *VAC* is somewhat stronger whereas the sensitivity of *RREg* to $GDPg_{t-1}$ is marginally lower. Recall from Chapter 7 on diagnostics that our rent growth model did not exhibit signs of parameter instability. The slope coefficients are statistically significant in the shorter sample period at the 5% level of significance. The explanatory power is 0.49 compared with 0.48 over the full sample period.

There are two key points to make relating to how we generate the forecasts for the period 2012 to 2017. First, will we use the actual values for the two regressors or forecasts we had at the time (that is in 2011)? Second, which coefficients should we use: the sub-sample (1985–2011) or full sample coefficients?

(ii)   *Perfect foresight in forecast evaluation*

Perfect foresight or information in forecasting refers to the use of the actual values of inputs in the forecast period. In our example, the forecast for *RREg* in 2012, 2013 and so on, we will use the realised values both of *VAC* and *GDPg*. Of course the actual values of *VAC* and *GDPg* are not known to us in 2011 when we make the 6-year forecast. We would have to use a forecast for *VAC* all the way through and a forecast for *GDPg* for 5 years since it is lagged 1 year in the model. This would have been a real-world forecast. By using the realised values of *VAC* and *GDPg*, it sounds as if we force the forecasts to be more accurate. The reason for using the actual values of the regressors and hence perfect foresight is a powerful one. By doing so we neutralise the error owing to inaccurate inputs. In the true real-world forecast, the error we would have made in the forecast period 2012–2017 would include errors in the forecasts for *VAC* and *GDPg*. If the forecast evaluation takes place with perfect foresight, the sources of the error originate in model specification and random effects. This is the reason for the popularity of perfect knowledge forecasts in forecast evaluation studies. Using the actual values of inputs in the hold out period is called an *ex-post* forecast. In general *ex-post* forecasts denote forecasts that are made with known input values during the forecasting period.

(iii)   Full sample or sub-sample coefficients

The second point relates to which model to use to generate the forecasts: equation (9.11) or (9.12)? That is, should we use the full or sub-sample coefficients? Again it sounds more realistic to use the model estimated with the smaller dataset (1985–2011). The other argument is that equation (9.12) will be used to forecast 2018, 2019, and so on, hence we are interested in the performance of the full-sample model. We would expect that forecasts based on coefficients (models) estimated over the full or a longer sample to win over forecasts from coefficients obtained over shorter samples. This is because in the former case, the model is trained with additional and more recent data and therefore the forecasts should be more accurate. However, it may not always be the case. If we use the full sample coefficients we obtain the fitted values we presented in the regression chapters earlier. This is an *in-sample* forecast. We calculate the forecasts for rent growth with both sets of coefficients. Table 9.1 contains the data and forecast calculations.

We consider the differences in the forecasts obtained from the two sets of coefficients rather small. This is not surprising since equations (9.11) and (9.12) do not differ much. A consistent pattern emerges, though. Forecasts obtained from the shorter model tend to be higher than the full sample model all along, but the difference between the two forecasts is no more than 1% in any year.

Table 9.2 shows the results of the forecast evaluation and calculation of the forecast metrics. It shows the calculations in detail for the full-sample coefficient forecasts and for comparison the corresponding values from the sub-sample model forecasts (shown in parentheses on the right-hand side of the forecast evaluation section). The Excel file for this chapter contains these calculations.

Looking at the metrics, we calculate a mean error of -1.6. Over the forecast horizon, the model on average predicts higher values than the realised *RREg*. The model, therefore, makes biased forecasts. In some years the error is virtually zero; in other years it is large as in 2014, when the forecast is too positive; whereas rent growth stalled resulting in an error of -7.3%. As we noted earlier, just a single large error (either positive or negative)

Table 9.1 Forecasts for Hong Kong office rent growth

	Data		*Forecast*	
	VAC (%)	*GDPg (%)*	*Full sample coefficients*	*Sub-sample coefficients*
2011	6.5	4.8		
2012	6.0	1.7	10.6	11.4
2013	7.0	3.1	3.9	4.8
2014	6.3	2.8	7.5	8.5
2015	8.0	2.4	3.0	3.6
2016	8.2	2.2	2.0	2.6
2017	9.5		-1.4	-1.0

Illustration of obtaining the forecast for 2012

Full sample coefficients:	$RREg_{2012} = 18.3 - 2.4 \times 6 + 1.4 \times 4.8 = 10.6$
Sub-sample coefficients:	$RREg_{2012} = 20.8 - 2.6 \times 6 + 1.3 \times 4.8 = 11.4$

Note: The forecasts differ from those using EViews due to rounding across the board.

Table 9.2 Forecast evaluation – full sample coefficients

	RREg, (%)		Errors (%)			A^2	F^2
	Actual (A)	Forecast (F)	Actual A – F	Absolute |A – F|	Squared (A – F)²		
(i)	(ii)	(iii)	(v)	(vii)	(ix)	(x)	(xi)
2011	*9.3*						
2012	6.5	10.6	−4.1	4.1	17.0	42.2	112.8
2013	3.9	3.9	0.0	0.0	0.0	15.0	15.1
2014	0.2	7.5	−7.3	7.3	53.3	0.0	56.6
2015	3.0	3.0	0.0	0.0	0.0	9.1	9.1
2016	0.1	2.0	−1.9	1.9	3.6	0.0	3.9
2017	2.6	−1.4	4.0	4.0	16.1	6.7	2.0
Sum			−9.3	17.3	89.9	73.1	199.4

Forecast evaluation

Metric:	Full sample coefficients	Sub-sample
Mean Forecast Error (ME)	−9.3/6 = −1.6	−2.3
Mean absolute error (MAE)	17.3/6 = 2.9	3.5
Mean squared error (MSE)	89.9/6 = 15.0	18.8
Root mean squared error (RMSE)	$\sqrt{15.0} = 3.9$	4.3
Theil's U1 inequality coefficient	$\dfrac{3.9}{\sqrt{73.1/6} + \sqrt{199.4/6}} = 0.42$	0.44

Note: The forecast evaluation results may differ from calculations in EViews due to rounding. These calculations are in the file 'Ch9_excel' and 'forecast evaluation' worksheet.

in a small sample can cause bias in the forecast. The mean absolute error is 2.9%. Since the unit of measurement is percentage we do not calculate the *MAPE*. The 6-year forecast obtained from equation (9.12) shows similar bias and a slightly higher error (the calculations are included in the chapter's Excel file).

The average of the square error values reported in column (ix) gives us the *MSE*. The value 15% is not informative to evaluate the variability of the forecast error. The *MSE* is a useful statistic to compare forecasts from alternative models and for the estimation of other statistics. The full sample model based forecast has a lower *MSE* than the forecast from the sub-sample model. Hence the variability of the forecast from the actual values is smaller for the former than the latter.

The *RMSE* value 3.9 can be compared to the standard error of the regression. Equation (9.11) has a standard error of 11.4 (full sample). It is a good sign that the *RMSE* value is lower than the standard error. As noted earlier, it is the convention to consider an *RMSE* value two times or over the standard error of the model a sign of a weak forecast. Since the *MSE* and *RMSE* values are mainly used to compare competing forecast models, we employ the scaled measure suggested by Theil (*U1* statistic) to further evaluate the model with respect to the quadratic loss function. The value of 0.42 does not indicate a strong forecast ability for (9.11), rather a moderate one. We mentioned earlier that a value of 0.20 or less would signify a good forecast. This finding is not surprising. Over the 6-step (year) horizon the model has made a large error in 3 out of 6 years that is penalised by the *MSE* and *U*1 metrics. The *U*1 statistic for the sub-sample forecasts indicates a marginally worse performance (0.44).

Let us keep in mind that with smaller forecast error samples (in our case consisting of just six observations) a large error in a particular period will affect significantly the tests that are based on the variance of the error, such as the *MSE* and the resulting *RMSE* and Theil's *U*1 coefficients.

(iv)　Comparison with benchmark or naïve forecasts

The forecast adequacy of the model can be illustrated with reference to a *naïve* model or a simpler methodology. It is also called the *benchmark* model. A naïve forecast could be the long-run average growth rate or the average growth rate in the 6 years prior to the year our forecast begins or last period's value. You can define the naïve forecast in any plausible way. Or one can make use of exponential smoothing methods available in all statistical packages and in Excel. We take the naïve forecast to be the value of *RREg* the previous year (this is the random walk model presented in Chapter 8). The relative performance of the two forecasts is assessed with Theil's *U*2 statistic. Table 9.3 shows the naïve forecast and the output of the forecast evaluation criteria for both sets of forecasts. The full calculations are in the chapter's accompanying Excel file.

On all criteria, with the exception of *MAE*, the naïve forecast outperforms the model-based forecast. We observe that the naïve forecast errors are less volatile or track the volatility of the actual series more closely. This is indicated by the lower values of *MSE* and *RMSE* for the naïve forecasts. The model-based forecasts achieved zero errors in a couple of years but there was a quite large error in 2014. The naïve forecast never manages to produce an error close to zero, but yet no large error is made. We have a small forecast sample: the large error the model makes works to its disadvantage.

These results should be seen to be specific to forecast horizon used. For example, we could replicate the test in another 6-year period. In the next section we elaborate further on the comparison with naïve models and show situations in which the naïve model underperforms the regression model. In any case, comparisons with naïve or benchmark methodologies is part of evaluating the forecast performance of any forecast methodology.

- **Online note #9.6 provides guidance to carry out and interpret forecast evaluation outputs in EViews**

Table 9.3 Model forecast versus naïve forecast

	Actual	Naïve Forecast	Error (naïve)	Forecast evaluation		
2011	9.3				Equation (9.11)	Naïve
2012	6.5	9.3	−2.8	*ME*	−1.6	−1.1
2013	3.9	6.5	−2.6	*MAE*	2.9	2.9
2014	0.2	3.9	−3.7	*MSE*	15.0	8.5
2015	3.0	0.2	2.8	*RMSE*	3.9	2.9
2016	0.1	3.0	−2.9	*U*1	0.42	0.34
2017	2.6	0.1	2.5	*U*2	1.33	

Table 9.4 Sensitivity of forecast evaluation to forecast period

	Actual	Forecast	Evaluation for forecast horizon:		
2011	9.3	11.9		2012–2017	2011–2016
2012	6.5	10.3	ME	−2.3	−2.4
2013	3.9	3.6	MAE	3.5	2.6
2014	0.2	7.3	MSE	15.8	12.3
2015	3.0	2.7	RMSE	4.3	3.5
2016	0.1	1.7	U1	0.44	0.28

In Table 9.4 we present another set of forecasts to draw attention to an important matter in forecast evaluation. The forecast evaluation results can be specific to the forecast period, in particular when the forecast period is short. A more complete analysis would require repeating the forecast evaluation over a different 6-year period(s), overlapping or discrete. The small sample of course limits the potential of this task. For example, evaluating the model forecast in the previous 6-year period 2006 to 2011 would leave us with 21 observations to estimate the model. Still though, we have some flexibility and we perform the forecast evaluation for the period 2011 to 2016. We estimate equation (9.11) with 26 observations to 2010. Table 9.4 shows the forecast and compares the evaluation metrics for the two periods.

The forecast is marginally more biased in the same direction in the period 2011–2016. The rest of the metrics though show a better performance with *MAE*, *MSE* and *RMSE* all taking smaller values. Theil's *U*1 statistic is notably lower, suggesting that the model has definitely moderate to good forecasting power. The second forecast period we selected is overlapping with the first one and only 1 year is different. Still the evaluation criteria showed sensitivity. With a larger sample, as we study in the next section, we will be able to repeat the forecast evaluation and get a good idea of the characteristics of the forecast error.

- **Online note #9.7: Further practice: Evaluation of rent level forecasts form different specifications**

9.7   Dynamic *ex-ante* forecasts and further testing – US property prices

In this section we show more practices to generate a sample of forecasts and forecast errors for evaluation. It is appropriate to work with a larger sample of data that will allow us to have a bigger set of forecasts to assess with additional tests. We also relax the assumption of perfect knowledge about the future values of inputs. Hence we will evaluate 'true' or 'real' world forecasts, also termed ex-ante forecasts, when we forecast beyond the sample period. Finally, this section illustrates how forecast assessment can feed back into the original model and necessitate adjustments.

9.7.1   The forecast model

We present the applications with a forecast model for property prices in the US. The data are drawn from RCA's database and its CPPI all property price index. From the monthly database we construct a quarterly series of all property capital growth (by taking the last month of the quarter as that quarter's value). The sample contains 72 observations (1Q2001 to 4Q2018). We adjust the data for inflation and construct a model of all property real prices using variables with leading properties, that is a series which precedes the movements in all property prices. Apparently such series are insightful for forecasting purposes. It means that a forecast for property prices a few periods ahead is obtained from current information and data for the leading variables. These variables can be macroeconomic, financial, survey data or other. Such series are interest rate spreads, the leading economic index for the US economy produced by the Federal Reserve Bank of St. Louis and similar series from the OECD and the Conference Board, ISM surveys on business and many more. These series are considered a precursor of developments in the economy. To the extent that such series lead the US economy, they could be useful for the real estate market too.

An initial investigation through cross-correlations shows that many of these series lead capital value growth, which makes them good candidate variables to forecast capital values. The objective in this section is not to come up with a complete and best model for all property capital values in the US using such variables but to illustrate rolling dynamic forecasts in larger samples. For this purpose we select the TED spread, a series the markets watch closely, as the variable in our forecasting regression model. The TED spread is the difference between the interest rate offered in the London interbank market for 3-month loans denominated in US dollars and the 3-month Treasury Bill rate. A high spread is an indication of strain in the financial markets and will impact on economic sentiment. The TED spread possesses strong leading properties for the US economy and real estate market. Table 9.5 shows the cross correlations of real price growth for all property in the US (*CPPIALLg*) and the TED spread (*TED*).

Cross correlations reveal a strong relationship between real price growth in all property capital values and the TED spread. The strongest correlation is at lag three. Price growth is also strongly correlated with TED at lags two and four. The negative sign is expected. A high TED spread indicates higher financial and economic risks that the real estate market is likely to reflect subsequently through say impact on sentiment and risk premia. Hence a higher spread has an adverse impact on real estate prices. Based on the cross-correlation results and some initial investigations we estimate equation (9.13).

$$CPPIALLg_t = 3.00 - 1.78TED_{t-2} - 1.850.00TED_{t-3} - 1.760.00TED_{t-4} \qquad (9.13)$$
$$(0.00)\ (0.00)\qquad\quad (0.00)\qquad\qquad\quad (0.00)$$

Adj. $\bar{R}^2 = 0.72$ DW = 0.82; Sample: 1Q2001–4Q2018 (72 obs); p-values in parentheses

Table 9.5 Correlation patterns for all property price growth and TED spread

Cross correlations			
CPPIALLg at time t and TED at t−i/t+i			
0	−0.41	0	−0.41
$t-1$	−0.50	$t+1$	−0.30
$t-2$	−0.72	$t+2$	−0.18
$t-3$	**−0.76**	$t+3$	−0.13
$t-4$	−0.71	$t+4$	0.01
$t-5$	−0.64	$t+5$	0.13
$t-6$	−0.59	$t+6$	0.20

Equation (9.13) is a model for *CPPIALLg$_t$* containing the strongest cross correlations with *TED*, which are all signed as expected and are statistically significant (*p*-values shown in brackets). There is evidence of first order serial correlation in this model, partly attributable to the high autocorrelation of *CPPIALLg*. For the purposes of this section, we do not attempt to resolve the issue and proceed to assess the forecast performance of this model.

9.7.2   Sample of rolling forecasts and evaluations

As we have a larger data sample than before, a forecast objective is to evaluate how well the equation (9.13) predicts at different horizons, for example say one, two, three, four or more quarters ahead. These forecasts can be perfect foresight forecast or real *out-of-sample* or *dynamic* forecasts. What we mean by out-of-sample forecasts is that we do not use any future information at the time of forecast, just data available at that time. This replicates a real world forecast. These dynamic forecasts are widely known as ex-ante forecasts. In general, ex-ante forecasts are those that use whatever information is available at the time of the forecast. Although ex-post forecasts are useful for exploring the accuracy of forecasting methodologies, it is ex-ante forecasts that are imperative.

Given the lag structure of equation (9.13) we can make predictions for *CPPIALLg* for one and two quarters without the need for future values of the TED spread. We initially illustrate the forecast performance of the model for one-step-ahead predictions. We generate these forecasts on a rolling basis over a 4-year period (16 quarters). More specifically, we estimate the model in the sample period 2001Q1 to 2014Q4 and forecast all property price growth for 2015Q1. Hence unlike the earlier example we do not use full sample coefficients. The forecast for 2015Q1 is our first one-step-ahead forecast. Now we roll forward the estimation of the model. We estimate the model for the period 2001Q1 to 2015Q1 and obtain the forecast for 2015Q2, which is our second one-step ahead forecast data point. We repeat the process until we exhaust the sample that is estimating the model for 2001Q1–2018Q3 and forecasting for 2018Q4. In this way we will get 16 one-quarter-ahead predictions. It is worth noting that all three terms in equation (9.13) retain their significance in all rolling estimations.

Table 9.6 presents the forecasts and their evaluation. We evaluate two sets of forecasts, the model based (equation (9.13)) forecast and a naïve forecast. The naïve forecast is the 'no change' from the previous quarter (random walk model). For example the forecast for 2Q2015 is the actual value in 1Q2015. Equation (9.13) does not outperform the naïve model. The forecasts of the former have a higher *MAE*. The average error is 0.66% in the model based forecast and 0.52% in the naïve forecasts. The forecast of the naïve method also have a lower *MSE* and *RMSE*. The *U*1 statistic points to a moderate forecast power for the model and a good forecast ability of the naïve method.

We present two other statistics which we use to compare the relative accuracy of forecasting models. Theil's *U*2 statistic is higher than one suggesting that the naïve forecast is superior to the model's forecast. Again we should note that these results are unique to the forecast period selected and the forecast horizon of one quarter. We look at this possibility next.

We examine the success of the TED spread for forecasting all property price growth over a longer horizon, which we take to be three quarters. Lead information of three quarters or 9 months is important in the real estate market. It gives a good window of time for portfolio

Table 9.6 Forecast evaluation for one-quarter-ahead forecasts for US all property price growth

	Actual (%)	Forecast model (%)	Error (%)	Forecast naïve (%)	Error (%)
1Q2015	3.15	1.78	1.37	3.78	−0.63
2Q2015	1.42	1.74	−0.32	3.15	−1.73
3Q2015	1.81	1.65	0.16	1.42	0.39
4Q2015	1.89	1.56	0.33	1.81	0.08
1Q2016	2.15	1.35	0.80	1.89	0.26
2Q2016	1.68	0.99	0.69	2.15	−0.48
3Q2016	1.71	0.73	0.98	1.68	0.03
4Q2016	0.71	0.66	0.05	1.71	−0.99
1Q2017	1.07	0.45	0.61	0.71	0.35
2Q2017	2.34	0.30	2.04	1.07	1.27
3Q2017	1.58	0.33	1.25	2.34	−0.76
4Q2017	1.11	0.85	0.26	1.58	−0.46
1Q2018	0.94	1.24	−0.30	1.11	−0.17
2Q2018	1.06	1.37	−0.31	0.94	0.12
3Q2018	1.15	0.79	0.36	1.06	0.10
4Q2018	0.92	0.49	0.43	1.15	−0.24

Evaluation	Model based	Naive
ME	0.53	−0.18
MAE	0.66	0.52
MSE	0.71	0.49
RMSE	0.84	0.70
U1	0.30	0.20
U2	1.20	

adjustments. Equation (9.13) is not suitable for a dynamic out-of-sample forecast of *CPPI-ALLg* three quarters ahead due to the TED lagged two quarters (TED_{t-2}) unless we predicate the value of TED for one quarter into the future in every round. We keep the analysis simple though and instead we estimate the model with the third and fourth lag of TED. After all we are testing the predictive ability of this indicator for different horizons. The estimated equation over the whole sample period is (9.14).

$$CPPIALLg_t = 2.67 - 2.68TED_{t-3} - 2.18TED_{t-4}$$

$$\quad\quad (0.00) \quad (0.00) \quad\quad\quad (0.00)$$

(9.14)

Adj. $\bar{R}^2 = 0.66$; DW = 1.11; Sample: 1Q2001–4Q2018 (72 obs); *p*-values in parentheses

We repeat the process to create the three-quarter-ahead forecasts and present the results in Table 9.7. For the first three-quarter-ahead forecast we estimate equation (9.14) to 2014Q2 and forecast three quarters ahead to 2015Q1 to obtain 1.62 (1.62%). Subsequently the model is estimated to 2014Q3 and we obtain the forecast for 2015Q2 (1.55%). In these

Table 9.7 Forecast evaluation for three-quarter-ahead forecasts for US all property price growth

	Actual	Forecast Model	Error (Model)	Forecast naïve	Error (Naïve)
	(%)				
2015Q1	3.15	1.62	1.53	2.14	1.01
2015Q2	1.42	1.55	−0.13	2.22	−0.80
2015Q3	1.81	1.54	0.27	2.65	−0.84
2015Q4	1.89	1.52	0.37	2.97	−1.08
2016Q1	2.15	1.39	0.76	2.78	−0.62
2016Q2	1.68	1.15	0.53	2.54	−0.87
2016Q3	1.71	0.70	1.00	2.07	−0.36
2016Q4	0.71	0.56	0.15	1.82	−1.11
2017Q1	1.07	0.72	0.35	1.88	−0.82
2017Q2	2.34	0.31	2.03	1.86	0.48
2017Q3	1.58	0.11	1.46	1.56	0.01
2017Q4	1.11	0.54	0.57	1.29	−0.18
2018Q1	0.94	1.09	−0.15	1.46	−0.52
2018Q2	1.06	1.35		1.42	
2018Q3	1.15	1.27		1.52	
2018Q4	0.92	0.42		1.49	
Evaluation					
ME		0.56		−0.43	
MAE		0.65		0.63	
MSE		0.75		0.50	
RMSE		0.87		0.71	
U1		0.31		0.19	
U2			1.23		

rolling estimations both TED_{t-3} and TED_{t-4} remain statistically significant all along. We also include a naïve forecast for comparison. We can use alternative naïve forecasts. Such a naïve forecast is to take the long-term average up to the time of the forecast. When we forecast for 2015Q1, our last actual observation is for 2014Q2. As the naïve forecast we can take the average over the period 2001Q1 to 2014Q2. The naïve forecast for 2015Q2 is the average for the period 2001Q1 to 2014Q3 and so forth. When evaluating the forecast the model outperforms the naïve (no change) forecast (results not shown in Table 9.7 but the chapter's Excel file contains the calculations). We also define the naïve forecast as the average of the past four quarters up to the time when we make the forecast. Hence the forecast for 2015Q1 is the average of the four quarter period 2013Q3 to 2014Q2 (that is 2.14 in Table 9.7).

The evaluation shows that the model under-predicts (0.56%) whereas the naïve model over-predicts (−0.43%). Both are biased, but they show different bias. The mean absolute error is pretty similar, however the *MSE* of the naïve model is lower than that of the regression model, 0.50 and 0.75, respectively. This is reflected in the *RMSE* (lower value for naïve forecast) and *U*1 statistic which favours the naïve model. Theil's *U*2 statistic takes a value above 1, hence the naïve forecast is better.

9.7.3   Implications for forecast comparisons of shifting trends

We should highlight the fact that the forecast evaluation presented in Table 9.7 takes place at a time of rising property values and persistent positive growth rates. The naïve method and other methodologies such as exponential smoothing, may capture that trend and owe their success to the rising trend in all property prices and ongoing positive growth. Further, *CPPIALLg* is strongly autocorrelated, hence the naïve method using past values when values are rising could be responsible for the good performance of the naïve model. Still, at times when there is greater volatility and in particular turning points equation (9.14) can generate signals about changing trends faster.

  We check this line of reasoning by repeating the forecast evaluation exercise for a different sample period. The sample period we select is the period that followed the collapse of Lehman Brothers in September 2008. We use the model with the TED spread to predict all property prices in the aftermath of the shock. The forecast is again a dynamic forecast three quarters ahead. We run equation (9.14) using data to 2008Q3 (the date of the event) and we forecast for three quarters to 2009Q2 seeking signals on market recovery (positive price growth). When we estimate (9.14) in this shorter period the term TED_{t-4} loses its significance. Hence at that time we would not include TED_{t-4} in our model. The equation is now reduced to equation (9.15). This is the model which we estimate on a rolling basis and obtain the three-step-ahead forecasts post the Lehman's collapse and into the global financial crisis.

$$CPPIALLg_t = 2.48 - 4.23TED_{t-3} \qquad\qquad\qquad\qquad (9.15)$$
$$\quad\;\; (0.00)\;\; (0.00)$$

  Adj. $\bar{R}^2$ = 0.45; DW = 0.94; Sample 2001Q1–2008Q3 (31 observations); *p*-values in parentheses.

We also use the previous naïve methodology (average of the four quarterly actual values to the date of the forecast) to get the naïve forecast. Hence the naïve forecast for 2009Q2 at the time of the forecast (2008Q3) is the average real price growth in the four-quarter period 2007Q4 to 2008Q3. Table 9.8 presents the results.

  Real price growth for all property in the US was negative until mid-2010 (2010Q2). The naïve forecast still predicts negative growth to the end of 2011. The forecast model using the TED spread (equation (9.15)) predicts positive growth through 2010Q1 (remember this forecast was made in 2009Q3). The model is wrong but not eventually. It predicts the positive growth about 6 months in advance of happening. In this sample, the model over-predicts whereas the naïve model under-predicts (forecasts more negative than actual values). The mean absolute value for the error of the naïve model is nearly twice that of the regression model.

  Similarly the *MSE* and *RMSE* values favour the regression model. This is illustrated by the *U*2 statistic which is lower than 1, hence the regression model is the preferred one. This example illustrates that when there is a turn in the trend of a series the naïve methods are likely to struggle to forecast accurately. External information such as the leading indicators

Table 9.8 Three-step ahead forecast evaluation post Lehman's event

	Actual	Forecast Model	Forecast naive
2009Q2	−7.98	−10.84	−4.05
2009Q3	−6.35	−2.85	−3.55
2009Q4	−4.30	−1.92	−4.47
2010Q1	−2.17	0.63	−5.29
2010Q2	−0.50	1.54	−5.81
2010Q3	0.28	1.35	−6.42
2010Q4	0.03	1.48	−5.20
2011Q1	−0.31	0.56	−3.33
2011Q2	0.56	1.36	−1.67
2011Q3	1.63	1.12	−0.59
2011Q4	1.77	0.98	−0.13
Evaluation			
ME		−1.16	2.13
MAE		1.83	3.47
MSE25		4.30	15.26
RMSE		2.07	3.91
U1		0.29	0.48
U2		0.53	

we referred to earlier, economic variables and other variables are appropriate to signal phase changes in the variable we predict.

- ***Online note #9.8: Calculation of the C-statistic***
- ***Online note #9.9: Practicing unbiasedness, efficiency and encompassing in forecast evaluation***

9.8　Directional forecast evaluation

The final task is an assessment of directional accuracy. We perform the tests over the two sample periods we forecast three steps ahead, 2015Q1–2018Q4 and 2009Q2–2011Q4. In our case if we define the directional change as a shift from positive growth to negative growth and vice versa the calculations are straightforward and details are given in Table 9.9 for the period 2009Q2–2011Q4.

The z_t variable takes the value 1 if the actual and forecast values are both positive or both negative, suggesting similar direction. The regression model correctly predicts 73% of the direction correctly compared with 55% of the naïve over the specific period. If we reiterate the analysis for the forecast sample period 2015Q1 to 2018Q4, both the model-based and naïve method forecasts have a success rate of 100%. This owes to the fact in this period all property prices showed uninterrupted positive growth.

Table 9.9 Evaluation of directional forecasts

	Actual	Forecast Model		z_t	Forecast naive		z_t
2009Q2	−7.98	−10.84	(−7.98) × (−10.84) = 86.5	1	−4.05	(−7.98) × (−4.05) = 32.3	1
2009Q3	−6.35	−2.85	(−7.35) × (−2.85) = 18.1	1	−3.55	22.5	1
2009Q4	−4.30	−1.92	8.3	1	−4.47	19.2	1
2010Q1	−2.17	0.63	−1.4	0	−5.29	11.5	1
2010Q2	−0.50	1.54	−0.8	0	−5.81	2.9	1
2010Q3	0.28	1.35	0.4	1	−6.42	−1.8	0
2010Q4	0.03	1.48	0.0	1	−5.20	−0.1	0
2011Q1	−0.31	0.56	−0.2	0	−3.33	1.0	1
2011Q2	0.56	1.36	0.8	1	−1.67	−0.9	0
2011Q3	1.63	1.12	1.8	1	−0.59	−1.0	0
2011Q4	1.77	0.98	1.7	1	−0.13	−0.2	0
Sum z_t				8			6
n (forecast sample				11			11
% correct predictions				8/11 = 73%			6/11 = 55%

We now define the directional change in a different way. We are interested in whether price growth will be stronger or weaker three quarters forward compared with the current period. We make the forecast for 2015Q1 three quarters earlier in 2014Q2 and examine whether our prediction for 2015Q1 is above or below 2.2%, the known growth in real prices in 2014Q2 (the time of the forecast). We show the calculations in Table 9.10.

Predicting the rate of growth of prices proves more challenging. For the first forecast data point both forecast methodologies predict a slower rate of growth than 2.2%, the naïve method just less. The calculations in Table 9.10 are self-explanatory and similar to those in Table 9.9. Interestingly, the correctly predicted growth rate is 69% for both forecast approaches.

In practice, the interest is more on forecasting negative and positive growth in rents and prices. When we predict variables such as yields/cap rates, it is important to know the direction a number of periods ahead compared to the current period. For example, we would like to know how successful a model is in predicting the direction in the yield four quarters ahead. This is important as lower yields result in capital gains, whereas if higher yields are predicted we should expect a hit on capital growth. Sign and direction is valuable in portfolio optimisation and management and as an input into market timing decisions. Numerous empirical studies focus on whether signs and directions can be predicted, which methodologies outperform and assess the forecast success (see Wilkens *et al.* (2005), Sinclair *et al.* (2010), Pönkä (2016, 2014), and Nyberg and Pönkä (2016)).

- **Online note #9.10: Selected forecast accuracy studies in real estate**

Table 9.10 Evaluation of directional forecasts for rate of growth

	Actual	Model						Naive	
		Forecast	$A_{t+3} - A_t$	$F_{t+3} - A_t$	$A_{t+3} - A_t \times F_{t+3} - A_t$	z_t		Forecast	z_t
2Q14	2.20								
3Q14	2.75								
4Q14	3.78								
1Q15	3.15	1.62	3.15−2.20 = 0.95	1.62−2.20 = −0.59	−0.55	0		2.14	0
2Q15	1.42	1.55	−1.33	−1.20	1.60	1		2.22	1
3Q15	1.81	1.54	−1.97	−2.24	4.41	1		2.65	1
4Q15	1.89	1.52	−1.26	−1.63	2.05	1		2.97	1
1Q16	2.15	1.39	0.73	−0.03	−0.02	0		2.78	1
2Q16	1.68	1.15	−0.14	−0.67	0.09	1		2.54	0
3Q16	1.71	0.70	−0.19	−1.19	0.22	1		2.07	0
4Q16	0.71	0.56	−1.44	−1.59	2.29	1		1.82	1
1Q17	1.07	0.72	−0.61	−0.96	0.58	1		1.88	0
2Q17	2.34	0.31	0.63	−1.40	−0.88	0		1.86	1
3Q17	1.58	0.11	0.86	−0.60	−0.52	0		1.56	1
4Q17	1.11	0.54	0.05	−0.52	−0.02	0		1.29	1
1Q18	0.94	1.09	−1.40	−1.25	1.75	1		1.46	1
2Q18	1.06	1.35	−0.52	−0.23	0.12	1		1.42	1
3Q18	1.15	1.27	0.04	0.16	0.01	1		1.52	1
4Q18	0.92	0.42	−0.02	−0.52	0.01	1		1.49	0
Sum z_t						11			11
n						16			16
% correct predictions						69%			69%

9.9   Qualitative forecasts and real estate forecasting in practice

Qualitative forecasts, which include judgmental views (see Figure 9.1), represent another approach to forecasting that has become popular across industries (see Lawrence *et al.*, 2006). In the online notes for this chapter we focus on judgmental forecasts usually offered by experts and individuals with significant market experience. Judgmental forecasting is common in real estate and it can be combined with quantitative or model based forecasts. Brooks and Tsolacos (2010, chapter 13) look into this topic in considerable detail. This section in the online version of the chapter draws upon their discussion. It covers justification for judgemental forecasting and limitations to judgemental forecasting. The section closes with reference to forecasting in practice in real estate.

- **Online note #9.11: Qualitative forecasts and real estate forecasting in practice**

9.10   Concluding remarks

The chapter begins with making the distinction between quantitative forecasting which is based on statistical models and qualitative forecasting that includes judgment and expert opinion. The chapter aims to provide sufficient background to assess the forecast adequacy of forecasting methodologies. Forecast evaluation is a key criterion in the adoption of an empirical model. The tests and analysis in this chapter represent the basic forecast evaluation statistics which are often used in real estate research. The discussion highlights the different objectives in forecasting and appropriate tests. It also shows different ways to construct a sample of forecasts and forecasts errors to evaluate.

We summarise the take home points in forecast evaluation that are stressed in this chapter.

- Forecasting has different objectives and appropriate forecasting statistics should be evaluated for these objectives.
- Understanding the sources of the forecast error rationalises the distinction between ex-post and ex-ante forecasting and evaluation as well as between *in-sample* and *out-of-sample* forecasting.
- The larger the sample of forecasts and forecast errors, the more reliable the forecast evaluation will be.
- The sample of the forecast and forecast errors can be compiled in different ways. It will much depend on the objectives of the forecast and availability of data.
- Throughout the chapter we assess the value added by the model, that is whether the model can outperform a naïve specification.
- The forecast evaluation is specific to the period conducted. The chapter points to the fact that in periods of volatility or turning points model-based forecasts are likely to forecast better than naïve models.
- The chapter presents an essential set of forecasting tests. It by no means covers the variety of tests in the large literature on forecast evaluation.

At the beginning of the chapter, we stated that forecast evaluation complements diagnostics checking and can loop back and inform the specification of the model. It should therefore be seen as part of developing robust models for the real estate market. Forecast evaluation informs our expectations about the behaviour of forecasts and errors from the chosen models. Finally, we refer to studies which focus entirely or partially on forecast evaluation. These studies give useful insight into approaches in forecast evaluation for our analysis.

In practice, there are constraints to quantitative analysis and situations that doubt the out-of-sample forecast of models, for example if a major event has just occurred that would impact market in the forecast period. In the online version of the chapter we point to the fact that judgmental forecasting can provide useful insights and inform the forecasts. We outline situations that call for judgmental input and discuss limitations to this type of forecasting. Judgmental forecasting is widespread in the real estate industry. The prevailing convention is to combine the two approaches to take advantage of their strengths. We refer to the 'house view' approach to arrive at the consensus forecasts within an organisation. The final remark is about forecast evaluation. Irrespective of forecast method adopted, this chapter gives a suite of tests to perform a thorough assessment. Samples of forecast errors are easy to construct from models. More work should be done to put together such samples for judgmental forecasting.

Our knowledge of forecasting benefits from the literature on how to improve the perception of the forecasts. This literature has not been reviewed in this chapter. One of the suggestions by Granger (1996) is to construct forecast uncertainty intervals at a 50% level of significance and not at 90% or 95%. Another suggestion by the same author is to use lagged forecast errors to allow for structural breaks. The topic is addressed in the context of real estate in Brooks and Tsolacos (2010).

Chapter 9 online resource

- EViews file: "ch9_eviews"
- Excel file: "ch9_excel"
- Chapter 9 accompanying notes

---

10 ARMA models

10.1 Introduction

Autoregressive moving average (ARMA) models exploit both information embedded in the autocorrelation pattern of the data and previous errors (the MA part) to model time series. They apparently represent a different structure to the regression models we studied in Chapters 5 and 6. We can generalise the ARMA (or ARIMA) specifications to include causal variables as we did in regression analysis.

ARMA and in general *time-series models* are *atheoretical* in the sense that they are not based on any theoretical model. These models are trained to capture the features of time-series data. This does not mean that the classical econometric modelling is abandoned. Rather we should consider the approaches complementary, aiming to inform us about the behaviour of time series.

ARMA models have a number of useful implications in real estate:

(i) ARMA models are considerably flexible to study stationary time series and describe the features of stationary series – for example, how relevant past shocks are to the current value of the variable we explain. The ARMA framework allows to capture and quantify the impact of unobservable random disturbances (shocks) and other phenomena determining a time series.

(ii) ARMA models enable us to model data series of higher frequency. In real estate modelling data series of monthly or higher frequency is in general constrained by data availability. For example in a study of the REIT and the underlying (private market) with monthly data may not be possible due to lack of monthly data for the private market.

(iii) A key use of ARMA models is short-term forecasting. ARMA models will tell us whether what happened in the past is relevant in the immediate future. ARMA forecasts can complement those from classical regression analysis. Forecasts from the latter are subject to revisions to reflect updates to the forecasts of inputs used in the model. This is not an issue in ARMAs unless the series has been revised historically (e.g. a revised rent series).

(iv) Forecasts from ARMA models represent the benchmark forecast. In the forecast evaluation Chapter 11 we used various naïve methods to compare the forecasts from regression models. ARMA represents the benchmark model which more sophisticated models are assessed against in terms of forecast accuracy. It would be expected that a more sophisticated model using variables suggested by real estate and economic theory would outperform the ARMA model.

In this section we outline the features of the ARMA model and practice with US property return data to specify an ARMA and forecast. A discussion of ARMA models and applications to real estate can be found in McGough and Tsolacos (1995), Tse (1997), Wilson *et al.* (2000), and Stevenson (2007).

10.2   AR and MA processes

ARMA models encompass two processes, the autoregressive (AR) process and the moving average (MA) process. We familiarised with the autoregressive process in Chapter 6 where we studied lagged effects in the real estate market and defined the autoregressive model.

10.2.1   *AR process*

A series that is time dependent can be written as:

$$y_t = \beta_0 + \beta_1 y_{t-1} + u_t \tag{10.1}$$

We defined this model (Chapter 6) as an autoregressive model. It simply says that the current level of y_t depends on its lagged value plus an error term. Equation (10.1) is an AR model of order 1, noted as AR(1). u_t is a white noise ($u_t \sim 0, \sigma^2$) error term. In the real estate market, an AR(1) model of total returns is:

$$TR_t = \beta_0 + \beta_1 TR_{t-1} + u_t \tag{10.2}$$

Total returns (TR) depend on the previous period, quarter or month. The coefficient β_1 gives the strength of the association. A positive sign suggests that if we observe high returns at time *t*, we should expect good returns next quarter (and perhaps in the next few quarters).

We can generalise equation (10.2). In the case of total returns, the more general form of the AR model is:

$$TR_t = \beta_0 + \beta_1 TR_{t-1} + \beta_2 TR_{t-2} + \ldots + \beta_p TR_{t-p} + u_t \tag{10.3}$$

This is an autoregressive model of order three *AR(p)* of total returns. In equation (10.3) the realisation of total returns at time *t* depend on total returns in the past *p* periods. It follows that future values of total returns are partially determined by the current and past returns. An AR process of total returns allows for momentum in the realisations of returns, slow adjustments and mean reversion effects.

The lag length *p* in (10.3) is determined by the calculation and minimisation of information criteria (most commonly the AIC and SIC; see Chapter 6). The lag length will depend on the how strongly the series (total returns) is autocorrelated and of course on the frequency of the data.

10.2.2   *The MA process*

A moving average sequence is a linear function of white noise disturbance terms, so that y_t depends on current and past errors. Thus y_t represents a linear combination of the current and previous independent error terms. The general form of a q-order moving average process (MA(q) process) can be expressed as:

$$y_t = \mu + u_t + \theta_1 u_{t-1} + \ldots + \theta_q u_{t-q}$$

$$(10.4)$$

The mean of the MA process is μ. The disturbance terms are independent and normally distributed $u_t \sim N(0, \sigma^2)$ and $Cov(u_t u_j) = 0$ for $i \neq j$.

The MA process also utilises autocorrelations.

Suppose an MA(1) model of total returns which is represented by a combination of two random and normally distributed disturbance terms u_t and u_{t-1}.

$$TR_t = \mu + u_t + \theta u_{t-1}$$

$$(10.5)$$

We can write

$$TR_t = \mu + u_t + \theta u_{t-1}$$
$$TR_{t-1} = \mu + u_{t-1} + \theta u_{t-2}$$

We observe that TR_t is determined by the current random outcome u_t but also partially (through parameter θ) by that of the previous period (u_{t-1}). Similarly TR_{t-1} is determined by the random impact u_{t-1} and partially by the past period's disturbance u_{t-2} (θu_{t-2}). The MA process also relies on the autocorrelation pattern of the series (total returns). If total returns were represented by an MA(2) process, then both random errors u_{t-1} and u_{t-2} would influence current total returns (TR_t). Even though the errors u_t are independent, total returns are autocorrelated. In the MA(1) case the correlation between TR_t and TR_{t-1} is not zero whereas the correlation between TR_t and TR_{t-2} or TR_t and TR_{t-3} and so forth is zero. In the case of an MA(2) representation, TR_t is correlated both with TR_{t-1} and TR_{t-2} but not with TR_{t-3}.

In order to have meaningful and not explosive outcomes, as per our discussion on stationarity, we restrict θ in equation (10.4) to be less than 1. More precisely the restriction is: $|\theta| \leq 1$. θ can take a positive or negative value.

Assume a shock in the real estate market at time t that has a large impact on returns, hence u_t is large. If total returns is an MA(1) series and θ is positive, we expect large impact on total returns in the period $t + 1$. Apparently, the size of the impact will be larger the higher the value of θ is. This discussion reminds us of the analysis of stationarity in Chapter 8.

MA is a process in which the forecast depends on past random errors (model mistakes). The errors from the past periods help to predict the series. Suppose we estimate the MA(1) model to monthly total return series (equation (10.5)) over the period January 2000 to December 2018, which is given by equation (10.6).

$$TR_t = 0.5\% + 0.4 u_{t-1}$$

$$(10.6)$$

We know that in December 2018 the model error was 3%. The expected value for January's total return (January 2019) is:

$$TR_{jan} = 0.5\% + 0.4 \times 3\% = 1.7\%$$

But the realised return in January will also include the random disturbance that month:

$$TR_{Jan} = 0.5\% + u_{Jan} + 0.4 \times 3\% \tag{10.7}$$

The error in January is:

$$u_{Jan} = TR_{Jan} - 0.5\% - 1.2\% \tag{10.8}$$

The prediction for February is:

$$TR_{Feb} = 0.5\% + u_{Feb} + 0.4 \times \left(TR_{Jan} - 0.5\% - 1.2\%\right) \tag{10.9}$$

10.2.3 *Invertibility*

An important property of the MA model is invertibility. The term invertibility relates to an MA process of finite order (say MA(4)) that can be converted into a stationary AR process of infinite order (AR(∞)). The condition of invertibility for MA processes is analogous to the stationarity condition for AR series.

Consider an MA(1) process without a constant to avoid complexity – hence the process has a zero mean. The following substitutions will remind you of those in the topic of stationarity.

$$
\begin{aligned}
y_t &= u_t + \theta u_{t-1} \\
&= u_t + \theta(\mu + y_{t-1} + \theta u_{t-1}) = u_t + \theta y_{t-1} + \theta^2 u_{t-1} \\
&= u_t + \theta y_{t-1} + \theta^2(y_{t-2} + \theta u_{t-3}) = u_t + \theta y_{t-1} + \theta^2 y_{t-2} + \theta^3 u_{t-3} \\
&= u_t + \theta y_{t-1} + \theta^2 y_{t-2} + \theta^3 y_{t-3} + \cdots + \theta^T u_0
\end{aligned}
\tag{10.10 and 10.11}
$$

The MA(1) process (equation (10.10)) is now expressed as an autoregressive sequence if the term $\theta^T u_0$ is zero. Innovations (errors) are inverted into a representation of past realised values. This can only happen if θ is less than 1 so that θ^T tends to zero if T (the sample size) is large enough. If the last term in equation is zero, we can estimate u_t through an autoregressive process with $\theta^T u_0$ being virtually zero. In equation (10.11) the more recent observation has a higher weight than a more remote observation. In the case where θ is higher than 1, then the more remote the observation the higher the impact on y_t. The AR process converges. Similarly if $\theta = 1$, the most recent weight is identical in size with distant weights. Both these cases, as we argued in stationarity, do not make sense and hence the preference for $|\theta| < 1$ that enables us to convert an MA(1) to an AR(∞) process.

Invertibility is a restriction programmed into econometric software for estimating MA coefficients.

10.3 **ARMA specification**

The combination of AR and MA processes results in the ARMA model. Hence we can model a stationary total return series using an ARMA representation:

$$TR_t = \mu + \beta_1 TR_{t-1} + \beta_2 TR_{t-2} + \ldots + \beta_p TR_{t-p} + \theta_1 u_{t-1} + \theta_2 u_{t-2} + \ldots + \theta_q u_{t-q} + u_t \tag{10.12}$$

Equation (10.12) is an ARMA model of order p and q (ARMA(p, q). u_t is white noise with zero mean. It has p lags of *TR* and q lags of the error or past innovations. If total returns were not stationary and we had to difference the series to achieve stationarity, we would have an ARIMA(p, d, q) model. The letter 'I' (which we used to denote integration) tells us we have differenced the series d times to make it stationary (integrated of order '0'). The estimation of (10.12) is based on maximum likelihood (see chapter's online resource for a brief discussion) and on the method of least squares.

Box and Jenkins (1978) proposed a systematic procedure to fit ARMAs to stationary series.

10.3.1 Identification

The first step is to establish the order of the ARMA, hence the order of order p and q. An alternative way (instead of using information criteria) is to observe the *acf* and *pacf* of the series (see online note #10.1 for autocorrelation patterns, significance and the correlogram). This method is discussed in several textbooks and in considerable detail in Brooks and Tsolacos (2010). The order of the AR terms is determined by the non-zero lags of the *pacf*. The *acf* of the series is declining geometrically. The non-zero lags of the *acf* dictate the order of the MA terms. The *pacf* of the series is declining geometrically. The use of information criteria (see Chapter 6) is however a more straightforward way to determine the order of the ARMA series. We choose the order that will minimise the chosen information criteria.

• **See online note #10.1 for using the correlogram to select the ARMA order**

10.3.2 Estimation

The technique commonly used to estimate ARMAs is the maximum likelihood.

• **See online note #10.2 for an explanation of the maximum likelihood estimator**

10.3.3 Diagnostics

In Chapter 7 we presented a host of criteria most of which apply to the ARMA models as well. In the Box-Jenkins approach the diagnostics recommended were overfitting and residual autocorrelation. Regarding overfitting the suggestion is to estimate a higher-order model. If the additional terms are not significant, then the original model is adequate. The test for linear independence of the residuals can be done through calculating the correlogram, using the Ljung-Box test or the LM test and the Jarque-Berra test. We also require the residuals in equation (10.12) to be white noise. We can test the residuals of (10.12) for unit roots. We require that the residuals are stationary. Also note that the order of p and q may be different.

10.4 Example

We specify an ARMA model for the NCREIF's national property returns index (NPI), which is of quarterly frequency. We deflate the index and take the lot of the real property return index. We obtain the real return series from the changes in the logs of the real return index (*dlnpir*). This series is stationary (see this chapter's Eviews file). We conduct the analysis in EViews. In Table 10.1 we show the autocorrelations from computing a correlogram in EViews. This is the autocorrelation pattern that ARMA will exploit.

Table 10.1 Autocorrelation pattern of real NPI returns (*dlnpir*)

Lags	Autocorrelation coefficient (AC)	Partial AC	Q-Statistic	p-value
1	0.67	0.67	73.8	0.00
2	0.56	0.20	125.6	0.00
3	0.46	0.04	160.5	0.00
4	0.39	0.03	185.3	0.00
5	0.18	−0.26	190.7	0.00
6	0.09	−0.05	192.1	0.00

Table 10.1 shows that total returns at quarterly frequency are strongly autocorrelated. Even after four quarters the correlation is moderate (0.39). The autocorrelation coefficients get close to zero after the 5th lag, hence MA terms of order 3,4 or 5 may be appropriate in the ARMA. For the *pacf* the coefficient gets close to zero after the second lag. There is a jump in the coefficient value at lag 5 but it seems to be transitory. The *pacf* suggests an AR order of 1 or 2.

To specify the ARMA in our example we will use two information criteria to select the order of AR and MA terms. We run several ARMAs and select those with the lowest information criteria. The ARMA order may differ by criterion.

The two information criteria select distinctly different ARMA specifications. The value of AIC is minimised for an ARMA(1,4), that is the model contains one AR term and four MA terms. The SIC selects two much more parsimonious models. We remarked in Chapter 6 that the AIC tends to select a more general model (model with more terms). The SIC value is similar for an ARMA(1,1) and an ARMA(2,0). Since the SIC does not distinguish between the two models we can choose the model to use for real NPI returns through an assessment of their forecast performance (which we illustrate later). The same applies (hence a second test) if we wish to do some further testing and assess the forecast success of ARMA models selected by AIC and SIC. It is assumed that the purpose of running the ARMAs is forecasting.

• **See online note #10.3 to specify an ARMA in EViews**

The preceding approach will be rather tedious, though, since we have to manually run lots of combinations of AR and MA terms. Think of monthly data with say a maximum order of 12 for AR and MA terms. Fortunately we can take advantage of the capacity of econometric software. EViews for example will automatically select the optimum order for us.

• **See online note #10.4 for automatic forecasting using ARMA (or ARIMA) in EViews**

The automatic ARIMA forecasting facility selects the model after any necessary transformation to the data to satisfy stationarity, and it further generates the forecasts from the model chosen on the pre-specified information criterion. It is a quick way to specify the ARMA model and obtain the forecast.

- **See online note #10.5 to obtain the forecast output from automatic ARIMA forecasting in EViews**

The forecast can also be obtained in the same way as we did in regression analysis. And similarly, we perform a forecast evaluation. We estimate the ARMA(2,0) model, obtain the forecast (as per online note #5) and as explained in online note #10.5 we further provide the forecast evaluation (online note #10.6).

- **Online note #10.6: ARMA forecast and forecast evaluation in EViews**

In Table 10.2 the ARMA(1,1) and ARMA(2,0) have similar SIC values whereas the model selected by AIC was an ARMA(1,4). In the online resource we perform a forecast evaluation for these three candidate specifications over two eight-quarter horizons. We further investigate forecast gains from combining the forecasts.

- **See online note #10.7 for a forecast evaluation of the different ARMA specifications**

Table 10.2 Determining the ARMA specification for real NPI returns (*dlnpir*)

Order of AR and MA terms	AIC	SIC
0,1	−5.277	−5.219
0,2	−5.424	−5.347
0,3	−5.433	−5.336
0,4	−5.584	−5.469
1,0	−5.533	−5.475
1,1	−5.561	**−5.484**
1,2	−5.551	−5.455
1,3	−5.556	−5.440
1,4	**−5.592**	−5.457
2,0	−5.561	**−5.484**
2,1	−5.550	−5.454
2,2	−5.549	−5.434
2,3	−5.559	−5.425
2,4	−5.583	−5.429
3,0	−5.551	−5.454
3,1	−5.538	−5.423
3,2	−5.567	−5.432
3,3	−5.568	−5.414
3,4	−5.571	−5.398
4,0	−5.539	−5.423
4,1	−5.546	−5.412
4,2	−5.564	−5.411
4,3	−5.576	−5.403
4,4	−5.564	−5.372

10.5   Concluding remarks

The second part of the chapter applied previous knowledge, such as stationarity, to present a pure time-series framework, which is popular in forecasting. We explained the uses of ARMA models and present a step by step guidance to specify these models in EViews and forecast. The frequency of real estate data also improves. For example we see more data series of monthly frequency. If the aim of the study is forecasting ARMA specifications will provide the benchmark model and the benchmark forecasts against which the contribution of other models will be assessed. Given the extensive use of ARMAs in financial markets, it is worth including ARMA models into our analysis if the context of course is appropriate.

Chapter 10 online resource

- Excel file: 'ch10_excel'
- EViews file: 'ch10_eviews'
- Chapter 10 accompanying notes

---

11 Vector autoregressions

11.1 Introduction

In previous chapters we used single regression models to study empirical relationships in the real estate market. These relationships were causal with no feedback effects. Each segment of the market was modelled independently, that is a separate model for vacancy and rents. Real estate market theory would determine choice of variables and causality (direction of impact) in these single regression models. The true dynamics of the market are more complex though. The simplification offered by single regression models to study empirical relationships in the market is necessary to illustrate principles in empirical modelling and is certainly useful in situations of limited data, that is data may not be available for all real estate market variables. Further, a single model regression analysis is easy to understand and communicate. It is seen as less of a black box and accepted by the wider real estate profession even among those practitioners who are not hands on with quantitative techniques.

The introduction to VAR analysis can be made with reference to multi-equation systems, recursive or simultaneous. These systems are not covered in the book, although there is an example in the book's website. For an application of such systems to real estate see Brooks and Tsolacos (2010). The multi-equation systems contain a number of equations describing the segments of the market, hence an equation for take up, another one for vacancy, new development, capital values and so forth. Such frameworks contain both endogenous and exogenous variables. The former determine and are determined by other variables in the system. The latter just determine variables in the system. Real estate market theory dictates (i) the specification of individual equations and (ii) causality imposed in the models and feedback effects.

The vector autoregression family of models offers an alternative statistical procedure to study the dynamic structure of variables. VARs have gained widespread use in macroeconomics and finance research as a framework to overcome the restrictions and exogeneity assumptions of the multivariate simultaneous models (see Gujarati, 2014 and online resource). Sims (1980) drops the distinction between endogenous and exogenous variables and proposes a system of equations in which all variables are endogenous. Hence, in such systems all variables determine each other. They offer an alternative framework to estimate multi-equation models.

This approach is relevant to modelling work in real estate markets. Take for example a model of commercial real estate construction, which includes vacancy, rents and prices with feedback effects. One way to model the relationship between these variables, either through a single regression model or a simultaneous system is to predetermine causality and exogeneity. We can therefore postulate that vacancy affects rent, rent is an input into prices and prices will partially determine the profitability of new construction. Construction in turn determines vacancy and rents. What if construction responds directly both to rents and vacancy? Simple theory may prove limited to determine the exact causality and interaction of variables. On the other hand, Pindyck and Rubinfeld (1998) state that theory may be too complicated to specify the model from the principles. A VAR overcomes this problem by allowing variables to affect each other. Vector autoregressions (VARs) are viewed as flexible modelling procedures providing a reliable benchmark in empirical analysis for alternative econometric representations such as complex models that are grounded in theory but impose more structure and restrictions on the linkages among variables.

VAR models are known to have good forecasting capabilities. The popularity of unrestricted VARs presented in this chapter and variants of VARs, such as Bayesian or structural VARs, is illustrated by their extensive adoption in macroeconomic forecasting and finance. Early studies highlight the advantages of the modelling flexibility they offer and their forecast capabilities. Hakkio and Morris (1984) report empirical findings that suggest VARs have superior forecast performance to misspecified structural models. Litterman (1984), Lupoletti and Webb (1986) and Webb (1984) offer empirical evidence on vector autoregressions achieving greater accuracy in macroeconomic forecasting. Bańbura *et al.* (2014) note the importance of VARs as benchmark forecasting models for alternative econometric specifications.

11.2   VAR specification

Vector autoregressions (VARs) are systems of equations in which all variables are endogenous that is they are determined by the lags of all variables in the system. It is a statistical framework to study linear interdependencies among multiple time-series variables. VARs are a generalisation of the ARMA model presented in Chapter 10 and a hybrid between autoregressive models and simultaneous equation models. The variables in the VAR are both dependent and independent – lags represent the independent variables. A system with two equations, hence with two variables modelled and interacting, denoted as VAR(2), and m number of lags will be written as:

$$y_t = a_1 + \beta_{11} y_{t-1} + \ldots + \beta_{1m} y_{t-m} + \gamma_{11} x_{t-1} + \ldots + \gamma_{1m} x_{t-m} + u_{1t} \tag{11.1a}$$

$$x_t = a_2 + \beta_{21} y_{t-1} + \ldots + \beta_{2m} y_{t-m} + \gamma_{21} x_{t-1} + \ldots + \gamma_{2m} x_{t-m} + u_{2t} \tag{11.1b}$$

In matrix form the VAR(2) is written as:

$$\begin{bmatrix} y_t \\ x_t \end{bmatrix} = \begin{bmatrix} a_1 \\ a_2 \end{bmatrix} + \begin{bmatrix} \beta_{11} & \cdots & \beta_{1m} & \gamma_{11} & \cdots & \gamma_{1m} \\ \beta_{21} & \cdots & \beta_{2m} & \gamma_{21} & \cdots & \gamma_{2m} \end{bmatrix} + \begin{bmatrix} u_{1t} \\ u_{2t} \end{bmatrix} \tag{11.2}$$

and more compactly,

$$\mathbf{Y}_t = \alpha + \beta \mathbf{Y}_{t-1} + \ldots + \beta \mathbf{Y}_{t-m} + u_t \tag{11.3}$$

where $\mathbf{Y}$ is the 2×1 vector of the two variables y and x, $\boldsymbol{\alpha}$ is the 2×1 vector of the intercepts (α_1 and α_2), $\boldsymbol{\beta}$ the 2×2 m vector of the coefficients β and γ and $\boldsymbol{u}_t$ the vector of the disturbances.

In this setting variable y (equation (11.1a)) is determined by own past values ($t - 1$ to $t - m$) and the lags of similar length ($t - 1$ to $t - m$) of the second variable x. The independent variables are similar in both equations of the system. We also note the lack of contemporaneous effects. This may not be too restrictive when we work with higher frequency data such as monthly and quarterly in real estate. Responses in the market and adjustments may occur with a lag due to inertia in the market. The lag length *m* is determined by the chosen information criterion or criteria, as theory is difficult to determine the optimal lag length. Since all variables on the right-hand side are predetermined, the VAR is said to be in a *reduced* form.

The VAR system of equations 11.1(a) and 11.1(b) is a *symmetric* VAR as it contains the same number of lags for each variable in each equation. There are no restrictions of any form imposed. This may result in a VAR with too many parameters and strong multicollinearity. In general multicollinearity is common in VARs rendering the interpretation of coefficients, their signs and statistical significance difficult. Restrictions can therefore be appropriate to address this problem leading to the so-called Bayesian and structural VARs which are not covered in this book. However, the principles and uses of unrestricted VARs presented in this chapter are applicable to VARs with restrictions especially in forecasting applications.

For a *symmetric* system, OLS (ordinary least squares) is used as the estimation method. Estimating the system with OLS is identical to estimating each equation with OLS. OLS estimates will be consistent. Efficiency will not be lost since both equations have identical regressors. For *asymmetric* systems, the more computationally demanding GLS (generalised least squares) estimator is used (not covered in this chapter).

It is further assumed that the VAR satisfies the following conditions adapted for our VAR(2) model:

(i)   The conditional expectation of the disturbances is zero: $E(u_{1t}) = E(u_{2t}) = 0$
(ii) The error variance is constant: $var(u_1) = \sigma_1^2$ and $var(u_2) = \sigma_2^2$
(iii) The autocorrelation in the errors is zero: $E(u_{1t}u_{1t-s}) = E(u_{2t}u_{2t-s}) = 0$ for $t \neq s$
(iv) The errors in the two equations are uncorrelated: $E(u_{1t}u_{2t}) = 0$.

In the absence of exogenous variables the disturbance variance-covariance matrix encompasses information from contemporaneous relationships among the variables in the system. Since contemporaneous effects are conveyed by the residuals, assumption (iv) may be difficult to hold in particular when contemporaneous effects are prominent. This will be a topic of further discussion in subsequent sections.

Traditionally VAR models included covariance stationary series. Therefore differencing or taking growth rates may be necessary to induce stationarity in *y* and *x* if one of them is I(1). If the two series are I(1) and cointegrated – they form a long-run relationship – we have to use the so-called vector error correction model, which resembles the VAR model and uses the error correction mechanism. In such a situation, the use of a VAR to study the dynamic relationship of cointegrated variables would be biased statistically. We examine this topic in the next chapter.

The issue of whether variables in a VAR should be stationary is debatable. Brooks (2014) highlights the loss of long-run information (information about co-movements among variables in the long run is thrown away) since variables need to be differenced to satisfy stationarity. Sims *et al.* (1990) also recommend against differencing or de-trending, even if the variables have a unit root. In the absence of a long-run relationship VARs are mostly estimated with stationary variables. In real estate markets where a high degree of smoothness characterises data anyway, the use of non-stationary variables is likely to result in long-lasting effects from shocks to the variables in the system, which may be unrealistic. Further, Granger causality analysis (studied later) requiring stationary series as inference is based on an asymptotic *F* distribution and to avert spurious causality.

The estimation of the VAR system involves the following stages:

(*i*) *Specifying the VAR:*
   Our objective in estimating a VAR may be the study of linkages among a number of variables in the real estate space such as rents, construction and capital values. Real estate market theory will guide the selection of variables (real estate, economic, other) in the system. Data availability may limit our choices. Defining the lag length or order of a VAR becomes straightforward with the use of information criteria. We should note that an over-parameterised VAR can result in a cumbersome system to estimate. Hence the first task will be to run unit root tests and consider stationary variables.

(*ii*) *Assessing the VAR:*
   A number of diagnostics, similar to those for single regressions, apply to VARs. Further, multivariate versions of these diagnostics tests are available to detect misspecification.

(*iii*) *Impulse responses and variance decompositions:*
   These are techniques to primarily assess the reaction of each variable to one-time shock in an impulse variable. They provide better insight of the system's dynamic behaviour as interpretations of individual parameters are of limited use for that purpose.

(*iv*) *Granger causality tests / Block exogeneity tests*:
   These tests examine the joint significance of lagged endogenous variables in each equation. Do these lags help explain and forecast a given variable in the presence of past values of that variable? These tests are used to establish whether some variables should be treated as exogenous. Pairwise tests can also determine unidirectional causality, bilateral causality (feedback effects) or lack of causality between variables.

(*v*) *Forecasting:*
   Dynamic forecasts are obtained from the unrestricted VAR, usually for short-term horizons. Forecasts can be conditional to future values of a variable. VAR forecasts are assessed with the forecast evaluation tests we presented in Chapter 9.

11.3 Specifying a VAR: an application to City of London office market

We now illustrate the concept of VAR and tasks in empirical investigations involving VARs with an example of the City of London office market. We adopt a VAR framework to study demand and rents in the City of London office market and their interaction with the economy. We are interested in the short-term dynamic relationship and transmission of shocks. Another objective is to forecast office demand and rents in the City of London. This example will provide the basis for generalisations and applications to other market contexts.

11.3.1 Building the city office market VAR – theoretical considerations

We start with setting the theoretical specification of the VAR. The relationship we examine is between office demand, office rents and the economy measured by UK GDP (national). We elaborate on this structure of the VAR in the online version of the chapter.

We also acknowledge the presence of other influences in our VAR system, for example construction and vacancy, which affect both rents and take up. For simplicity and the purpose of this case study we do not include further real estate variables. We have just over 100 quarterly observations in our sample which may prove too small if we include too many variables.

We can of course test alternative VAR specifications especially if one of the target variables is for example vacancy or new construction.

The general form of the three variable VAR in our example is:

$$TU_t = a_1 + \beta_{11}TU_{t-1} + \ldots + \beta_{1m}TU_{t-m} + \gamma_{11}RRg_{t-1} + \ldots + \gamma_{1m}RRg_{t-m} \qquad (11.4a)$$
$$+\delta_{11}GDPg_{t-1} + \ldots + \delta_{1m}GDPg_{t-m} + u_{1t}$$

$$RRg_t = a_2 + \beta_{21}TU_{t-1} + \ldots + \beta_{2m}TU_{t-m} + \gamma_{21}RRg_{t-1} + \ldots + \gamma_{2m}RRg_{t-m} \qquad (11.4b)$$
$$+\delta_{21}GDPg_{t-1} + \ldots + \delta_{2m}GDPg_{t-m} + u_{2t}$$

$$GDPg_t = a_3 + \beta_{31}TU_{t-1} + \ldots + \beta_{3m}TU_{t-m} + \gamma_{31}RRg_{t-1} + \ldots + \gamma_{3m}RRg_{t-m} \qquad (11.4c)$$
$$+\delta_{31}GDPg_{t-1} + \ldots + \delta_{3m}GDPg_{t-m} + u_{3t}$$

where *TU* is take up of office space in the City of London, *RRg* is real rent growth (prime headline real rent growth), and *GDPg* is GDP growth in the UK as a whole. The disturbances u_1, u_2 and u_3 satisfy the conditions presented earlier.

We observe that our theoretical VAR (equations 11.4a–11.4c) contains identical terms on the right-hand side for the three equations. The lag length, which will be similar in all three components (equations) of the VAR, will be determined later. There are no contemporaneous effects in the equations of the system. The impact of any contemporaneous effects are picked up by the disturbances in the equations. In the presence of contemporaneous effects u_1, u_2 and u_3 will be correlated and will make inferences from impulse responses and variance decompositions unreliable. We will not be able to distinguish the effects of the impulse variable on the system. Hence this point needs attention and we cover it later.

An issue we may have in the real estate market, particularly in the direct market, is data availability. The unrestricted VAR consumes degrees of freedom fast (too many parameters to be estimated). The number of estimated parameters in a VAR is given by $g + kg^2$, where g is the number of equations and k is the number of lags. Hence for three equations ($g = 3$) with two lags ($k = 2$), the number of parameters to estimate is $3 + 2 \times 3^2 = 21$ parameters. Three equations and three lags will result in the estimation of 30 parameters. Apparently, the consumption of degrees of freedom by the VAR makes it challenging to apply it to the direct real estate market when long enough series are not available. Of course data constraints in the securitised market (REITS) are not an issue.

11.3.2   *City of London office market VAR: estimation*

In our example we will use a dataset of just over 100 quarterly observations. The unrestricted VAR will contain stationary (covariance stationary) variables. The null hypothesis of a unit root is strongly rejected for all three variables. Since the series are covariance stationary we proceed to establish the number of lags in the VAR or determine the VAR order.

The information criteria we discussed in Chapter 6 guide our decision on the lag length of the VAR. The VAR order is determined by minimising the value(s) of our chosen criterion (or criteria if we use more than one to inform our decision). It is really the choice of the researcher which specific criterion or criteria to use. In the case of VARs the Akaike could select a high-order VAR, that is a VAR with several lags that will limit the degrees of freedom if our sample is small. On the other hand, a criterion such as the Schwarz criterion may select too few lags and useful information about the dynamics of the relationships may be lost due to a short lag structure in the VAR.

For our City of London VAR, we consider both the Akaike information criterion (AIC) and the Schwarz information criterion (SIC). We select the number of lags that will minimise each of these two criteria for the system as a whole. In Table 11.1 we also present the values of these criteria for the individual real estate equations.

As we work with quarterly data, it is conventional to set the maximum lag in the VAR at four quarters. Both AIC and SIC for the system select a VAR(1) specification, the length that minimises the value of these criteria. Since our target is the office market, the selection could follow the minimisation of AIC and SIC of the TU or RRg equation. By coincidence, these criteria take their lowest value both in the TU and RRg equations at lag one. Thus, there is strong evidence for estimating the City office VAR with one lag.

Statistical software could have in-built facilities to determine the order of VAR automatically. As an example, Table 11.2 reports relevant summary results by information criterion and lag length from EViews (see online note #11.1).

Table 11.1 Lag order selection

Lag Order		AIC	SIC
VAR(1)	System	**7.97**	**8.28**
	TU equation	1.51	1.61
	RRg equation	**5.29**	**5.39**
VAR(2)	System	8.15	8.70
	TU equation	1.57	1.75
	RRg equation	5.35	5.53
VAR(3)	System	8.24	9.02
	TU equation	1.64	1.90
	RRg equation	5.40	5.66
VAR(4)	System	8.37	9.39
	TU equation	1.66	2.00
	RRg equation	5.46	5.80

Table 11.2 Defining the lag order of the VAR in EViews

Lag	LogL	LR	FPE	AIC	SIC	HQ
0	−432.21	NA	1.32	8.79	8.87	8.82
1	−383.76	92.98*	0.60*	8.00*	8.31*	8.12*
2	−382.67	2.03	0.70	8.15	8.71	8.38
3	−377.57	9.16	0.76	8.23	9.02	8.55
4	−375.36	3.84	0.87	8.37	9.39	8.78
		LR test statistic	Final Prediction Criterion	Akaike Information Criterion	Schwarz Information Criterion	Hannan–Quinn Information Criterion

Sample: 1992Q2 – 2017Q3
Lags with asterisk denote chosen optimal lag length by criterion

All five criteria in Table 11.2 select one lag for the City office market VAR. The values for the AIC and SIC marginally differ from those in Table 11.1 due to a sample difference (starting date). In Table 11.2 the sample is constant for all lag lengths used (as determined by VAR(4)). For VAR(4) the AIC and SIC values are identical in the two tables.

In our example we have a straightforward case in which all criteria select one lag. This may not always be the case. What if we had a situation in which the AIC would select four lags and SIC two lags? The decision should take into consideration the number of observations and the number of variables in the system. In our sample we have 102 quarterly observations. A system with three variables and four lags would require the estimation of $3 + 4 \times 3^2 = 39$ parameters. The system will still run, but we need to estimate a large number of parameters, reducing notably the degrees of freedom. Hence estimating the system with two lags may be more appropriate in this case. Of course both models can be estimated and assessed further. In particular, if the objective is forecasting, the forecast capacity of the models should be examined.

Table 11.3 presents the results of the VAR(1) model for the City office market. The system is estimated with OLS as all variables on the right-hand side are predetermined variables. We pointed out that the interpretation of coefficients in a VAR is difficult due to multicollinearity. The coefficients may not get the expected sign or be statistically significant. Consider for example the rent growth equation. The term TU_{t-1} is positive (which is expected as stronger demand tends to push rent higher and foster rent growth) and statistically significant. GDP growth, although statistically significant, takes a negative sign, which is not expected. This is the unrestricted VAR.

The adjusted R-squared values denote that this VAR has rather weak capability to explain real rent growth and office take up. It explains 21% of take-up variation and 27% of real rent growth. The highest R^2 value is obtained for the GDP growth equation. A closer look at the results suggests that the key variable driving GDP growth is its previous value ($GDPg_{t-1}$) which is highly significant (t-ratio 7.6), the result of an autocorrelated series despite being stationary.

• ***Online note #11.1: VAR estimation and selection in EViews***

Table 11.3 VAR output for City office market

	TU_t *(take up)*	RRg_t *(real rent growth)*	$GDPg_t$ *(GDP growth)*
C	1.01	−4.61	0.15
	(6.0)	(−4.1)	(1.1)
TU_{t-1}	0.35	2.35	0.04
	(3.8)	(3.8)	(0.5)
RRg_{t-1}	−0.01	−0.09	−0.01
	(1.8)	(2.3)	(1.1)
$GDPg_{t-1}$	−0.09	−0.61	−0.08
	(1.2)	(2.1)	(7.6)
Adj. R-squared	0.21	0.27	0.40
Akaike IC	1.51	5.29	1.20
Schwarz IC	1.62	5.40	1.31
System			
Akaike IC	7.97		
Schwarz IC	8.28		

Note: Numbers in parentheses are t-ratios. Sample period: 1992Q2 – 2017Q3.

11.4   VAR diagnostics

A number of diagnostics can be performed to the residuals of the VAR. We discuss two groups of tests. The first set refers to the simple residual based tests we presented in Chapter 8. The residual series of each equation in the model are individually tested for normality, autocorrelation and heteroscedasticity. Further the residuals can be examined for unit roots. The second group of diagnostics encompasses multivariate tests that apply to the vector of errors in the VAR or the group of the residuals, the so-called multivariate diagnostics tests. In a VAR with three variables the group of residuals will contain three arrays of residuals.

In this section we summarise the output the multivariate tests. The complete presentation both of the univariate and multivariate tests along with guidance how to perform the tests in EViews is contained in the online version of the chapter.

11.4.1   Multivariate residual autocorrelation tests

The multivariate test for autocorrelations proposed by Portmanteau (see Lütkepohl, 1991) for up to fourth order autocorrelation in the system rejects the null hypothesis if no residual autocorrelation of order two, three or four is not rejected at the 90%, 95% or 99% significance level. The VAR system residuals do not exhibit autocorrelation. Similarly, the multivariate LM test described in Johansen (1995) gives no evidence of autocorrelation in the VAR residuals up to fourth order.

11.4.2   Multivariate tests for heteroscedasticity

The VAR system heteroscedasticity LM test (no cross terms) shows no heteroscedasticity in the VAR system overall, although there is evidence of heteroscedasticity in one of the six test equations (see online chapter).

11.4.3   Multivariate normality

A generalisation of the simple Jarque-Bera test we studied in Chapter 8 allows us to estimate multivariate tests for normality in the vector of VAR residuals. From the joint multivariate test results the null hypothesis of joint normality of the residuals is rejected. It seems that equations for *TU* and *RRg* fail normality due to skewness problems whereas the *GDPg* equation fails normality due to kurtosis problems. Online note #11.2 provides a detailed coverage of these tests in EViews.

• **Online note #11.2: VAR diagnostics in EViews**

11.5   Impulse response functions – City of London office market VAR

Impulse responses track the movement (response) of a variable to one-time shocks or innovations to other variables in the system. This helps to study the dynamic behaviour of the system to shocks in a variable (impulse variable). Impulse response functions is a plot of the trajectory of the responses. In our example a shock to GDP growth will affect both take up and rent growth. We want to know how take up and rent growth respond through time after the time of the GDP growth shock. Similarly, a shock to *TU* will affect rent growth and GDP growth (since GDP growth is considered as endogenous in our VAR). It may be hard to rationalise how a shock to *TU* in City of London will affect UK GDP growth. Of course it can be argued that a shock due to take up is due to an economic event that will be reflected in national GDP growth too. In any case the impulse response function will help us understand and assess this debatable impact.

A shock or innovation is an extreme value that a macroeconomic, monetary or real estate series can take. For example, the UK economy with an average growth of 2.5% in the post-war period contracted by nearly 5% in 2009 in the aftermath of Lehman Brothers' collapse. How do we incorporate these shocks in the VAR system? We apply the shock to the residuals of the specific equation. A common practice to illustrate a shock is to increase (or decrease) the error term in the impulse variable equation by one or two standard deviations or to increase the error by the unit of measurement of the dependent variable in that equation (in our VAR, say GDP by 4% or take up by 500,000 sq ft).

- **Online note #11.3: Example to illustrate impulse responses**

In our example we will calculate nine responses since we have three variables in the system – a shock to each variable and response of three variables to that shock. Hence:

(i)   Innovation to *TU* and response of *TU*, *RRg* and *GDPg*
(ii) Innovation to *RRg* and response of *TU*, *RRg* and *GDPg*
(iii) Innovation to *GDPg* and response to *TU*, *RRg* and *GDPg*.

Table 11.4 shows the impulse responses along with standard errors. Statistical software will usually estimate impulse responses and the standard errors accompanying them. The impulse responses shown in Table 11.4 are based on output obtained from EViews.

Table 11.4 Impulse responses

Period (quarter)	Response of TU			Response of RRg			Response of GDPg		
	Innovation in:			Innovation in:			Innovation in:		
	TU	RRg	GDPg	TU	RRg	GDPg	TU	RRg	GDPg
1	0.506	0.0	0.0	0.455	3.316	0.0	−0.050	0.023	0.430
	0.035	*0.0*	*0.0*	*0.330*	*0.232*	*0.0*	*0.043*	*0.043*	*0.030*
2	0.185	0.087	0.048	1.224	0.741	0.555	−0.003	0.059	0.260
	0.049	*0.047*	*0.040*	*0.329*	*0.316*	*0.266*	*0.047*	*0.047*	*0.039*
3	0.096	0.056	0.060	0.693	0.440	0.569	0.022	0.050	0.167
	0.036	*0.027*	*0.038*	*0.241*	*0.190*	*0.247*	*0.041*	*0.035*	*0.042*
4	0.054	0.037	0.055	0.403	0.291	0.480	0.027	0.038	0.112
	0.027	*0.020*	*0.033*	*0.191*	*0.147*	*0.221*	*0.034*	*0.027*	*0.039*
5	0.032	0.025	0.044	0.248	0.198	0.375	0.024	0.029	0.076
	0.020	*0.016*	*0.027*	*0.149*	*0.116*	*0.191*	*0.027*	*0.021*	*0.034*
6	0.020	0.017	0.034	0.160	0.138	0.283	0.019	0.021	0.053
	0.015	*0.012*	*0.022*	*0.116*	*0.092*	*0.161*	*0.021*	*0.017*	*0.029*
7	0.013	0.012	0.025	0.107	0.097	0.209	0.015	0.015	0.037
	0.012	*0.010*	*0.018*	*0.090*	*0.073*	*0.133*	*0.016*	*0.013*	*0.024*
8	0.009	0.008	0.018	0.074	0.069	0.152	0.011	0.011	0.027
	0.009	*0.007*	*0.014*	*0.069*	*0.058*	*0.108*	*0.012*	*0.010*	*0.019*

Note: Standard errors in italics. Cholesky decomposition.

The ordering of the variables is *TU, RRg, GDPg*, which is the order of setting up and estimating the equations in the system. The first response we examine is that of *TU* to shocks to *TU, RRg* and *GDPg*. We start with the shock or innovation to *TU*. The innovation takes the form of a one-standard-deviation rise in the residuals of the *TU* equation. As expected, *TU* responds immediately (the same period) and take up (*TU*) rises by 0.5 million sq ft, which is the standard deviation in the residuals of the *TU* equation. At the time of the *TU* shock there is no impact on either *RRg* or *GDPg* because this is the first innovation to consider assuming no other shocks. The impact of the own shock (*TU* on *TU*) declines, and after eight quarters it eventually dissipates.

There is a small positive impact from innovations in *RRg* and *GDPg* on *TU*. A standard deviation shock to the residuals of the *RRg* equation (3.316%) results in about 0.087 million sq ft or 87,000 sq ft response of take up in the period after the shock. Hence if the shock to rent growth occurred in the first quarter of the year, take up responded by increasing by 87,000 sq ft in the second quarter. Space became more expensive in Q1 as rent growth accelerated but take up picked up as well, the likely result of good market conditions (hence the rent growth shock). At period $t + 2$ there is a further impact of 56,000 sq ft with the effect eventually dying away. If we add up these responses for eight quarters (hence 0.087 + 0.056 + . . . + 0.08), we will get a cumulative response of 0.241 or 241,000 sq ft after 2 years out of the one-time shock. Given average quarterly take up in this market of 1.6 million sq ft, the impact of a rise in rent by 3.3% is apparently small.

The response path of *TU* to national *GDP* growth differs slightly. A one-standard-deviation positive shock to *GDP* growth, that is 0.43%, roughly 1.7% annualised, will trigger a *TU* rise of 0.048 (48,000 sq ft). At time $t + 3$ the impact gets stronger and demand rises by a further 0.060 (60,000 sq ft). Hence there is a lag in the response of *TU* to *GDP* growth to reach the peak impact. Such lagged responses in the real estate should be expected. In quarter four (three quarters after the event at $t = 1$) the response eases to 0.055 (55,000 sq ft) and continues to decline. The cumulative impact after eight quarters is 0.284 (284,000 sq ft). As in the case of rent growth we would not consider this impact strong. How about if we ignore the information criteria on the basis that the lag length of one quarter is too short and estimate a VAR(2). The respective cumulative impact will be 0.254 (254,000 sq ft). Estimating a higher-order VAR does not translate into stronger impulse responses. You are encouraged to examine this for yourself.

Turning to *RRg* and *GDPg* the own responses at time $t = 1$ (3.316% in the *RRg* equation and 0.43% in the *GDPg* equation) are the one-standard-deviation shocks. Let us study the response of *RRg* to a take-up shock. When take up is shocked by 560,000 sq ft (one standard deviation), real rent growth will accelerate by 0.455 (0.46%) at time $t + 1$ and by 1.224 (1.2%) at time $t + 2$ when the impact peaks. The cumulative impact on rent growth is 4.8% after 1 year and 5.4% after 2 years. The respective response from a shock to national GDP growth is more moderate accounting to 1.7% after 1 year and 2.8% after 2 years. The national economy has an impact on City office rents but certainly not as large as take up.

Finally the response of *GDPg* to take up is unstable (two negative impacts followed by a positive impact) and of minor severity. Shock to rent growth does not seem to move GDP growth either (0.06% at peak time $t + 2$). The results are relevant to the question of whether City office market variables impact on *GDPg* as assumed in this VAR setting.

Figure 11.1 presents a graphical analysis of the data presented in Table 11.5. Selected impact responses are plotted (solid lines) with two-standard-error boundaries calculated by the statistical software. Traditionally impulse responses are shown in graphic form, and it is a standard feature of statistical software to produce the following impulse response function

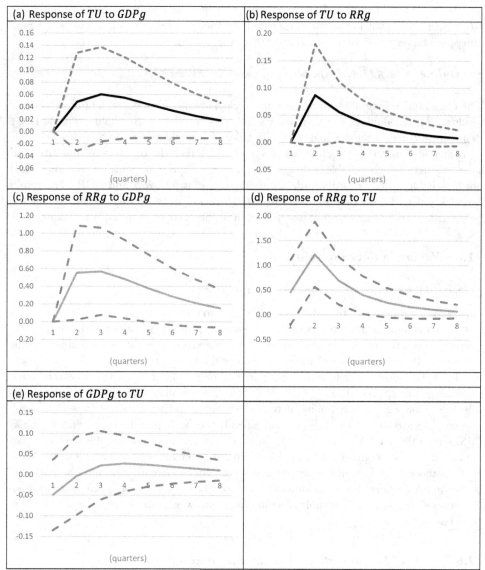

Dotted lines are two standard errors

Figure 11.1 Graphs of impulse responses

graphs. The vertical axis is in the units of the variable that responds to the shock. The horizontal axis shows time.

A practical rule to assess the stability of impulse responses is to observe the standard error boundaries. A stable impulse response outcome is indicated by both an upper and lower boundary that are positive or negative depending on the relationship. A positive upper limit and a negative lower boundary denote instability. An impulse response function close to the zero line indicates no relationship.

Panels (a) to (d) represent stable impulse responses. In panel (e) the upper band is positive and the lower band negative. This is indicative of a questionable if any impact from *TU* to *GDPg*, an issue that will further be investigated in subsequent tests. In online note #11.3 summary impulse responses for comparisons are also presented.

*   ***Online note #11.4: Impulse responses in EViews***

The calculation of impulse responses must be carried out with independent or orthogonal residuals. The residuals of the VAR show small cross-correlations (see online note #11.3), a desirable outcome. If the residuals of the equations in the system are moderately or strongly correlated (e.g. correlation coefficients of, say, over 0.3) then the ordering of innovations or shocks matters (which variable is shocked first) and the impulse response calculations would not be reliable. In our case the ordering of the shocks is *TU*, *RRg* and *GDPg*. The results would be different if the ordering was *RRg*, *GDPg* and *TU*. In our example some bias can still be present, but the situation is not worrying with strongest correlation coefficient being 0.13.

11.6   Variance decompositions

11.6.1   *Background to variance decomposition*

An alternative method to study the effects of shocks to the variables in the system and therefore their dynamic relationship is variance decomposition. It provides information on the relative importance of each innovation (shock to the error of equations) in explaining the variance of the forecast error of the variables over the time horizon we specify. Own shocks usually explain most of the forecast error variation of the variables with the effects of other variables getting stronger through time. Weak effects from other variables after several periods denote low explanatory ability of these variables in the VAR and perhaps lack of relevance. Forecasts for each variable are obtained by the VAR and the mean squared errors (MSEs) are calculated. Variance decomposition determines how much of the MSE of a particular variable is accounted for by the innovation in the errors of each variable in the VAR. Again orthogonalised residuals are required so the discussion earlier on impulse responses is relevant to variance decomposition analysis too. The calculation of variance decomposition sounds complicated. Technology would help, however, as these methods are inbuilt in econometric software.

11.6.2   *City of London office VAR: variance decomposition*

We compute forecast error variance decompositions in our example. We examine the relative importance of innovations in each of the three equations in our VAR to explain the variance of the forecast error obtained from forecasting the three variables *TU*, *RRg* and *GDPg* within the VAR. Table 11.5 reports the results obtained in EViews.

Table 11.5 shows variance decompositions for up to eight quarters. The 'S.E.' column provides the forecast error by forecast horizon. The remaining three columns give the percentage of the forecast error variance explained by each innovation (the shock to the residuals of each equation). Each row adds up to 100 (100%). *TU* comes first in the ordering, hence the only source of the one-quarter-ahead forecast variation is its own shock. In the second period, 96.7% is explained by its own shock and 2.5% is attributable to *RRg* shock. As the horizon lengthens, *RRg* and *GDPg* explain more and after eight quarters they jointly account for about 8%. This is an indication of *RRg* and *GDPg* not explaining adequately in our VAR. The results are different for the rent growth equation. After three quarters *TU* innovations are responsible for 15% of the rent growth error variance with no further impact recorded. After eight quarters 15.8% is accounted by innovations in *TU* and 7.4% in *GDPg*. For *GDPg* earlier results pointing to an absence of impact from both *TU* and *RRg* are confirmed. *TU* and *RRg* combined are responsible for a mere 5% of the forecast error variation in *GDPg* after eight quarters. It appears that rent growth is the best explained variable in this VAR.

If the order in the Cholesky decomposition alters, some differences will appear in the results, but they are not major in our example. In Table 11.6 we show the results for *TU* and *RRg* when the order is *GDPg, RRg, TU* (see previous section).

After eight quarters now *RRg* and *GDPg* explain 11.5% of the forecast error variation in *TU* compared with 8% in the original ordering of the variables. For *RRg* the new ordering results in an impact of less than 20% explained by *TU* and *GDPg* compared with over 23% in the previous ordering. This sensitivity of the results to ordering owes to the presence of

Table 11.5 Forecast error variance decomposition for London City office VAR

Period	TU				RRg				GDPg			
	S.E.	TU	RRg	GDPg	S.E.	TU	RRg	GDPg	S.E.	TU	RRg	GDPg
1	0.51	100	0	0	3.35	1.85	98.15	0	0.43	1.32	0.29	98.39
2	0.55	96.71	2.51	0.78	3.68	12.58	85.15	2.27	0.51	0.96	1.58	97.46
3	0.56	94.73	3.39	1.89	3.82	15.02	80.64	4.34	0.54	1.03	2.26	96.71
4	0.57	93.51	3.72	2.77	3.88	15.62	78.64	5.73	0.55	1.22	2.64	96.14
5	0.57	92.80	3.87	3.33	3.91	15.78	77.65	6.57	0.56	1.38	2.84	95.78
6	0.57	92.41	3.94	3.66	3.92	15.82	77.15	7.03	0.56	1.48	2.95	95.57
7	0.57	92.20	3.97	3.84	3.93	15.83	76.89	7.28	0.56	1.54	3.01	95.45
8	0.57	92.09	3.98	3.93	3.94	15.83	76.75	7.42	0.56	1.57	3.04	95.39

Note: Cholesky decomposition used: Ordering: *TU, RRg, GDPg*. Column labelled 'S.E.' contains the forecast error of the variable at the given forecast horizon. As stated in EViews, the source of this forecast error is the variation in the current and future values of the innovations to each endogenous variable in the VAR.

Table 11.6 Sensitivity of variance decomposition results to re-ordering

Period	TU				RRg			
	S.E.	TU	RRg	GDPg	S.E.	TU	RRg	GDPg
1	0.51	96.70	1.97	1.32	3.35	0	99.86	0.14
2	0.55	92.82	5.72	1.45	3.68	10.11	88.28	1.61
3	0.56	90.92	6.85	2.23	3.82	12.70	84.03	3.27
4	0.57	89.82	7.23	2.95	3.88	13.44	.82.07	4.48
5	0.57	89.18	7.38	3.44	3.91	13.68	81.08	5.24
6	0.57	88.83	7.44	3.73	3.92	13.77	80.57	5.67
7	0.57	88.64	7.47	3.89	3.93	13.80	80.30	5.90
8	0.57	88.54	7.48	3.98	3.94	13.82	80.16	6.02

Note: Cholesky decomposition used: Ordering: *DPg, RRg, TU*.

some cross correlation among residuals. The difference in the results would have been more prominent if cross correlations were stronger. Overall in our VAR, *GDPg* cannot adequately be explained by the *TU* and *RRg*, but also *TU* is not explained well either. What values would indicate good performance by the VAR in explaining the forecast error variance of variables? The more of the forecast variance is explained, the better is the answer. Several real estate data series display high autocorrelation (especially in the direct market), hence own shocks are prominent and persist restricting the impact of other variables. A value close to 50% for joint accountability of the rest of the variables after a few periods would represent good performance, but really the higher this value the better.

• **Online note #11.5: Variance decompositions in EViews**

11.7   Granger causality tests

11.7.1   *Theoretical setting*

The VAR framework can be used as a means of running Granger causality tests to establish whether movements in one variable precede movements in another. These tests do not necessarily suggest that one variable causes the other. However, if real estate theory indicates causality between time series variables then these tests can be seen as 'true' causality tests generating empirical evidence to support or challenge a priori causality hypothesis.

Consider the following bivariate VAR(3) model:

$$y_t = a_1 + \beta_{11} y_{t-1} + \beta_{12} y_{t-2} + \beta_{13} y_{t-3} + \gamma_{11} x_{t-1} + \gamma_{12} x_{t-2} + \gamma_{13} x_{t-3} + u_{1t} \tag{11.5a}$$

$$x_t = a_2 + \beta_{21} y_{t-1} + \beta_{22} y_{t-2} + \beta_{23} y_{t-3} + \gamma_{21} x_{t-1} + \gamma_{22} x_{t-2} + \gamma_{23} x_{t-3} + u_{2t} \tag{11.5b}$$

We would like to test whether *x* causes *y* or *y* causes *x*. We are investigating four options and outcomes:

(i)   *x* causes *y* but *y* does not cause *x*. In this case we have a unidirectional relationship. Movements of *x* can predict movements of *y* but not the other way round.

In the context of equations (11.5(a)) and (11.5(b)) we test the following hypothesis:

In equation (11.5(a)): H_0: $\gamma_{11} = \gamma_{12} = \gamma_{13} = 0$
against H_1 that one of $\gamma_{11}, \gamma_{12}, \gamma_{13}$ coefficients is not zero.

and

In equation (11.5(b)): H_0: $\beta_{21} = \beta_{22} = \beta_{23} = 0$
against H_1 that one of $\beta_{21}, \beta_{22}, \beta_{23}$ coefficients is not zero.

In equation (11.5(a)) we therefore test if lags of variable *x* can predict variable *y* (or precede movements of *y*) in the presence of lags of *y*. Hence this is a test of zero restrictions on the coefficients $\gamma_{11}, \gamma_{12}, \gamma_{13}$. Inferences can be made from an *F*-test or a chi-square test we studied in earlier chapters. In equation (11.5(b)) we examine if lags of *y* can predict *x* (or precede movements of *x*) in the presence of lags of *x*.

If H_0 in equation (11.5(a)) is rejected (the set of γs is statistically different from zero) but not in equation (11.5(b)) (the set of βs is not statistically different from zero) there is Granger causality from x to y only.

(ii) x does not cause y but y causes x. Again a unidirectional relationship exists which now runs from y to x. Movements of y can predict movements of x but not the other way round.

Following the process described under (i), if the set of coefficients γ_{11}, γ_{12} and γ_{13} are not zero in equation (11.5(a)) but β_{21}, β_{22} and β_{23} are not statistically different from zero, there is Granger causality from y to x.

(iii) x causes y and y causes x. This is a situation of bidirectional or bilateral association and mutual dependence between the two variables. There are feedback effects between x and y, a common situation in the real estate market.

The sets of γ_{11}, γ_{12} and γ_{13} coefficients in equation (11.5(a)) are statistically significantly different from zero and so are coefficients β_{21}, β_{22} and β_{23} as a group in equation (11.5(b)).

(iv) Neither x causes y, nor does y cause x. There is no relationship between the two variables, they are independent.

Coefficients γ_{11}, γ_{12} and γ_{13} are not jointly significant in equation (11.5(a)) and coefficients β_{21}, β_{22} and β_{23} are not jointly significant in equation (11.5(b)).

Test results for these options are subject to the lag length in equations 11.14(a) and 11.5(b). This is important in real estate market analysis where lagged responses and lagged feedback effects are common. For example, a bidirectional relationship may be found after three or four lags. A longer lag length allows more time to detect feedback effects. The lag length in the bivariate VAR 11.5(a)–11.5(b) can be determined by the information criteria for selecting the optimal VAR order. It is worth though running the Granger causality test with different lag lengths (e.g. 2 and 4 lags for quarterly data – see Pindyck and Rubinfeld (1998).

11.7.2 *Application of Granger causality tests in City of London office market*

The calculation of Granger causality tests in our example will reveal the causal relationships among the three variables. We had evidence from impulse responses and variance decompositions suggesting the GDP is not explained by the other two variables. We can further examine this point with Granger causality tests. Initially we run pairwise tests estimating systems as described by Equations 11.5(a) and 11.5(b).

Table 11.7 gives the output of a bivariate VAR system in the context of equations 11.5(a) and 11.5(b). Three sets of pairwise tests are run: *RRg* and *TU*, *GDPg* and *TU* and *GDPg* and *RRg*. Although the optimum lag for the VAR system is one lag, we run the pairwise tests with 1, 2 and 4 lags in case causality exists at longer lag length.

There is clearly a strong bilateral relationship between *TU* and *RRg* at lag 1. According to the *p*-values, we reject both null hypotheses. At lags 2 and 4 causality runs only from *TU* to *RRg*. The results for the relationship between *TU* and *GDPg* do not point to any

Table 11.7 Pairwise causality tests at different lags – City of London offices

Null Hypothesis:	1 lag			2 lags			4 lags		
	Obs	F-Stat	Prob.	Obs	F-Stat	Prob.	Obs	F-Stat	Prob.
RRg does not Granger cause TU	102	4.61	0.03	101	1.47	0.23	99	1.06	0.38
TU does not Granger cause RRg		14.04	0.00		6.11	0.00		2.81	0.03
GDPg does not Granger cause TU	102	2.80	0.10	101	1.15	0.32	99	0.72	0.58
TU does not Granger cause GDPg		0.97	0.33		0.32	0.73		0.74	0.56
GDPg does not Granger cause RRg	110	5.52	0.02	109	2.51	0.09	107	1.31	0.27
RRg does not Granger cause GDPg		4.48	0.04		2.22	0.11		1.06	0.38

causality. There is some evidence of *GDPg* causing *TU* at the 10% level of significance at lag 1. This result is also consistent with the findings of the variance decompositions and impulse responses.

In the third bivariate relationship *GDPg* causes *RRg* with feedback effects from *RRg* to *GDPg* at lag 1. Rents may convey influences which are reflected in *GDPg*, e.g. reflecting an external influence on both *GDPg* and *RRg*. *GDPg* Granger causes *RRg* after two quarters but not the other way round, and when 4 lags are considered there is no causal relationship. Overall there is evidence (considering the results for both 1 and 2 lags) that *GDPg* causes *RRg*. If we consider the results in Tables 11.5 and 11.6, we can further conclude that the influence of *GDPg* on *RRg* is not particularly strong. In the online chapter we generalise the discussion to block exogeneity tests.

- **Online note #11.6: Block exogeneity tests**
- **Online note #11.7: Granger causality/Block exogeneity tests in EViews**

11.8   VAR forecasting – London office market

We now use our estimated VAR for the London City office market (Table 11.3) to forecast the three variables *TU*, *RRg* and *GDPg* with our interest of course being on the two office market series. The VAR will generate and reproduce forecasts through its lag structure. Our VAR is estimated up to 2017Q3. We generate a four quarter forecast that is 2017Q4 to 2018Q3. Since we only have one lag we use the values of the variables in 2017Q3 to forecast their values for 2017Q4. The forecasts for 2017Q4 will subsequently be used to obtain the forecast for 2018Q1 and so forth. Table 11.8 shows the forecasts and how they are calculated.

The forecasts in Table 11.8 are dynamic and unconditional in the sense that no future values of any of the three variables are used. The VAR forecasts stronger GDP growth, a small in increase in take up and slightly stronger real rent growth. These forecasts are consistent with our expectations.

There was evidence in our earlier analysis that *GDPg* was exogenous in the VAR system. On such signs, we can condition the VAR forecasts for *TU* and *RRg* to *GDPg* forecasts obtained from another source if available and not from the VAR. Or we can conduct scenario analysis: what if *GDPg* was to be greater or weaker than the prediction from the VAR? Investors are interested in downside shocks. Events such as Brexit introduce uncertainty. Market participants would like to have estimates of downside scenarios, e.g. adverse developments to GDP growth.

Table 11.8 Unconditional out-of-sample VAR forecasts

Actual			
	TU		1.577 (million sq ft)
2017Q3	*RRg*		−0.985 (qoq %)
	GDPg		0.436 (qoq %)
Forecasts			
	TU	**1.007 + 0.353***1.577 + **0.025***(−0.985) + **0.112***0.436 =	1.588
2017Q4	*RRg*	**−4.614 + 2.354***1.577 + **0.215***(−0.985) + **1.292***0.436 =	−0.550
	GDPg	**0.152 + 0.041***1.577 + **0.014***(−0.985) + **0.606***0.436 =	0.467
	TU	**1.007 + 0.353***1.588 + **0.025***(−0.550) + **0.112***0.467 =	1.606
2018Q1	*RRg*	**−4.614 + 2.354***1.588 + **0.215***(−0.550) + **1.292***0.467 =	−0.391
	GDPg	**0.152 + 0.041***1.588 + **0.014***(−0.550) + **0.606***0.467 =	0.492
	TU	**1.007 + 0.353***1.606 + **0.025***(−0.391) + **0.112***0.492 =	1.619
2018Q2	*RRg*	**−4.614 + 2.354***1.606 + **0.215***(−0.391) + **1.292***0.492 =	−0.282
	GDPg	**0.152 + 0.041***1.606 + **0.014***(−0.391) + **0.606***0.92 =	0.511
	TU	**1.007 + 0.353***1.619 + **0.025***(−0.282) + **0.112***0.511 =	1.629
2018Q3	*RRg*	**−4.614 + 2.354***1.619 + **0.215***(−0.282) + **1.292***0.511 =	−0.203
	GDPg	**0.152 + 0.041***1.619 + **0.014***(−0.282) + **0.606***0.511 =	0.524

In order to introduce external *GDPg* forecasts we just replace the VAR based *GDPg* forecasts numbers of 0.467 in 2017Q4, 0.492 in 2018Q1 and 0.511 in 2018Q2 with the desired *GDPg* numbers. The value of 2018Q3 is not needed since it the lag of *GDPg* that is used in the VAR forecast. All coefficients remain as they are in Table 11.8. These calculations are not shown here, but they can be found in the accompanying Excel file.

11.8.1 *Forecasting from VAR with exogenous variables*

How about if it is the contemporaneous growth in GDP that matters in the VAR model of the office market? In this case the value of *TU* and *RRg* in 2017Q4 is determined by the values of these variables in 2017Q3 and *GDPg* growth in 2017Q4. In this case we need to re-estimate the VAR with *GDPg* specified as an exogenous (hence the VAR has two exogenous terms in the estimation dialog box, the intercept *c* and *GDPg*). The new VAR output with *GDPg* as an exogenous variable is shown. We now have two endogenous variables and hence two equations:

$$TU_t = 1.073 + 0.355 TU_{t-1} + 0.030 RRg_{t-1} - 0.015 GDPg_t \tag{11.6a}$$

Adj. $R^2 = 0.20$

$$RR_t = -4.395 + 2.322 TU_{t-1} + 0.227 RRg_{t-1} + 0.977 GDPg_t \tag{11.6b}$$

Adj. $R^2 = 0.26$

A comparison with Table 11.3 shows that the size of coefficients and explanatory power has little changed. The optimal lag is again 1. This owes to the fact that *GDPg* should have been treated as exogenous influence – we have lots of evidence about it. The forecasts from this model are given in Table 11.9. In our example, quarterly forecasts for GDP growth are provided by an economics forecasting company.

Table 11.9 VAR forecasts conditional on predetermined *GDPg*

Actual			
	TU		1.577 (million sq ft)
2017Q3	*RRg*		−0.985 (qoq %)
	GDPg		0.436 (qoq %)
Forecasts			
	TU	**1.073 + 0.355***1.577 **+ 0.03***(−0.985) **− 0.015***0.43 =	1.597
2017Q4	*RRg*	**−4.395 + 2.322***1.577 **+ 0.227***(−0.985) **+ 0.977***0.43 =	−0.531
	GDPg		0.430
	TU	**1.073 + 0.355***1.597 **+ 0.03***(−0.531) **− 0.015***0.400 =	1.618
2018Q1	*RRg*	**−4.395 + 2.322***1.597 **+ 0.227***(−0.531) **+ 0.977***0.400 =	−0.463
	GDPg		0.400
	TU	**1.073 + 0.355***1.618 **+ 0.03***(−0.463) **− 0.015***0.400 =	1.628
2018Q2	*RRg*	**−4.395 + 2.322***1.618 **+ 0.227***(−0.463) **+ 0.977***0.400 =	−0.352
	GDPg		0.400
	TU	**1.073 + 0.355***1.628 **+ 0.03***(−0.352) **− 0.015***0.38 =	1.635
2018Q3	*RRg*	**−4.395 + 2.322***1.628 **+ 0.227***(−0.352) **+ 0.977***0.38 =	−0.323
	GDPg		0.380

The third-party forecasts for GDP growth are less optimistic than those obtained from the VAR. Both the rent growth and take up forecasts are pretty similar to those of the original VAR (VAR *GDPg* endogenous). Given the weaker GDP growth (externally predicted), real rent growth is slightly more negative than previously. Take up is somewhat more positive, the result of a negative *GDPg* coefficient in the *TU* equation. The *TU* forecast is counterintuitive. Perhaps lags of *GDPg* are more appropriate, however it is a strong sign that we need to do more work on measuring economic influences in our VAR.

In all the preceding cases, if we think that the information criteria were too strict we can repeat the forecast evaluation using a VAR with 2 lags.

11.8.2 *VAR forecast evaluation*

The forecast performance of our original VAR is now illustrated over the latest eight-quarter horizon. We estimate the model for the period 1992Q2 to 2015Q3 and we forecast the variables for remaining eight quarters. We obtain a dynamic forecast; no actual values in the last eight quarters are used. The estimated VAR model over the shorter period is shown.

$$TU_t = 1.000 + 0.361TU_{t-1} + 0.025RRg_{t-1} + 0.110GDPg_{t-1} \qquad (11.7a)$$

Adj. $R^2 = 0.22$

$$RRg_t = -4.656 + 2.382TU_{t-1} + 0.213RRg_{t-1} + 1.303GDPg_{t-1} \qquad (11.7b)$$

Adj. $R^2 = 0.27$

$$GDPg_t = 0.145 + 0.046TU_{t-1} + 0.015RRg_{t-1} + 0.610GDPg_{t-1} \qquad (11.7c)$$

Adj. $R^2 = 0.41$

A comparison of the coefficients of the shorter VAR with the full time VAR shows little variation in the coefficient values. This is a welcome result, a sign of coefficient stability. We generate forecasts for 2015Q4 to 2017Q3 and a basic forecast evaluation is performed. The actual values in 2015Q3 are used to obtain forecasts for 2015Q4 and we roll the forecasts forward as Table 11.10 illustrates. Table 11.11 contains the forecasts for the entire forecast horizon.

From just observing the forecasts it becomes apparent that the unrestricted VAR(1) we estimated does not replicate the quarterly variation in real estate series. It forecasts a slight fall both in take up and real rent growth. The last row shows averages for actual and forecast values over the forecast horizon. The averages do not differ much though. A formal forecast evaluation is presented in Tables 11.12 and 11.13.

Table 11.10 Calculating the VAR in-sample forecasts

Actual			
	TU		1.940 (million sq ft)
2015Q3	*RRg*		1.992 (qoq %)
	GDPg		0.452 (qoq %)
Forecasts			
	TU	**0.996 + 0.361***1.940 + **0.025***1.992 + **0.110***0.452 =	1.797
2015Q4	*RRg*	**−4.656 + 2.382***1.940 + **0.213***1.992 + **1.303***0.452 =	0.979
	GDPg	**0.145 + 0.046***1.940 + **0.015***1.992 + **0.610***0.452 =	0.540
	TU	**0.996 + 0.361***1.796 + **0.025***0.978 + **0.110***0.540 =	1.729
2016Q1	*RRg*	**−4.656 + 2.382***1.796 + **0.213***0.978 + **1.303***0.540 =	0.537
	GDPg	**0.145 + 0.046***1.796 + **0.015***0.978 + **0.610***0.540 =	0.572
. . .	. . .		. . .

Table 11.11 VAR in-sample dynamic forecasts

Period	*Actual*			*Forecast*		
	TU	*RRg*	*GDPg*	*TU*	*RRg*	*GDPg*
2015Q3	*1.940*	*1.992*	*0.452*	–	–	–
2015Q4	1.973	4.980	0.721	1.797	0.979	0.540
2016Q1	1.774	0.000	0.179	1.729	0.537	0.572
2016Q2	1.216	−0.839	0.536	1.697	0.323	0.582
2016Q3	1.239	−0.757	0.533	1.681	0.214	0.583
2016Q4	2.011	−0.602	0.707	1.673	0.154	0.581
2017Q1	1.616	−0.710	0.351	1.668	0.119	0.579
2017Q2	1.907	−1.400	0.262	1.665	0.097	0.577
2017Q3	1.577	−0.985	0.436	1.663	0.083	0.575
Average	1.66	−0.04	0.47	1.70	0.31	0.57

Note: Output from a software package may differ slightly to the numbers in this table, as in these calculations we round numbers to three decimal points at each round in the calculations.

Table 11.12 Forecast evaluation: take up

	Actual	Forecast	Error	Abs error	% abs error	Squ. error
2015Q4	1.973	1.797	0.176	0.176	8.9%	0.031
2016Q1	1.774	1.729	0.045	0.045	2.5%	0.002
2016Q2	1.216	1.697	−0.481	0.481	39.6%	0.231
2016Q3	1.239	1.681	−0.442	0.442	35.7%	0.195
2016Q4	2.011	1.673	0.338	0.338	16.8%	0.114
2017Q1	1.616	1.668	−0.052	0.052	3.2%	0.003
2017Q2	1.907	1.665	0.242	0.242	12.7%	0.059
2017Q3	1.577	1.663	−0.086	0.086	5.5%	0.007
			ME	MAE	MAPE	MSE
			−0.033	0.233	15.6%	0.080
						RMSE
						0.283

Note: ME: Mean error; MAE: Mean absolute error; MAPE: Mean absolute percentage error; MSE: Means squared error; RMSE: root mean squared error.

Table 11.13 Forecast evaluation: real rent growth

	Actual	Forecast	Error	Abs error	Squ. error
2015Q4	4.980	0.979	4.001	4.001	16.006
2016Q1	0.000	0.537	−0.537	0.537	0.288
2016Q2	−0.839	0.323	−1.162	1.162	1.350
2016Q3	−0.757	0.214	−0.971	0.971	0.943
2016Q4	−0.602	0.154	−0.756	0.756	0.571
2017Q1	−0.710	0.119	−0.829	0.829	0.687
2017Q2	−1.400	0.097	−1.497	1.497	2.240
2017Q3	−0.985	0.083	−1.068	1.068	1.140
			ME	MAE	MSE
			−0.352	1.352	2.903
					RMSE
					1.704

Table 11.12 provides the forecast evaluation for take up. According to the mean error there is a small overprediction as on average the forecast values are higher than the actual values. The mean absolute error is 0.233 or 233.000 sq ft. This translates into a 15.6% mean absolute error. The RMSE is lower than the standard error of the TU equation in the VAR – a desired result.

The evaluation of RRg forecasts is presented in Table 11.13. The VAR misses the strong growth in real rents in 2015Q4 and predicts positive real rent growth all the way at a declining rate. Actual real rent growth was negative after 2015Q4. Hence on average the VAR overpredicts real rent growth by 0.35%. The mean absolute error is 1.35%. This is not bad performance in terms of error size and direction although the negative growth is not predicted in 2015Q3. The RMSE of the forecasts is lower than the standard error of the

RRg regression (3.45%). On these statistics and sample we forecast the VAR does a good job predicting *TU* and *RRg*.

It would interesting to compare the preceding forecast evaluation statistics with those from alternative VARs. For example a VAR in which *GDPg* is treated as exogenous. As the realised values of *GDPg* are used in the forecasts we should expect better performance.

As we noted in Chapter 9, we may wish to assess whether the forecast performance is sensitive to the chosen forecast horizon. We can therefore focus on another eight-quarter period, e.g. 2012Q1 to 2013Q4). Further VAR forecasts for one, two, three quarters and so forth can be made recursively and evaluated. For example, estimating the VAR until 2013Q4 and obtain a forecast four quarters ahead, that is for 2014Q4. Roll the VAR estimation one quarter on to 2014Q1 and forecast for 2015Q1. In this way we generate a series of four-quarter-ahead forecasts which we evaluate in the same way.

- ***Online note #11.8: VAR forecasting in EViews***

11.9 VAR advantages and limitations

This chapter has illustrated the usefulness of VAR for studying relationships in the real estate market and for forecasting. Next we summarise the advantages of unrestricted VARs and highlight shortcomings.

11.9.1 Advantages of VARs

(i) Flexible and easy to use to examine relationships in the real estate market and interactions between its three main segments, occupier, investment and development. These sectors in reality are interrelated and the VAR will allow for feedback effects.
(ii) Theory guides the selection of variables in the VAR. There is no need to determine which variables are exogenous or determine a priori causality. All variables are endogenous: they explain and are determined by other variables. We can include exogenous variables as well, either on the basis of theory or as a result of relevant tests.
(iii) Easy to estimate with OLS since there are no contemporaneous terms of the endogenous variables in the system.
(iv) VARs have good forecasting capabilities. Even if they are *atheoretical*, unrestricted VAR forecasts can be used as a benchmark.
(v) VAR present a framework to illustrate the dynamics of relationships and the impact of shocks through impulse response functions and variance decompositions.
(vi) Block exogeneity and causality tests are useful tools to study relative movements even if theory is not clear. For example examining causality between returns in New York and London office markets. Do we have a strong theory that links the two markets? We can still examine whether movements of returns in one market precede movements in the other or whether these two markets are independent.
(vii) Functional form is not a worry since the equations in the unrestricted VAR are of linear form.

11.9.2 Disadvantages

(i) There could be too many parameters to estimate in the VAR, as we indicated earlier. Although this is not a major issue for financial data, it is in the real estate market. In the direct market, monthly data are not available everywhere and for all series. Historic data of whatever frequency can be short. Hence there may be sample issues for VARs with several endogenous variables and lags (say three and over endogenous variables and over 3 lags).

(ii) Difficult direct interpretation of the coefficients. If the VAR has too many parameters, multicollinearity arises making inference about statistical significance problematic.

(iii) Contemporaneous relationships are not accounted for.

(iv) Lag length criteria select different lag lengths that can pose difficulties about which model to adopt. Since a key objective is forecasting, one can assess the forecast capacity of alternative models and select the model the best forecasts our target variables. One can check what different it makes to impulse responses and variance decompositions.

- **Online note #11.9: Use of VARs in real estate – selected studies**

11.10   Concluding remarks

In this chapter, we have illustrated the uses and workings of VAR modelling with an application to real estate. We focused on unrestricted VARS but the principles are similar with restricted versions of VARs. One of the VAR criticisms is its atheoretical nature. In our example of the City of London office market, we did not establish a strong relationship between take up and the general economy. In one version of the VAR, a weaker GDP growth was associated with stronger take up. This is counterintuitive. In such situations, theory should be used to refine the VAR. Hence, in our case we should seek other economic variables for consideration.

Another point relates to the number of variables in the VAR. In our example, we could include more variables such as availability and office construction. We should be careful though not to over-parameterise our VAR especially if the data sample is not large. Not all these variables may be needed in the VAR. Too many variables and lags will result in a complex structure and perhaps with no forecast gains. For example we may not need both take up and vacancy or both vacancy and construction. When it comes to forecasting one should consider a simple VAR first containing the target variable(s) and subsequently assess the gains from complicating the VAR structure.

Chapter 11 online resource

- EViews file: "ch11_eviews"
- Excel file: "ch11_excel"
- Chapter 11 accompanying notes

12 Epilogue

Most real estate students are taught methods of quantitative analysis at various levels, from elementary statistical analysis to advanced methods depending on the nature of the programme. Quantitative techniques are valuable in tackling the complex problems of modern real estate and essential in many functions of the industry. The motivation of this book is the increasing need for quantitative analysis in real estate reflecting the evolution of the profession. The aim of this book is to explain quantitative analysis in real estate by adopting a practical rather than theoretical approach.

- The world of real estate is getting more sophisticated all the time. Graduates of real estate programmes are expected to have at least basic skills in quantitative analysis as the profession advances.
- As a main asset class now, analytical frameworks in real estate should reflect practices in mainstream and niche asset classes that can be 'quants' heavy to facilitate comparable analysis.
- Datasets in real estate are expanding. The exploitation of the potential of these databases requires good quantitative skills.
- Improvements in technology support quantitative analyses. The plurality of software (econometric and spreadsheet packages) and in-built methods open up many options for analysis.
- Real estate students and analysts in the profession are asking multifaceted questions necessitating a more intense interrogation of data and study of relationships. This is manifested by the topics investigated and analysis contained in student research projects and industry research papers.
- There have been external pressures on any asset class for an objective assessment of investment risks. Quantitative analysis is an important instrument to satisfy these regulatory demands.

The central theme has been to promote the acquisition and improvement of quantitative skills of real estate students and analysts. The book covers the needs of undergraduate and postgraduate real estate students, those studying for professional qualifications and practitioners.

We believe that our book prepares the reader for undertaking applied real estate market investigations by focusing on three areas:

(i) Preparation of data for analysis. This task requires you to understand the information recorded in the data before undertaking manipulations and graphing the data. During this process you can expect to extract useful information from the data. Data transformations may be necessary not only for statistical analysis but also to make visible trends inherent in the data and for comparisons with other real estate or asset class data.

(ii) Understanding the fundamental concepts underlying the application of statistical analysis in real estate. The book introduces and applies basic and intermediate level statistical calculations accompanied by explanations of the mathematical and statistical concepts underlying data analysis such as probability distributions, correlation analysis and hypothesis testing.

(iii) Ability to carry out an empirical investigation of relationships. The book takes the reader through the various stages in the process, from defining the empirical relationship in the form of a testable 'model' that makes use of the data available to choosing the right statistical methodology, estimation and validating the model. The interpretation of the results and implications for future trends (predictions) are the final fundamental tasks in the process.

Quantitative analysis not only complements other subjects studied on a real estate degree programme but also with business practices in the real estate industry. Quantitative analysis in the industry is instructive. It is informed by market knowledge and enables empirical investigations of importance to a business to be carried out. The links with key subjects including real estate economics, investment and portfolio analysis, valuations and development appraisals are apparent. Economic and financial theories alongside market knowledge provide guidance on important testable propositions and the subsequent results can be used in rent and yield analysis to assist with valuations, appraisals and portfolio management.

The quantitative methods presented in this book is widely applied in real estate, and every student of real estate will come across them somewhere in their course and later in the workplace. The structure of the book makes it clear that a basic analysis of real estate data is important before proceeding in applying more sophisticated techniques. Hence the book contains two chapters that extend from the preparation and transformation of data, to methods which summarise the information contained in the data, descriptive statistics, use of distributions, methods identifying possible linear relationships among data and hypothesis testing.

We devote five chapters (of which two are online) on regression analysis and evaluation of regression models (with time-series or cross-section data). Regression analysis (single and multiple) will remain the backbone of modelling work in real estate for a number of reasons. It is easy to understand and to communicate empirical findings. It is the base upon which to apply more sophisticated methodologies. Another reason for its popularity in real estate is the nature of the real estate data, that is rather short time-series databases characterised by low frequency. Of course, this applies to the private and not the public market. In our view, regression analysis is an essential skill for real estate analysts. It gives the analyst more flexibility and options to objectively study the market. The material in this book covered model building based on theory, model validation and interpretation of results.

The book also covers selective advanced topics. We include a separate chapter on the subject of stationarity which is relevant to time-series data. This allows us to study the implications of trended data, random walks and white noise processes in the statistical analysis we undertake. Random walks and white noise processes are common concepts in finance. Using this chapter as background, we construct ARMA models to model time-series data and generate forecasts. Vector autoregressions (VARs) allow for interactions within the real estate market and with the broader investment and economic environment. In the online resource we also cover long-term relationships in real estate (cointegration) and panel analysis.

This book has placed emphasis on forecasting – a core application of quantitative analysis in real estate. Forecasts from regression models are the easiest to obtain and explain and are considered to be the benchmark forecasts. The book also takes you through each step in obtaining forecasts from more advanced techniques. Our step-by-step exposition makes the process transparent and helps the reader acquire a better understanding of the process which should prove to be useful in communicating forecasts to non-cognitive audiences. We highlight the importance of taking into consideration qualitative forecasting during the evaluation to complement the analysis. We believe that after reading the forecast section of this book the reader will be able to understand the process, execute a forecast using an econometric model and combine both types of forecasts in an analysis.

Real estate students often find quantitative analysis difficult. Typically, they are not interested in mathematical abstraction, proofs and derivations, but are more concerned with understanding the principles of methodologies and interpretation of results. We have made extensive use of real estate data to illustrate statistical concepts. In each chapter we provide detailed practical examples to illustrate tests and estimations using a range of real estate data. In the examples we pay particular attention to the interpretation and practical implications of the statistical methods which we hope will be beneficial both to students and real estate professionals to draw robust conclusions. We have made significant use of Excel and EViews, spreadsheet and econometric software commonly used by real estate researchers to illustrate the execution of quantitative methods. Quantitative analysts should have knowledge of the underlying theoretical concepts and derivations. An interpretation of the results of a procedure or a test must be accompanied by a good understanding of the purpose and nature of the test and what the procedure is about.

The book has not covered all the methods that can be used to tackle real estate problems and relationships. For example, discrete choice models such as probit and logit models designed to explain categorical outcomes (e.g. explaining credit default rates or housing tenure choices) or simultaneous equations in real estate. Such methods are not included in the book but they are planned for the online resource.

Our closing remark is a reminder that we can only capitalise on the benefits of quantitative techniques if we combine an understanding of statistical methods with good real estate market knowledge and a strong theoretical background. Such a combination will ensure that we arrive at valid and robust conclusions. And remember that in the workplace your audience will not always be statistically literate. Irrespectively, it is your task to communicate the results of quantitative analysis in a simple and practical way. Be assured that a proficient theoretical background, market knowledge that includes a comprehension of raw and transformed data and competence in applying statistical methods will assist you to carry out influential investigations which have a business impact.

Appendix A

Statistical tables

Table A1 Standard normal table

Area under the normal curve
For example:

If $z = 1.96$, $P(0 \leq Z \leq 1.96) = 0.4750$.

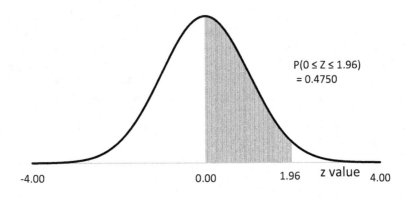

z	0.0000	0.0100	0.0200	0.0300	0.0400	0.0500	0.0600	0.0700	0.0800	0.0900
0.0	0.0000	0.0040	0.0080	0.0120	0.0160	0.0199	0.0239	0.0279	0.0319	0.0359
0.1	0.0398	0.0438	0.0478	0.0517	0.0557	0.0596	0.0636	0.0675	0.0714	0.0753
0.2	0.0793	0.0832	0.0871	0.0910	0.0948	0.0987	0.1026	0.1064	0.1103	0.1141
0.3	0.1179	0.1217	0.1255	0.1293	0.1331	0.1368	0.1406	0.1443	0.1480	0.1517
0.4	0.1554	0.1591	0.1628	0.1664	0.1700	0.1736	0.1772	0.1808	0.1844	0.1879
0.5	0.1915	0.1950	0.1985	0.2019	0.2054	0.2088	0.2123	0.2157	0.2190	0.2224
0.6	0.2257	0.2291	0.2324	0.2357	0.2389	0.2422	0.2454	0.2486	0.2517	0.2549
0.7	0.2580	0.2611	0.2642	0.2673	0.2704	0.2734	0.2764	0.2794	0.2823	0.2852
0.8	0.2881	0.2910	0.2939	0.2967	0.2995	0.3023	0.3051	0.3078	0.3106	0.3133
0.9	0.3159	0.3186	0.3212	0.3238	0.3264	0.3289	0.3315	0.3340	0.3365	0.3389
1.0	0.3413	0.3438	0.3461	0.3485	0.3508	0.3531	0.3554	0.3577	0.3599	0.3621
1.1	0.3643	0.3665	0.3686	0.3708	0.3729	0.3749	0.3770	0.3790	0.3810	0.3830
1.2	0.3849	0.3869	0.3888	0.3907	0.3925	0.3944	0.3962	0.3980	0.3997	0.4015
1.3	0.4032	0.4049	0.4066	0.4082	0.4099	0.4115	0.4131	0.4147	0.4162	0.4177
1.4	0.4192	0.4207	0.4222	0.4236	0.4251	0.4265	0.4279	0.4292	0.4306	0.4319

1.5	0.4332	0.4345	0.4357	0.4370	0.4382	0.4394	0.4406	0.4418	0.4429	0.4441
1.6	0.4452	0.4463	0.4474	0.4484	0.4495	0.4505	0.4515	0.4525	0.4535	0.4545
1.7	0.4554	0.4564	0.4573	0.4582	0.4591	0.4599	0.4608	0.4616	0.4625	0.4633
1.8	0.4641	0.4649	0.4656	0.4664	0.4671	0.4678	0.4686	0.4693	0.4699	0.4706
1.9	0.4713	0.4719	0.4726	0.4732	0.4738	0.4744	0.4750	0.4756	0.4761	0.4767
2.0	0.4772	0.4778	0.4783	0.4788	0.4793	0.4798	0.4803	0.4808	0.4812	0.4817
2.1	0.4821	0.4826	0.4830	0.4834	0.4838	0.4842	0.4846	0.4850	0.4854	0.4857
2.2	0.4861	0.4864	0.4868	0.4871	0.4875	0.4878	0.4881	0.4884	0.4887	0.4890
2.3	0.4893	0.4896	0.4898	0.4901	0.4904	0.4906	0.4909	0.4911	0.4913	0.4916
2.4	0.4918	0.4920	0.4922	0.4925	0.4927	0.4929	0.4931	0.4932	0.4934	0.4936
2.5	0.4938	0.4940	0.4941	0.4943	0.4945	0.4946	0.4948	0.4949	0.4951	0.4952
2.6	0.4953	0.4955	0.4956	0.4957	0.4959	0.4960	0.4961	0.4962	0.4963	0.4964
2.7	0.4965	0.4966	0.4967	0.4968	0.4969	0.4970	0.4971	0.4972	0.4973	0.4974
2.8	0.4974	0.4975	0.4976	0.4977	0.4977	0.4978	0.4979	0.4979	0.4980	0.4981
2.9	0.4981	0.4982	0.4982	0.4983	0.4984	0.4984	0.4985	0.4985	0.4986	0.4986
3.0	0.4987	0.4987	0.4987	0.4988	0.4988	0.4989	0.4989	0.4989	0.4990	0.4990

Table A2 The *t* distribution table

	Confidence intervals					
	80%	90%	95%	98%	99%	100%
	Level of Significance for one tailed Test, a					
df	0.100	0.050	0.025	0.010	0.005	0.0005
	Level of Significance for two tailed Test, a					
	0.20	0.10	0.05	0.02	0.01	0.001
1	3.078	6.314	12.706	31.821	63.657	636.619
2	1.886	2.920	4.303	6.965	9.925	31.599
3	1.638	2.353	3.182	4.541	5.841	12.924
4	1.533	2.132	2.776	3.747	4.604	8.610
5	1.476	2.015	2.571	3.365	4.032	6.869
6	1.440	1.943	2.447	3.143	3.707	5.959
7	1.415	1.895	2.365	2.998	3.499	5.408
8	1.397	1.860	2.306	2.896	3.355	5.041
9	1.383	1.833	2.262	2.821	3.250	4.781
10	1.372	1.812	2.228	2.764	3.169	4.587
11	1.363	1.796	2.201	2.718	3.106	4.437
12	1.356	1.782	2.179	2.681	3.055	4.318
13	1.350	1.771	2.160	2.650	3.012	4.221
14	1.345	1.761	2.145	2.624	2.977	4.140
15	1.341	1.753	2.131	2.602	2.947	4.073
16	1.337	1.746	2.120	2.583	2.921	4.015
17	1.333	1.740	2.110	2.567	2.898	3.965
18	1.330	1.734	2.101	2.552	2.878	3.922
19	1.328	1.729	2.093	2.539	2.861	3.883

(*Continued*)

Table A2 (Continued)

20	1.325	1.725	2.086	2.528	2.845	3.850
21	1.323	1.721	2.080	2.518	2.831	3.819
22	1.321	1.717	2.074	2.508	2.819	3.792
23	1.319	1.714	2.069	2.500	2.807	3.768
24	1.318	1.711	2.064	2.492	2.797	3.745
25	1.316	1.708	2.060	2.485	2.787	3.725
26	1.315	1.706	2.056	2.479	2.779	3.707
27	1.314	1.703	2.052	2.473	2.771	3.690
28	1.313	1.701	2.048	2.467	2.763	3.674
29	1.311	1.699	2.045	2.462	2.756	3.659
30	1.310	1.697	2.042	2.457	2.750	3.646
40	1.303	1.684	2.021	2.423	2.704	3.551
50	1.299	1.676	2.009	2.403	2.678	3.496
60	1.296	1.671	2.000	2.390	2.660	3.460
100	1.290	1.660	1.984	2.364	2.626	3.390
120	1.289	1.658	1.980	2.358	2.617	3.373
Infinity	1.282	1.645	1.960	2.326	2.576	3.291

Table A3 *F* distribution critical values

F – distribution critical values for 5% level of significance ($p = 0.05$)
(df = degrees of freedom)

Denominator $(T - K)$

df	Numerator df (m)						
	1	*2*	*3*	*4*	*5*	*7*	*10*
5	6.61	5.79	5.41	5.19	5.05	4.88	4.74
10	4.96	4.10	3.71	3.48	3.33	3.14	2.98
15	4.54	3.68	3.29	3.06	2.90	2.71	2.54
20	4.35	3.49	3.10	2.87	2.71	2.51	2.35
21	4.32	3.47	3.07	2.84	2.68	2.49	2.32
22	4.30	3.44	3.05	2.82	2.66	2.46	2.30
23	4.28	3.42	3.03	2.80	2.64	2.44	2.27
24	4.26	3.40	3.01	2.78	2.62	2.42	2.25
25	4.24	3.39	2.99	2.76	2.60	2.40	2.24
30	4.17	3.32	2.92	2.69	2.53	2.33	2.16
35	4.12	3.27	2.87	2.64	2.49	2.29	2.11
40	4.08	3.23	2.84	2.61	2.45	2.25	2.08

Table A4 Critical values for χ^2 distribution

(*df* = degrees of freedom)

Denominator (T − K)			
df	*Level of significance (p-value)*		
	0.10	*0.05*	*0.01*
1	2.71	3.84	6.64
2	4.61	5.99	9.21
3	6.25	7.82	11.35
4	7.78	9.49	13.28
5	9.24	11.07	15.09
6	10.65	12.59	16.81
7	12.02	14.07	18.48
8	13.36	15.51	20.09
9	14.68	16.92	21.67
10	15.99	18.31	23.21

See Gujarati and Porter (2009) Appendix D for detailed statistical tables

References

Akaike, H. (1973) Information theory and an extension of the maximum likelihood principle, in Petrov, B. and Csaki, F. (Eds.), *Second International Symposium on Information Theory*, Akademai Kiado, Budapest, 267–281.

Akarim, Y. and Sevim, S. (2013) The impact of mean reversion model on portfolio investment strategies: Empirical evidence from emerging markets, *Economic Modelling*, 31, 453–459.

Armstrong, J. (2001) Evaluating forecasting methods, in Armstrong, J. (Ed.), *Principles of Forecasting: A Handbook for Researchers and Practitioners*, Chapter 14, Kluwer Academic Publishers, Norwell, MA.

Armstrong, J. and Collopy, F. (1992) Error measures for generalising about forecasting methods: Empirical comparisons, *International Journal of Forecasting*, 8, 69–80.

Bai, J. and Perron, P. (2004) Multiple structural change models: A simulation analysis, in Corbea, D., Durlauf, S. and Hansen, B. E. (Eds.), *Econometric Essays*, Cambridge University Press, Cambridge.

Bai, J. and Perron, P. (2003) Computation and analysis of multiple structural change models, *Journal of Applied Econometrics*, 18, 1–22.

Bai, J. and Perron, P. (1998) Estimating and testing linear models with multiple structural changes, *Econometrica*, 66, 47–78.

Ball, M., Lizieri, C. and MacGregor, B. (1998) *The Economics of Commercial Property Markets*, Routledge, London, UK.

Ball, M. and Tsolacos, S. (2002) UK commercial property forecasting, the devil is in the data, *Journal of Property Research*, 19(1), 13–38.

Bańbura, M., Giannone, D. and Lenza, M. (2014) Conditional forecasts and scenario analysis with vector autoregressions for large cross-sections, *Working Paper Series*, No 1733, European Central Bank.

Baum, A. and Crosby, N. (2014) *Property Investment Appraisal*, Blackwell Publishing, Oxford.

Blaskowitz, O. and Herwartz, H. (2011) On economic evaluation of directional forecasts, *International Journal of Forecasting*, 27(4), 1058–1065.

Blundell, G. and Ward, C. (1987) Property portfolio allocation: A multifactor model, *Land Development Studies*, 4, 145–156.

Box, G. and Jenkins, G. (1978) *Time Series Analysis: Forecasting and Control*, 3rd edition, Holden Day, San Francisco.

Brooks, C. (2014) *Introductory Econometrics for Finance*, Cambridge University Press, Cambridge.

Brooks, C. and Tsolacos, S. (2010) *Real Estate Modelling and Forecasting*, Cambridge University Press, Cambridge.

Brooks, C. and Tsolacos, S. (2001) Forecasting real estate returns using financial spreads, *Journal of Property Research*, 18(3), 235–248.

Brooks, C. and Tsolacos, S. (2000) Forecasting models of retail rents, *Environment and Planning A*, 32, 1825–1839.

Brown, G. and Matysiak, G. (2000) *Real Estate Investment: A Capital Market Approach: A Capital Market*, London: Prentice Hall.

Brown, R., Durbin, J. and Evans, J. (1975) Techniques for testing the constancy of regression relationships over time, *Journal of the Royal Statistical Society*, Series B, 35, 149–192.

Campbell, J. and Perron, P. (1991) Pitfalls and opportunities: What macroeconomists should know about unit roots, *NBER Macroeconomics Annual*, 6, 141–201.

Caner, M. and Kilian, L. (2001) Size distortions of tests of the null hypothesis of stationarity: Evidence and implications for the PPP debate, *Journal of International Money and Finance*, 20, 639–657.

CBRE (2019) *A UK Four Quadrants Pricing Model, Viewpoint*, UK Research, January, CBRE, London.

Chatfield, C. (1988) Apples, oranges and mean square error, *International Journal of Forecasting*, 4, 515–518.

Chauvet, M. and Piger, J. (2008) A comparison of the real time performance of business cycle dating methods, *Journal of Business and Economic Statistics*, 26(1), 42–49.

Chow, G. (1960) Tests of equality between sets of coefficients in two linear regressions, *Econometrica*, 28(3), 591–605.

Clapp, J. (1993) *Dynamics of Office Markets, AREUEA Monograph*, October, The Urban Land Institute Press, Washington.

Clements, M. and Hendry, D. (2011) Forecasting from misspecified models in the presence of unanticipated location shifts, in Clements, M. and Hendry, S. (Eds.), *The Oxford Handbook in Economic Forecasting*, Oxford University Press, Oxford.

Clements, M. and Hendry, D. (1998) *Forecasting Economic Time Series*, Cambridge University Press, Cambridge.

Collopy, F. and Armstrong, S. (1992) Rule-based forecasting: Development and validation of an expert systems approach to combining time series extrapolations, *Management Science*, 38(10), 1394–1414.

D'Arcy, E., McGough, T. and Tsolacos, S. (1997) National economic trends, market size and city growth effects on European office rents, *Journal of Property Research*, 14(4), 297–308.

Davidson, R. and MacKinnon, J. (1993) *Estimation and Inference in Econometrics*, OUP Catalogue, Oxford University Press, Oxford.

Devaney, S., McAllister, P. and Nanda, A. (2017) Which factors determine transaction activity across U.S. metropolitan office markets?, *Journal of Portfolio Management*, 43(6), 90–104.

Devaney, S., Scofield, D. and Zhang, F. (2019) Only the best? Exploring cross-border investor preferences in US gateway cities, *The Journal of Real Estate Finance and Economics*, 59, 490–513.

Dickey, D. A. and Fuller, W. A. (1979) Distribution of the estimators for autoregressive time series with a unit root, *Journal of the American Statistical Association*, 74, 427–431.

Diebold, F. X. (1998) *Elements of Forecasting*, South-Western College Publishing, Cincinnati.

Diebold, F. X. (1993) On the limitations of comparing mean square forecast errors: Comment, *Journal of Forecasting*, 12, 641–642.

Diebold, F. X. and Lopez, J. (1996) Forecast evaluation and combination, in Maddala, G. S. and Rao, C. R. (Eds.), *Handbook of Statistics*, North Holland, Amsterdam.

DiPasquale, D. and Wheaton, W. C. (1992) The markets for real estate assets and space: A conceptual framework, *Journal of the American Real Estate and Urban Economics Association*, 20(2), 181–198.

Durbin, J. and Watson, G. (1951) Testing for serial correlation in least squares regression, *Biometrika*, 38, 159–171.

Elliott, G., Rothenberg, T. and Stock, J. (1996) Efficient tests for an autoregressive unit root, *Econometrica*, 64, 813–836.

FRSBF (2011) Natural vacancy rates in commercial real estate markets, *Federal Reserve Bank of San Francisco (FRBSF) Economic Letter*, 2001–27, October.

Fuerst, F., Milcheva, S. and Baum, A. (2015) Cross-border capital flows into real estate, *Real Estate Finance*, 31(3), 103–122.

Fuller, W. A. (1976) *Introduction to Statistical Time Series*, John Wiley & Sons, New York.

Füss, R., Stein, M. and Zietz, J. (2012) A regime-switching approach to modelling rental prices of U.K. real estate sectors, *Real Estate Economics*, 40(2), 317–350.

Gerlow, M. E., Irwin, S. H. and Liu, T.-R. (1993) Economic evaluation of commodity price forecasting models, *International Journal of Forecasting*, 9, 387–397.

Godfrey, L. G. (1988) *Specification Tests in Econometrics*, Cambridge University Press, Cambridge.

Godfrey, L. G. (1978b) Testing for higher order serial correlation in regression equations when the regressors include lagged dependent variables, *Econometrica*, 46(6), 1303–1310.

Goyal, A. and Welch, I. (2008) A comprehensive look at the empirical performance of equity premium prediction, *Review of Financial Studies*, 21, 1455–1508.

Granger, C. (1996) Can we improve the perceived quality of economic forecasts?, *Journal of Applied Econometrics*, 11(5), 455–473.

Gujarati, D. (2014) *Econometrics by Example*, Palgrave Macmillan, New York.

Gujarati, D. and Porter, D. (2010) *Essentials of Econometrics*, McGraw-Hill/Irwin, The University of Michigan, New York.

Hakkio, C. and Morris, C. (1984) Vector autoregressions: A user's guide, *Research Working Paper*, 84(10) Research Division, Federal Reserve Bank of Kansas City.

Hannan, E. and Quinn, B. (1979) The determination of the order of autoregression, *Journal of the Royal Statistical Society*, Series B, 41(2), 190–195.

Hendershott, P. H. (1996) Rental adjustment and valuation in overbuilt markets: Evidence from the Sydney office market, *Journal of Urban Economics*, 39, 51–67.

Hendershott, P. H. (1995) Real effective rent determination: Evidence from the Sydney office market, *Journal of Property Research*, 12(2), 127–135.

Hill, C., Griffiths, W. and Judge, G. (2001) *Undergraduate Econometrics*, 2nd edition, John Wiley & Sons, Hoboken, NJ.

Home and Communities Agency (2015) *Employment Density Guide*, 3rd edition, November, Home and Communities Agency, London.

Home and Communities Agency (2010) *Employment Density Guide*, 2nd edition. Home and Communities Agency, London.

INREV (2018) *Calculation Guide for INREV Indices*, INREV, Amsterdam.

Jarque, M. and Bera, A. (1980) Efficient tests for normality, homoscedasticity and serial independence of regression residuals, *Economic Letters*, 6, 255–259.

Johansen, S. (1995) *Likelihood-Based Inference in Cointegrated Vector Autoregressive Models*, Oxford University Press, Oxford.

Jouini, J. and Boutahar, M. (2005) Evidence on structural changes in U.S. time series, *Economic Modelling*, 22, 391–422.

Jowsey, E. (2011) *Real Estate Economics*, Palgrave Macmillan, Basingstoke.

Kelly, H. (1983) Forecasting office space demand in Urban Area, *Real Estate Review*, 13, 87–95.

Keogh, G. (1994) Use and investment markets in British real estate, *Journal of Property Valuation and Investment*, 12(4), 58–72.

Key, T. *et al.* (1994) *Understanding the Property Cycle, Report on Economic and Property Cycles*, Royal Institution of Chartered Surveyors, London, May.

Kim, M., Nelson, R. and Startz, R. (1991) Mean reversion in stock prices? A reappraisal of the empirical evidence, *Review of Economic Studies*, 58(3), 515–528.

Krystalogianni, A., Matysiak, G. and Tsolacos, S. (2004) Forecasting UK commercial real estate cycle phases with leading indicators: A probit approach, *Applied Economics*, 36(20), 2347–2356.

Kwiatkowski, D., Phillips, P., Schmidt, P. and Shin, Y. (1992) Testing the null hypothesis of stationary against the alternative of a unit root, *Journal of Econometrics*, 54, 159–178.

Lawrence, M., Goodwin, P., O'Connor, M. and Onkal, D. (2006) Judgemental forecasting: A review of progress over the last 25 years, *International Journal of Forecasting*, 22(3), 493–518.

Leitch, G. and Tanner, J. E. (1991) Economic forecast evaluation: Profit versus the conventional error measures, *American Economic Review*, 81(3), 580–590.

Litterman, R. (1984) Forecasting with Bayesian vector autoregressions: Four years of experience, *Working Paper 259*. Federal Reserve Bank of Minneapolis, August.

Lizieri, C. and Pain, K. (2014) International office investment in global cities: The production of financial space and systemic risk, *Regional Studies*, 48(3), 439–455.

Lupoletti, W. M. and Webb, R. H. (1986) Defining and improving the accuracy of macroeconomic forecasts: Contributions from a VAR model, *Journal of Business*, 59(2), 263–285.

Lütkepohl, H. (1991) *Introduction to Multiple Time Series*, Springer-Verlag, Berlin.

MacKinnon, J. G. (1996) Numerical distribution functions for unit root and cointegration tests, *Journal of Applied Econometrics*, 11, 601–618.

Markowitz, H. (1952) Portfolio selection, *Journal of Finance*, 7(1), 77–91.

Mauck, N. and Price, S. M. (2017) Determinants of foreign versus domestic real estate investment: Property level evidence from listed real estate investment firms, *Journal of Real Estate Finance and Economics*, 54(1), 17–57.

McAllister, P. and Nanda, A. (2016a) Does real estate defy gravity? An analysis of foreign real estate investment flows, *Review of International Economics*, 25(4), 924–948.

McAllister, P. and Nanda, A. (2016b) Do foreign buyers compress office real estate cap rates?, *Journal of Real Estate Research*, 38(4), 569–594.

McAllister, P. and Nanda, A. (2015) Does foreign investment affect U.S. office real estate prices?, *Journal of Portfolio Management*, 41(6), 38–47.

McGough, T. and Tsolacos, S. (1995) Forecasting commercial rental values using ARIMA models, *Journal of Property Valuation and Investment*, 13(5), 6–22.

Newey, W. K. and West, K. D. (1987) A simple, positive semi-definite, heteroskedasticity and autocorrelation consistent covariance matrix, *Econometrica*, 55, 703–708.

Ng, S. and Perron, P. (2001) Lag length selection and the construction of unit root tests with good size and power, *Econometrica*, 69, 1519–1554.

Nyberg, H. and Pönkä, H. (2016) International sign predictability of stock returns: The role of the United States, *Economic Modelling*, 58, 323–338.

Oikarinen, E. and Falkenbach, H. (2017) Foreign investors' influence on the real estate market capitalization rate: Evidence from a small open economy, *Applied Economics*, 49(32), 3141–3155.

Pastor, L. and Stambaugh, R. (2012) Are stocks really less volatile in the long run?, *Journal of Finance*, 67(2), 431–478.

Paye, B. and Timmermann, A. (2006) Instability of return prediction models, *Journal of Empirical Finance*, 13(3), 274–315.

Perron, P. (2006) Dealing with structural breaks, in Patterson, K. and Mills, T. C. (Eds.), *Palgrave Handbook of Econometrics, Vol. 1: Econometric Theory*, Palgrave Macmillan, New York, 278–352.

Perron, P. and Ng, S. (1996) Useful modifications to some unit root tests with dependent errors and their local asymptotic properties, *The Review of Economic Studies*, 63(3), 435–463.

Pesaran, H., Smith, R. and Yeo, J. (1985) Testing for structural stability and predictive failure: A review, *Manchester School*, 53, 280–295.

Pesaran, M. H. and Timmerman, A. (1992) A simple non-parametric test of predictive performance, *Journal of Business and Economic Statistics*, 10(4), 461–465.

Phillips, P. C. B. and Perron, P. (1988) Testing for a unit root in time series regression, *Biometrika*, 75, 335–346.

Pindyck, R. and Rubinfeld, D. (1998) *Econometric Models and Economic Forecasts*, 4th edition, Irwin/McGraw Hill, New York.

Pirounakis, S. (2013) *Real Estate Economics*, Routledge, Abingdon, UK.

Pönkä, H. (2016) Real oil prices and the international sign predictability of stock returns, *Finance Research Letters*, 17, 79–87.

Pönkä, H. (2014) Predicting the direction of US stock markets using industry returns, *Discussion Paper*, No. 385, Helsinki Center of Economic Research, September.

Pons, J. (2000) The accuracy of IMF and OECD forecasts for G7 countries, *Journal of Forecasting*, 19(1), 53–63.

Ramsey, J. B. (1969) Tests for specification errors in a classical linear least-squares analysis, *Journal of the Royal Statistical Society*, Series B, 71, 350–371.

Rapach, D., Strauss, J. and Zhou, G. (2010) Out-of-sample equity premium prediction: Combination forecasts and links to the real economy, *Review of Financial Studies*, 23(2), 821–862.

Rapach, D. and Wohar, M. (2006) In-sample vs. out-of-sample tests of stock return predictability in the context of data mining, *Journal of Empirical Finance*, 13(2), 231–247.

Rees, D. and Wood, M. (2007) The four quadrant investment model, *Journal of Investment Strategy*, 2(1), 67–71.

Refenes, A.-P. (1995) *Neural Networks in the Capital Markets*, JohnWiley, Chichester, England.

RICS (2015) Code of measuring practice, in *RICS Professional Standards and Guidance, Global*, 6th edition, May, Royal Institution of Chartered Surveyors, London.

RICS (2013) Analysis of commercial lease transactions, in *RICS Guidance Note*, September, Royal Institution of Chartered Surveyors, Hong Kong.

Schwarz, G. (1978) Estimating the dimension of a model, *The Annals of Statistics*, 6, 461–464.

Schwert, G. (1989) Test for unit roots: A Monte Carlo investigation, *Journal of Business and Economic Statistics*, 7(2), 147–159.

Sims, C. A. (1980) Macroeconomics and reality, *Econometrica*, 48, 1–48.

Sims, C. A., Stock, J. and Watson, M. (1990) Inference in linear time series models with some unit roots, *Econometrica*, 58(1), 113–144.

Sinclair, T., Stekler, H. and Kitzinger, L. (2010) Directional forecasts of gdp and inflation: A joint evaluation with an application to federal reserve predictions, *Applied Economics*, 42(18), 2289–2297.

Spierdijk, L., Bikker, J. and Hoek, P. (2012) Mean reversion in international stock markets: An empirical analysis of the 20th century, *Journal of Inter Journal of International Money and Finance*, 31(2), 228–249.

Stevenson, S. (2007) A comparison of the forecasting ability of ARIMA models, *Journal of Property Investment & Finance*, 25(3), 223–240.

Theil, H. (1971) *Principles of Econometrics*, North Holland, Amsterdam.

Theil, H. (1966) *Applied Economic Forecasting*, North-Holland, Amsterdam.

Thompson, P. (1990) An MSE statistic for comparing forecast accuracy across series, *Journal of International Forecasting*, 6(2), 219–227.

Tonelli, M., Cowley, M. and Boyd, T. (2004) Forecasting office building rental growth: Using a dynamic approach, *Pacific Rim Real Estate Journal*, 10(3), 283–304.

Tse, R. (1997) An application of the ARIMA model to real-estate prices in Hong Kong, *Journal of Property Finance*, 8(2), 152–163.

Tsolacos, S., Keogh, G. and McGough, T. (1998) Modelling use, investment and development in the British office market, *Environment and Planning A*, 30, 1409–1427.

Turner, P. (2010) Power properties of the CUSUM and CUSUMSQ tests for parameter instability, *Applied Economics Letters*, 17(11), 1049–1053.

Voith, R. and Crone, T. (1988) National vacancy rates and the persistence of shocks in U.S. office markets, *AREUEA Journal*, 16, 437–458.

Webb, R. H. (1984) Vector autoregressions as a tool for forecast evaluation, *Economic Review*, Federal Reserve Bank of Richmond, January/February, 3–11.

Wheaton, W. C. (1990) Vacancy, search, and prices in a housing market matching model, *Journal of Political Economy*, 98(6), 1270–1293.

White, H. (1980) A heteroskedasticity-consistent covariance matrix estimator and a direct test for heteroskedasticity, *Econometrica*, 48, 817–838.

Wilkens, K., Heck, J. and Cochran, S. (2005) Risk and return properties of portfolios based on directional forecasts, *Managerial Finance*, 31(8), 58–76.

Wilson, P., Okunev, J., Ellis, C. and Higgins, D. (2000) Comparing univariate forecasting techniques in property markets, *Journal of Real Estate Portfolio Management*, 6(3), 283–306.

Winker, P. (2002) Vektor autoregressive modelle, in Schröder, M. (Ed.), *Finanzmarkt-Ökonometrie*, Schäffer-Poeschel Verlag, Stuttgart, 213–262.

Wooldridge, J. (2012) *Introductory Econometrics: A Modern Approach*, 5th edition, South-Western Cengage Learning, Mason, OH.

Wu, D.-M. (1974) Alternative tests of independence between stochastic regressors and disturbances: Finite sample results, *Econometrica*, 42(3), 529–546.

Wyatt, P. (2013) *Property Valuation*, 2nd edition, Wiley-Blackwell, Oxford.

Index

Printed in the United States
By Bookmasters